STABILITY AND CONSTANCY IN VISUAL PERCEPTION

WILEY SERIES IN BEHAVIOR

KENNETH MacCORQUODALE, Editor

University of Minnesota

A Temperament Theory of Personality Development
ARNOLD H. BUSS AND ROBERT PLOMIN

Serial Learning and Paralearning
E. RAE HARCUM

Increasing Leadership Effectiveness
CHRIS ARGYRIS

Stability and Constancy in Visual Perception:
Mechanisms and Processes
WILLIAM EPSTEIN

STABILITY AND CONSTANCY IN VISUAL PERCEPTION: Mechanisms and Processes

WILLIAM EPSTEIN, Editor
University of Wisconsin—Madison

A WILEY-INTERSCIENCE PUBLICATION

JOHN WILEY & SONS, New York • London • Sydney • Toronto

1-20-78

Library of Congress Cataloging in Publication Data:

Main entry under title:

Stability and constancy in visual perception.

(Wiley series in behavior)
"A Wiley-Interscience publication."
Includes bibliographies.
1. Visual perception. 1. Epstein, William

1931- [DNLM: 1. Visual perception. WW103 S775]
BF241.S75 152.1'4 76-28769
ISBN 0-471-24355-8

Printed in the United States of America

10 9 8 7 6 5 4 3 2 1

CONTRIBUTORS

V. R. Carlson, Laboratory of Psychology and Psychopathology, NIMH, Bethesda, Maryland

James Comerford, Department of Psychology, York University, Toronto, Ontario, Canada

Stanley Coren, Department of Psychology, University of British Columbia, Vancouver, British Columbia, Canada

R. H. Day, Department of Psychology, Monash University, Clayton, Victoria, Australia

Sheldon Ebenholtz, Department of Psychology, University of Wisconsin, Madison, Wisconsin

William Epstein, Department of Psychology, University of Wisconsin, Madison, Wisconsin

Joan Girgus, Department of Psychology, City College, New York, New York

Walter C. Gogel, Department of Psychology, University of California, Santa Barbara, California

Gunner Johansson, Department of Psychology, University of Uppsala, Uppsala, Sweden

B. E. McKenzie, Department of Psychology, La Trobe University, Bundoora, Victoria, Australia

Hiroshi Ono, Department of Psychology, York University, Toronto, Ontario, Canada

Tadasu Oyama, Department of Psychology, Chiba University, Yayoi-cho, Chiba-shi, Japan

Whitman Richards, Department of Psychology, Massachusetts Institute of Technology, Cambridge, Massachusetts

Irvin Rock, Department of Psychology, Rutgers University, Newark, New Jersey

Wayne Shebilske, Department of Psychology, University of Virginia, Charlottesville, Virginia

SERIES PREFACE

Psychology is one of the lively sciences. Its foci of research and theoretical concentration are diverse among us, and always on the move, sometimes into unexploited areas of scholarship, sometimes back for second thoughts about familiar problems, often into other disciplines for problems and for problem-solving techniques. We are always trying to apply what we have learned, and we have had some great successes at it. The Wiley Series in Behavior reflects this liveliness.

The series can accommodate monographic publication of purely theoretical advances, purely empirical ones, or any mixture in between. It welcomes books that span the interfaces within the behavioral sciences and between the behavioral sciences and their neighboring disciplines. The series is, then, a forum for the discussion of those advanced, technical, and innovative developments in the behavior sciences that keep its frontiers flexible and expanding.

KENNETH MACCORQUODALE

Minneapolis, Minnesota
December 1974

PREFACE

With the exception of the first chapter, which is primarily historical, this book presents a set of original essays concerned with contemporary research and theory. Although the reader will find much that is new in these essays, he will recognize many of the same concerns that dominated the early history of the experimental study of perception. Certainly Helmholtz, returning to the scene after an 80-year leave, would find himself on familiar terrain. To some extent this state of affairs may be attributed to the refractory nature of the problems under consideration. To a larger extent it is due to the development of new techniques and the discovery of new information, one consequence of which has been a revival of interest in problems once considered resolved. Consider two examples: the perception of visual direction and the ontogenesis of the perceptual constancies. It is understandable that Helmholtz's masterly treatment of the questions of visual direction induced the feeling that the determinants of perceived visual direction and its components, for example, registered eye position, were perfectly understood. Yet, as Shebilske's chapter shows, new experimental procedures and new findings challenge some of Helmholtz's basic assumptions such as emphasis on outflow information. It is plain that the matter is much more complicated than Helmholtz supposed. Concerning the ontogenesis of the constancies, there was until recently a tendency to regard the problem as intractable because of the allegedly insuperable difficulties in securing reliable data. Another widely prevalent view was that infants do not experience a stable perceptual world. This opinion was nurtured by the empiristic dogma that characterized much of American psychology. The chapter by Day and McKenzie shows that neither the pessimism nor the dogma was warranted. The development and application of new experimental procedures—the stimulus preference method, habituation measures, and operant conditions procedures—have generated new and exciting data that will lead to a rebirth of interest in the question.

In a number of instances, although the considerations taken up in the essays have a familiar ring, the analyses have achieved a level of detail that was not possible in earlier treatments. Compare the treatment of instructional (attitudinal) effects in Carlson's contribution to this book with Brunswik's or Koffka's. Comparable advances are apparent in Ebenholtz's analysis of orientation constancy, Gogel's examination of space perception, Ono and Comeford's discussion of stereoscopic depth, and Coren and Girgus' treatment of the illusions. Oyama's contribution, which is another example of a familiar question, utilizes a new method of analysis that promises more exact characterizations of the causal relations so frequently implicit in explanations of constancy.

Turning to the contributions of Johannson and Richards, we find presentations that display less continuity with past approaches. Johannson's focus on kinetic displays and his theoretical assertions that draw on modern developments in geometry would seem strange to the ghost of Helmholtz, and Richards' exposition, resting as it does on the new discoveries in the field of visual physiology, can hardly have been anticipated by him.

Rock's essay explicitly links itself to the past, but it is not a rehearsal of old arguments and data. In fact, the title "In Defense of Unconscious Inference" is misleading if it suggests to the reader that the essay is an attempt to defend Helmholtz's doctrine. Instead it is a long overdue examination and explication of the process of unconscious inference in perception. The thesis and its implications are subjected to an exacting scrutiny which Helmholtz surely would have appreciated but which unhappily he would not have or could not have performed himself. Of course, Rock's analysis is facilitated by contemporary findings not available to Helmholtz and his immediate successors, but the principal reason is that Rock has set out to ask the questions that earlier writers have largely ignored.

This book concludes with a chapter aimed at bringing into focus a variety of methodological, empirical, and theoretical questions that recur in the various essays. The book is directed to graduate students and other investigators in the field of perception. A less prepared reader who is interested primarily in one of the substantive topics treated in any single essay might do well to supplement his selection with a review of the initial and final chapters.

"On behalf of the contributors and myself I wish to acknowledge with thanks the various publishers and authors who allowed us to reproduce materials originally appearing in their publications."

WILLIAM EPSTEIN

Madison, Wisconsin
September 1976

CONTENTS

STABILITY AND CONSTANCY IN VISUAL PERCEPTION

of scientific observation: the method of observation ought not to affect the object of observation. Starting with the theoretical assumption that the true sensory facts are local phenomena that depend on local stimulation, observational procedures were tailored to reveal these facts. The observer was instructed to adopt an analytic (reductionist) attitude and aids, such as an artificial pupil, were introduced to facilitate the prescribed mode of observation. Under such conditions a careful observer may report that the shape of an object appears to change as orientation of the object is changed. If asked to report the whiteness of a gray surface, the observer may report that whiteness changes with every change of illuminance. According to the introspectionist, these reports reflect the true sensory core of perception. The fact of ordinary uninstructed viewing, that is, an object appears almost constant in shape as orientation varies and almost constant in whiteness as illuminance changes, reflects the accretions of experience.

As is well known, the Gestalt psychologists were determined to redirect attention from artificially isolated sense data to immediate spontaneous perceptions. For Gestalt theory the phenomena available to everyday experience were the facts that ought to be emphasized in a theory of perception. The pure sensations of the introspectionist were artifacts of a method, mere curiosities of little theoretical interest. Nowhere was this clearer than in the case of the constancies. Whereas the introspectionist tried to strip away the constancies so that the genuine sensory core could be observed, the Gestalt psychologist considered the constancies in perception to be the primary data.

The arguments advanced by the Gestalt psychologists were convincing. The core-context theory and analytic introspectionism, weakened by problems of reliability and the challenge of radical Behaviorism, were totally undermined by the theoretical analysis of Gestalt psychology. But Gestalt theory may have been too successful. The compelling polemic against sensationalism seems to have foreclosed further examination within experimental psychology of the status of the experiences studied by the introspectionist. Notwithstanding agreement that the collection of sensations does not constitute the inventory out of which the perceptual world is constructed, useful questions remain concerning this other domain of sensory experience. Consider the following as one illustration: How is it that adoption of a pictorial or reductionist attitude results in the kind of experience that Gibson (1950) has called the "visual field" and what is seen when one sees the visual field? Not the world of physical objects that constitutes the environment. Not the retinal image, for even if a percipient could scan his own image it would be found to be as different from the "visual field" as it is from the distal world. Boring (1952) has suggested that Gibson's visual field may be equated with the appearance of objects under reduced viewing conditions. I agree with Gibson (1950) in disputing this equation. At least one difference between the visual field and experience under reduced viewing is plain. The visual field may be voluntarily relinquished in favor of the more familiar experience that Gibson calls the "visual world." Thus the visual field and the visual world are alternatives that may be selected by the observer. In contrast, under reduced viewing conditions, the experience is mandatory; for example, if all stimuli for distance have been eliminated, the observer is not free to choose between perceiving a size that is proportional to visual angle and perceiving a size that corresponds to objective size. However, even if we were to accept Boring's identification, it ought to be recognized that important questions may be raised concerning the interpretation of perception under reduced conditions. As one example, does the covariation of perceived size and retinal size reflect direct one-to-one determination of perceived size by retinal size or is the relationship between perceived size and retinal size mediated by an assumption or guess about distance?

CONSTANCY AND KEY RELATIONSHIPS IN PERCEPTION

Setting aside the polemical motivations of the Gestalt psychologist, we may attribute the continued interest in the constancies to the fact that examination of them directs attention to the key relationships in the analysis of perception: (1) between distal and proximal stimuli, (2) between proximal stimulation and perception, and (3) between distal stimuli and perception.

Distal Stimuli and Proximal Stimuli

Agreement is widespread regarding the relationship between distal and proximal stimuli. The conventional view, already expressed clearly in the early works by Descartes (1637) in *La Dioptrique* and by Berkeley (1709) in *An Essay Toward a New Theory of Vision* is that the distal-proximal relationship is intractably equivocal. At the turn of the eighteenth century this premise was so widely accepted that Berkeley (1709, Sec. 2) commenced his inquiry with a statement of the premise unelaborated by comment or defense. Moreover, although much of Berkeley's theory was later disputed, the correctness of this premise was not questioned. Indeed the claim that proximal stimulation cannot specify the properties of the distal situation is an assumption held by almost all twentieth century theoreticians, regardless of their differences. For Ames and the Transactionalists (e.g., Ittleson, 1960, Chapter 4) the assumption is expressed in the concept of "equivalent configurations," the family of physical arrangements for which impingement is invariant. For Brunswik (1947) the assumption of equivocality in the distal-proximal relationship is the starting point for the development of his probabilistic learning theory of space perception. Nor is it only the empiricistic theories that insist on the equivocality of stimulation. In his elegant analysis of "why things look as they do," Koffka (1935) argues that the

> confusion of distant and proximal stimuli can have a fatal effect on psychological theory. The danger . . . lies in the fact that for each distant stimulus there exists a practically infinite number of proximal stimuli; thus, the "same stimulus" in the distant sense may not be the same stimulus in the proximal sense; as a matter of fact it seldom is (p. 80).

This view of the distal-proximal relationship has had a profound effect on theories of perception. It has led to accounts like Berkeley's in which optical variables function like words in a language; the optical variables are arbitrary signs whose meanings are learned. In Berkeley's opinion there is no more reason to believe that a variable of stimulation will contain information about the environment than there is to think that the auditory event "table" will inform a listener who knows no English. In general, acceptance of the doctrine of equivocality has encouraged theories that emphasize the contributions of extrastimulation factors such as selective principles of organization or cognitivelike algorithms that structure the perceptual world. In other words, the equivocality of the distal-proximal relationship has seemed to call for the identification of mechanisms that disambiguate, supplement, or organize the proximal input.

J. J. Gibson's (1950) theory has been a notable exception. Gibson has been singular in questioning the original premise. According to this author, earlier writers have been misled by faulty specifications of both the distal and the proximal stimulation: although it is unimpeachable that the relationship between an isolated bit of the environment and its isolated proximal counterpart may be equivocal, the relationship between the total visual scene and the total pattern of proximal stimulation is always unique. The distal-proximal relationship properly defined is unambiguous. Although secondhand accounts of Gibson's theory frequently focus on Gibson's psychophysical hypothesis, it is my view that this claim concerning the specificity of the distal-proximal relation constitutes Gibson's special contribution. Obviously, without the psychophysical hypothesis Gibson would not have a theory of perception; nevertheless, the psychophysical hypothesis in different form was anticipated by others including Descartes (1637). In fact, Epstein and Park (1963) suggested that the hypothesis may be considered a reconstructed "constancy thesis" with the stimulus redefined and percepts replacing sensations. In contrast, the Gibsonian view of the distal-proximal relationship represents a daring break with a centuries-old assumption and makes it possible for Gibson to offer a stimulation-based theory of veridicality.

Proximal Stimuli and Perception

No perceptual phenomena are better suited than the constancies for drawing attention to the necessity for careful scrutiny of the proximal-perceptual relationship. If, in agreement with the practice of the conventional analysis, we consider only the local retinal counterpart of the distal property, for example, the retinal subtense associated with a particular sized object, we find that the percept displays a notable degree of independence from the retinal counterpart. Two examples will suffice:

1. Displacement of the image over the retina will be accompanied by perceived object movement if the eye is stationary, whereas the same retinal displacement will be associated by no perceived object motion if the observer has moved the eyes by an amount equal to the retinal displacement. Also, absence of retinal displacement will be the occasion of perceived movement if the eyes have been moved, but no perceived movement will be reported if absence of retinal displacement is accompanied by absence of eye movement.
2. The retinal orientation that is associated with an object's orientation in the frontal plane will change each time the observer changes his viewing position, but under typical conditions of observation perceived orientation will remain the same despite the retinal rotations. Also, two objects with different orientations in space will have identical retinal orientations if they are viewed successively with different head positions, but under typical conditions two different orientations will be perceived corresponding closely to the orientations of the objects.

The independence of perception from local retinal events exemplified above is characteristic of all perceptual discriminations. Although this fact may have been recognized from a very early time, it does not seem to have troubled students of perception before the seventeenth century.* Hamlyn (1961) has provided a use-

*There are, of course, exceptions to this statement; for example, in an influential treatise Porterfield (1759) writing about the perception of velocity claimed strict dependence of perceived absolute velocity on angular velocity or the rate of retinal velocity: "When the Body moves transversely in a Plane parallel to our eyes, we judge of its Motion from the Motion of its Image on the Retina . . . " (p. 421). (Consequently), "if two Objects, unequally distant from the Eye, move with equal Velocity, the more remote one will appear slower; or if their Celerities (speeds) be proportional to their Distances, they will appear equally swift" (p. 422, parenthesized words not in original). A more recent example is provided by W. H. Lawrence writing on the laws of perspective in the fourteenth edition of *Encyclopedia Britannica* (1929). Thus "Axiom 2. Parallel lines appear to approach one another as they recede from the eye. . . ." Lawrence has converted the prescriptions of the geometry of perspective into descriptions of perceptions.

ful survey of the notions of sensation and perception in the history of philosophy beginning with the classical Greek philosophers and it appears that until the seventeenth century neither philosophers concerned with epistemology nor those concerned with developing a causal theory of perception took explicit cognizance of the special nature of the proximal-perceptual relationship. Several factors may have contributed to full recognition of the problematic nature of the proximal-perceptual relationship (Hatfield & Epstein, 1976). One may have been the discovery of the reinal image and the growth of understanding of the dioptrics of the eye which were provided by the work of Descartes, Kepler, and Scheiner early in the seventeenth century. Another factor may have been the growth of Empiristic philosophy which stressed that knowledge originated in perception and led to consideration of the relationship between perception and the physical world. Finally, there was the wish among some philosophers and scientists to develop a scietific causal theory of perception. In charting the sequence of events or stages of processing culminating in conscious perception the contribution of the retinal image came under scrutiny.

Only one theorist demurs from the consensus about the independence of perception from stimulation. In the view of J. J. Gibson perception is directly a function of retinal stimulation. Whereas the fact of perceptual constancy coupled with the presumed fact of variability of retinal stimulation suggested to other theorists that to explain perception we are required to take into account more than stimulation, Gibson draws a different inference. The facts of constancy suggest to him that there must be invariant properties of stimulation that control perception, thereby ensuring constancy. Perception is directly dependent on stimulation and the facts of the constancies point to the existence of a particular useful property of stimulation: higher order invariants.

Distal Stimulation and Perception

An adaptive perceptual system is geared to extract useful information about the environment. The constancies seem to exemplify this attainment. Despite the continuous transformations of optical input, the mechanisms of the perceptual system operate to guarantee constancy. The adaptive utility of this achievement cannot be exaggerated. The perceptual world without constancies would be a bewildering experience that would make the acquisition of adaptive behavior extremely difficult. Their adaptive significance is another reason for the sustained theoretical interest in the constancies. Nevertheless, it would be incorrect to attribute great prominence to considerations of adaptive significance.* Most in-

*Emphasis on the adaptive utility of the constancies can lead to admission of teleological explanations. In *The Analysis of Sensations* Mach (1959, pp. 80-82) considers the potential benefits and risks associated with a teleological approach to understanding perception.

vestigators of the constancies are interested more in constancy mechanisms or processes than in constancy as a product. Nowhere is this more evident than in the frequent attempts to confirm hypotheses about the basis of constancy through experiments on illusions. The general opinion is that although constancy and veridicality may be distinguished from illusion on the basis of extraperceptual criteria, that is, conventional assessments of correspondence between reports and physical measurement, veridicality and illusion differ neither in the character of the experience nor in the underlying process. The illusory size appearance in a Distorted Room is experienced as directly as the veridical size appearance of this book and the mechanism of size perception is the same in both cases. Thus it is plain that it is constancy and not veridicality that is the focus of concern and it is the mechanism and not the achievement that commands interest. Koffka reflects this position in the following statement:

> . . . the constancy problem should be reformulated in this way: What shape, size, brightness will correspond to a certain local stimulus pattern under various total external and internal conditions? Once we have answered this question we shall know when to expect constancy, when not. Indeed some effects of non-constancy are just as striking as the effects of constancy . . ." (1935, p. 227).

The tendency to attribute illusions and veridicality to the same process is common to theoreticians of widely divergent views. It is also an old practice. Berkeley's treatment of the Barrovian Case (1709, Sections 29-30) is an illustration. In analyzing this case, Berkeley attributes an illusory perception of distance to image blur, the same stimulus that in Berkeley's theory typically suggests accurate estimation of distance. This tendency is also clearly present in Helmholtz's (1890) writings as well as in the work of the modern-day empiricists.

Two exceptions are worth noting, however. Again Gibson separates himself from the rest by insisting that illusions are peculiarities that occur only under special conditions having little in common with normal conditions of stimulation. Accordingly, the principles that determine illusory perception are not the principles that govern veridical perception. Veridicality is an important feature of perception in Gibson's analysis and constancy is an obvious exemplification of this tendency toward veridicality.

The other exception is Brunswik (1947) who attributed great theoretical significance to the capacity of the visual system for discriminating the properties of the environment. Emphasis on perceptual achievement or "distal focusing" is the principal theme of Brunswik's psychology. It follows that he found the constancies to be of paramount theoretical interest. His emphasis on constancies as achievement is manifest in his advocacy of correlational techniques and in his development of a quantitative expression for the degree of constancy. The Brunswik Ratio expresses the accomplishment of the visual system by locating the percept between two poles designating the proximal and distal values of the to-be-discriminated property. The closer the percept approaches the distal value, the greater the degree of constancy yielded by the Brunswik Ratio. It is fair to say that Brunswik exhibited greater concern with assessment of the constancy achievement and establishment of the generality of constancy, for example, his treatment of perceived weight density, volume, and value, than in elucidation of the underlying process.

THEORIES OF PERCEPTUAL CONSTANCY

The Learning Approach

One approach to explaining the constancies proposes that constancy is the product of learning that begins in infancy. Historically, the favorite mechanism of learning has been association. One learning theory clearly implied by Berkeley (1709) presumed that the constancies are learned through a process of association between visual and haptic-kinesthetic stimulation. Ostensibly, associations between constant tangible properties and visual stimulations are formed so that the optical input comes to "suggest" the tangible property. It may be said that visible objects are not constant; they only suggest the tangible properties that are.

This account may be difficult to accept because neither the associative process nor the terms in the association (the optical variable and tangible associate) are present in consciousness. Berkeley was aware of this reservation and he used a linguistic analogy to demonstrate the plausibility of this state of affairs: he argued that the circumstances are similar to the experience of reading. Ordinarily the reader is unaware of the written word as an orthographic structure per se. Nor is the reader usually aware of the mediating events that originally created the meaning of the words. Instead, it is as if he perceived the meanings directly. It is the same with sense data (Berkeley's "proper objects of sight"), the tangible associates, and visual experience. In both cases the great frequency of occurrence of the association coupled with the fact that the percipient is distally oriented serves to inhibit awareness of the associative process. Helmholtz, whose views are similar to Berkeley's, offers the same explanation of the absence of awareness of the visual sense data and the mediating tactual events:

> The connection between the sensations and external objects may interfere very much with the perception of their simplest relations . . . the practice of associating them with things outside of us actually prevents us from being distinctly conscious of the pure sensations (1890, p. 9).

Although this argument may suffice to allay the doubts to which it is addressed, there is an objection that neither Berkeley nor Helmholtz dealt with satisfactorily. The thesis that constancies are learned by association with touch implies that haptic sensory input is not subject to the ambiguities and shortcomings that set the problem for visual perception. This assumption certainly is questionable. Tactual sensory input, to be a guarantor of constancy, requires as much processing as optical input. The tactual sense is no more capable of providing direct access to object properties than the visual sense. To this objection we should add that contemporary experimental evidence has given little support to Berkeley's thesis [for a review see Rock, 1975, pp. 377-382; for a particularly pertinent study see Bower, Broughton, & Moore's (1970) experiment with young infants].

A view similar to Berkeley's was favored by Helmholtz who proposed that the percepts of vision are acquired on the basis of association between optical input and nonvisual inputs; for example, touch, inflow from, and outflow to muscles. In fact, contemporaries of Helmholtz (e.g., Sully, 1878) considered his theory in direct line of descent from Berkeley. The thesis also has contemporary exponents. Probably the most radical expression is to be found in Taylor's (1962) behavioristic theory. With appropriate allowances for notational preferences and the vernacular of conditioning, Chapter 6 of Taylor's book which describes the acquisition of the constancies could have been written by Berkeley.

Although he differs in important respects from proponents of a strict empiricistic approach, Piaget (Piaget & Inhelder, 1969) also subscribes to the doctrine that touch educates vision:

> Hence (size) constancy makes its appearance . . . after the coordination of vision and prehension . . . the size of an object is variable to the sight but constant to the touch . . . . It is no accident, then, that constancy of size makes its appearance after and not before the coordination of vision and prehension; although of a perceptual nature, constancy thus depends upon the overall sensori-motor schemes (Piaget & Inhelder, 1969).

A different type of learning theory was proposed by James (1890). The theory begins with the assumption that there is a "normal" viewing position and that the appearance of an object in this position has "extraordinary pre-eminece." A memory trace of the appearance of the object in this position is stored. All retinal transformations serve as stimuli to elicit this memory trace, which, once activated, is so strong that it replaces the sensation that might otherwise result from the present input. Consider the case of shape constancy that James develops in some detail (1890, Vol. 2, pp. 237-240). On different occasions a circle will project very different shapes. In this family of projections the circular shape is preeminent because it is the projection that obtains when the form is viewed under optimal conditions, that is, in the frontal plane. In this position,

> we believe we see the object as it *is*; elsewhere only as it seems . . . we must learn that when we get one of the former (other, projective transformation) appearances to translate it into the appropriate one of the latter class (normal appearance or circle); we must learn of what optical "reality" it is one of the optical signs. Having learned this . . . we attend exclusively to the "reality" and ignore . . . the sign by which we came to apprehend it (p. 239).

A similar argument is developed by James for size constancy. The apparent size of the object at arm's length becomes the preeminent size in memory into which all other sensations of size are transmuted:

> Out of all the visual magnitudes of each known object we have selected one as the REAL one to think of, and degraded all the others to serve as its signs. This "real" magnitude is determined by aesthetic and practical interests. It is that which we get when the object is at the distance most propitious for exact visual discrimination of its details. This is the distance at which we hold anything we are examining. Farther than this we see it too small, nearer too large. And the larger and the smaller feeling vanish in the act of suggesting this one, their more important meaning. As I look along the dining table I overlook the fact that the farther plates and glasses *feel* so much smaller than my own, for I *know* that they are all equal in size; and the feeling of them, which is a present sensation, is eclipsed in the glare of the knowledge, which is a merely imagined one (James, 1890, Vol. 2, pp. 179-180).

It appears that James' hypothesis has not been taken up seriously by subsequent writers, although the various elements of his theory can be identified in other works. Contemporary attempts to demonstrate the effectiveness of the cues known size for distance and known shape for slant may reflect implicit acceptance of the premise that there is a memory trace of a preeminent apparent size or shape. There is also considerable reliance on the hypothesis of a memory trace of image size-distance combination in Rock's (1975, pp. 71-73) analysis of size perception and in Rock's discussion of adaptation to a miniaturized scene (1966, pp. 159-164; 1975, pp. 73-75). Finally, Hering's (1920) inclusion of "memory colors" as one of the determinants of constancy is an obvious case in point.

The notion that there is a "normal" condition for viewing is a necessary feature of Helmholtz's rule: ". . . such objects are always imagined as being present . . . as would have to be there in order to produce the same impression on the nervous mechanism, the eyes being used *under ordinary normal condi-*

tions" (Helmholtz, 1890, Vol. 3, p. 2, my italics). The assumption of "normal" conditions for observations of distal properties also figures prominently in Katz's (1935) classical analysis of the perception of color:

> We believe that we apprehend the genuine colour of an object only under certain particular conditions of illumination. Neither twilight nor direct sunlight presents the genuine colour of an object. We must rather choose an intensity of illumination such as there is in the open air when the sky is slightly clouded. We shall call such intensity of illumination *normal* illumination (Katz, 1935, p. 83).*

In the case of shape constancy James's assertion that current input is translated into the normal appearance implies a mental operation by which shapes are rotated or displaced in mental space. Precisely this claim is made in the contemporary studies of the reaction time to rotated shapes by Shepard and his associates (Cooper & Shepard, 1973; Shepard, 1975; Shepard & Metzler, 1971). Although Shepard does not apply his findings to the analysis of shape constancy, the results are interpreted as evidence that man can execute mental rotations in imaginary space. Under certain conditions these operations of mental rotation are executed to judge whether two shapes that are oriented differently are the same or to determine whether a single rotated form matches the stored representation of a familiar form in long-term memory.

The Algorithm Approach

The algorithm approach to the constancies proposes that the visual system operates according to rules of processing that combine variables to generate constancy of perception. One set, which may be called the intrinsic variables, corresponds to the local retinal correlate of the to-be-discriminated distal property; for example, the retinal orientation or the retinal subtense. The other variables may be called the extrinsic variables and correspond to information about the organismic or environmental conditions that contribute to the variability of the intrinsic variables; for example, position of the head or distance of the object

*The "normalcy" assumption also is employed by the Gestalt psychologists but in a manner very different than the abovementioned writers. For the latter normalcy is determined by frequency of occurrence and optimization of the conditions of viewing. Gestalt theory explicitly rejects this notion (Koffka, 1935, pp. 221-22) in favor of a dynamic definition of normalcy. Thus, in discussion shape constancy, the frontal plane is designated as the normal orientation because this plane "is well balanced within itself so that special forces are required to dislocate it. In such a plane stimulus patterns would produce perceptual patterns according to the most simple laws" (Koffka, 1935, p. 231).

from the observer. Consider the case of perceived size. That distance is taken into account in perceiving size is implicit in Ptolemy's (ca. A.D. 150) explanation of the moon illusion and is given fuller expression by Descartes (1637) in his *La Dioptrique*:

> . . . concerning the way in which we see the size and shape of objects, there is no need for me to say anything in particular, other than that these characteristics are presumed from what we see of the distance and arrangements of the parts. In other words, their size is estimated by the knowledge or opinion we have of their distance as compared with the size of the image they cast at the back of the eye, and not absolutely by the size of these images, as is fairly evident from the fact that, whereas [the images of] objects [in the eye] may be a hundred times bigger than when they were [for objects] ten times further away, they still do not seem one hundred times bigger, but practically equal, as long as we are not deceived as to their distances (p. 000).

The size-distance invariance hypothesis is the contemporary version of this theory: a visual angle determines a unique ratio of perceived size to perceived distance. Analogous formations have been advanced for the other perceptual properties (see Epstein, 1973, for a review); for example, Helmholtz (1890) in his analysis of visual direction proposed that egocentric visual location depends on integration of retinal location and the felt position of the eye. Identical visual directions result if the algebraic sum of the retinal location and the eye position are the same.

The widespread acceptance of this type of explanation is due in large measure to its empirical successes and its broad applicability (Epstein, 1973; Epstein & Park, 1963; Hochberg, 1971; Rock, 1975). But there is probably another reason. The taking-into-account process operates as if under the control of an internalized logic which is isomorphic to the structure of the world of light and objects. This assertion is easiest to comprehend in the perception of size at a distance and shape at a slant. It is as if an invisible geometer were at work, following rules of a natural geometry that correspond to Euclidian geometry. Of course, no contemporary investigator maintains this view. Nevertheless, part of the appeal of the algorithm approach is that it imports into the visual system a set of familiar and intuitively reasonable rules.

It is of historical interest that the notion of "natural geometry" was widely advocated without reservation by many writers before Berkeley; for example, Kepler (1604), in discussing the basis of distance perception, spoke of the operation of a "distance measuring triangle," an internal surveyor's triangulator. In the same vein Descartes (1637) asserted that perception involves "intricate reasoning similar to that done by geometers." Berkeley (1709) objected to the doctrine of natural geometry for two reasons:

1. The doctrine was based on the mistaken reification of geometrical or mathematical fictions; for example, light rays: " . . . those lines and angles have no real existence in nature, being only an hypothesis framed by mathematicians . . . " (Section 14).

2. The doctrine implied that a percept was computed on the basis of inputs which were themselves never experienced, for example, the direction of light rays from the borders of objects, but "no idea which is not itself perceived can be to me the means of perceiving any other idea" (Berkeley, 1709, Section 10). Berkeley's objections notwithstanding, the notion of "natural geometry" continued to appeal to certain writers. For example, Leibnitz, Berkeley's contemporary, proposed that we see "by dint of reasoning about rays according to the laws of optics."

Although the classical doctrine of natural geometry was eschewed by the investigators of the last century, Helmholtz's characterization of perception as unconscious conclusions based on unconscious inferences, as well as later variants of this characterization, has strong kinship with the mode of processing proposed by Kepler, Descartes, and Leibnitz. Because of its seemingly paradoxical implications, the term "unconscious inference" has been subjected to frequent criticism, but the peculiar state of affairs which led Helmholtz to introduce the term cannot be ignored.

Although the algorithm theory proposes that the perceptual system operates according to ratiomorphic rules, the operations are not spontaneously present in consciousness. Indeed they often cannot be experienced even when attention is directed to them. Compare the experience of computing the length of one of the legs of a right angle, given a particular acute angle, and length of the other leg with seeing the size of an object at a distance. It may be objected that the perceptual process is so overpracticed and rapid that observation of the process is made difficult. The geometric analogy on this account is not entirely appropriate. Even were this to be admitted, we must recognize that the perceptual situation is more peculiar than we have noted so far. Not only is it unnecessary for the operations to be conscious and deliberate, it is even unnecessary for the information that constitutes the raw material of the algorithm, the premises in Helmholtz's inferential process, to be conscious. That consciousness is not required is obvious in the willingness of advocates of the algorithm approach (e.g., Eissler, 1933; Holaday, 1933, Klimpfinger, 1933) to distinguish between registered (unconscious) and perceived values of extrinsic variables and to aver that it is the former that enter into the algorithm. In this case the conscious percept is attributed to an unconscious process operating on inputs that have no conscious correspondents. In itself this is not a manifestly unpalatable offering nor is it without precedent. What causes discomfort is the clear disposition among proponents of algorithm theory to refer to the processes *as if* they were conscious, cognitive and like reasoning. The ontological nature of the perceptual algorithm is a matter requiring clarification.

The Psychophysical Approach

According to the psychophysical approach, perception is the direct result of invariant retinal stimulation. Neither learning nor ratiomorphic processing are necessary. It is difficult to find exact historical precursors of this view. If by perception we intended "sensation," then it could be said that the psychophysical hypothesis was anticipated by the constancy thesis. But this would disregard the fact that it is constancy in immediate perceptual experience that concerns us.

This approach owes its development largely to the work of contemporary investigators (Gibson, 1950; Wallach, 1939, 1948, 1959). Their thesis is that despite transformations of the extrinsic variable (e.g., positon, illuminance) an aspect of the retinal stimulation remains invariant. Obviously this aspect is not the local retinal counterpart of the to-be-discriminated property. Gibson and Wallach do not dispute the contention of the conventional analysis that the local proximal counterpart is unreliable, but they argue that if the relevant stimuli are construed differently then invariant reliable aspects of stimulation can be found. These are higher order variables such as ratios and gradients. They hypothesize that any transformation of position or other extrinsic variable that does not alter the higher order variable will leave perception unaltered. Constancy prevails because the stimulus is constant.

Thus whiteness constancy does not result because illumination is registered and taken into account, as the algorithm approach suggests or because of past experience, as Helmholtz advocates. Whiteness constancy results because transformation of the illumination of the entire field affects local luminance but it does not affect ratios of luminance (Wallach, 1948). Luminance ratios are the determinants of perceived whiteness so that if the luminance ratio is constant apparent whiteness will also be constant. A strictly analogous explanation has been offered for the constancy of perceived velocity (Wallach, 1939, 1959). This constancy may be attributed to the fact that alterations of viewing distance transform not only the angular extent of the movement path, and thus angular velocity, but the angular extent of the framework of movement as well and both are transformed in the same proportion. As a consequence, relative angular velocity remains constant, ensuring speed constancy. The approach may also be applied to size constancy. The perceived size of an object is determined by its retinal size in relation to the field of retinal stimulation (Gibson, 1950). Consider a chair viewed from various distances. The size of the image of the chair will vary inversely with distance but so will the density of the gradient of stimulation associated with the floor. Because the proportional variation with distance will be the same, the extent of occlusion of the retinal field provided by the chair will remain constant and so will perceived size. No recourse need be made to learning or a size-distance algorithm.

One question occurs immmediately. What does the psychophysical hypothesis have to say about the types of covariance between percepts that have usually been offered as evidence for the algorithm approach? How is lawful interdependence between judgments of the focal distal property and judgments of the extrinsic variable to be understood? Why, for example, are overestimations of size accompanied by overestimations of distance. In the case of perceived whiteness why does viewing the target in "concealed shadow" lead to misjudgments of illuminance and whiteness, that is, overestimation of illuminance and underestimation of whiteness?

The answer is that correlational relationships do not necessarily imply causal relationships. Thus size and distance perception may covary because the percepts depend on partly correlated populations of retinal variables. In this case size and distance perception would not be causally related but only related through the mediation of partly over-lapping sets of stimuli. It is startling to find precisely this argument in Berkeley's analysis of perceived size:

> . . . the same extension at a near distance shall subtend a greater angle, and at a further distance a lesser angle. And by this principle, we are told, the mind estimates the magnitude of an object, comparing the angle under which it is seen with its distance, and thence inferring the magnitude thereof. What inclines men to this mistake (beside the humour of making one see by geometry) is that the same perceptions or ideas which suggest distance, do also suggest magnitude. But if we examine it, we shall find they suggest the latter, as immediately as the former. I say they do not first suggest distance, and then leave it to the judgment to use that as a medium, whereby to collect the magnitude, but they have as close and immediate a connexion with the magnitude, as wtih the distance; and suggest magnitude as independently of distance as they do distance independently of magitude (1709, Section L111).

The concealed shadow effects have two independent consequences and may be explained along similar lines:

1. Because the shadow is confined to the target, rather than shading the entire field, a new luminance ratio is created, thereby eliciting a new whiteness percept, that is, a darker appearance.

2. Because the shadow is deceptively exposed, S assumes that the illuminance level is uniform over the field and at the level of the background, thereby resulting in overestimation of target illuminance. Accordingly, although this pair of judgments seems to reflect interdependence in conformity with the algorithm theory, the judgments may also be handled within the framwork of the psychophysical hypothesis.

It may seem paradoxical to find Berkeley, the archempiricist, offering the same argument that has been presented on behalf of the advocates of psychophysical theory. The paradox vanishes, however, if a misunderstanding of Berkeley, which is common in secondary sources (e.g., Boring, 1942), is corrected. Although Berkeley emphasized learning, it was not his opinion that the process underlying perception was active or ratiomorphic. Instead he conceptualized it as strictly passive and associative. The relationship between sensory stimuli and perception, according to Berkeley, was "suggestion," a passive elicitation based on association. Because of this conceptualization, Berkeley felt impelled to reject the causal construction of the size-distance relationship with its ratiomorphic implications. In large measure, although not entirely, passivity characterizes the perceptual process according to the psychophysical approach. Thus the psychophysical theory and Berkeley's theory agree on this point and this makes sense of the concordance of their views on the size-distance question.

There is another set of observations that may pose even greater difficulties for the psychophysical approach to the constancies. The case of reversible perspective figures illustrates the problem. These figures, brought to the attention of the scientific community by Necker (1832), appear to undergo reversals in depth during the course of uninterrupted inspection. It is obvious that such occurrences cause problems for perceptual theories that stress exclusive dependence on retinal stimulation. In fact, the Gestalt psychologists (Koffka, 1935; Kohler, 1929) were quick to seize the opportunity to criticize such theories on these grounds. Here I wish to draw attention to another aspect of these phenomena. As a rule, *both* depth and size covary in a lawful manner when reversals occur. It is easy to see how a constancy mechanism postulated by the algorithm theory accounts for this fact, but it is not so evident how the psychophysical theory of size perception would manage. To recast this difficulty in more general terms, the psychophysical theory has difficulty in accounting for covariation of percepts that occur in the absence of alterations of optical input. An excellent discussion of this difficulty is presented by Hochberg (1974) in his evaluation of Gibson's theory.

THE EARLY EXPERIMENTS

The brief excerpts from the writings of Descartes and Berkeley make it clear that some of the constancy phenomena were known long before the advent of experimental psychology. In particular, the phenomenon of size constancy was noted often by the early writers. (For a pre-Cartesian description, see Peckham's *Perspectiva communis*, ca. 1480, which is available in English translation.) Despite awareness of the constancies as perceptual phenomena, there was little or no experimental assessment of the phenomena. The first systematic studies of constancy did not occur until the latter half of the nineteenth century when Martius (1889) examined size constancy. Systematic studies of whiteness constancy were first reported by Katz in 1911, of shape constancy by Thouless in 1931, and of speed constancy by Dembitz in 1927. These investigators set out to measure constancy under controlled conditions of observation, using the psychophysical procedures of the day. In a sense, the early experiments supplied operational definitions of the constancy phenomena. Rather than relying on naturalistic variation and observation, these investigators deliberately introduced controlled variations of the extrinsic variables, for example, viewing distance, orientation, and level of illumination, and assessed the degree of constancy of the focal percept as the extrinsic variable was altered. In the following paragraphs the principal achievements of the early research are summarized.

Just as a number of the constancy phenomena were familiar to students in the preexperimental period so was there an awareness that the degree of constancy depended on the conditions of viewing. One of the objectives of the early experiments was an examination of this dependence. Attention quickly focused on two types of determinant: environmental and subject. The prinicpal environmental determinants were the cues that determine the phenomenal correlates of the extrinsic variables; for example, the cues for perceived distance. Two experimental procedures were introduced to study the effects of cues. In the older procedure cues are gradually added, starting with one and accumulating others on successive tests. The original studies of the "alley problem" illustrate this procedure. In the later studies (e.g., Holaday, 1933, and Eissler, 1933) the procedure was reversed. On the initial test the full complement of cues is present and successive tests are conducted under conditions that omitted cues one by one. The Holway-Boring (1941) study of size perception is a familiar example. It was Brunswik's (1947, p. 25) contention that the difference between the accumulation and omission procedures represents an important difference in theoretical orientation. Although this assertion is repeated by Postman and Tolman (1958), the basis for the claim is unclear. There is no evidence that the results of the two procedures differ. In any event, both procedures have been superseded by counterbalancing the order of presentation of the various conditions. Two principal findings emerged from the early studies: (a) Addition of cues enhances constancy and omission of cues reduces constancy. (b) In their effect on perception of the focal property there is a tendency for the cues to be interchangeable; that is, as long as a cue or complement of cues reliably determines perception of the extrinsic variable (e.g., distance, orientation), the effect on perception of the focal property (e.g., size, shape) will be the same for different cues or constellations of cues. Although this tendency can be discerned in the results of the early experiments, and was remarked on by Brunswik (1947) experiments explicitly designed to examine the intersubstitutability of cues have never been executed nor were the implications of this result explored.

Although the dependence of constancy on environmental variables was established early, the dependence of constancy on the perceptual (conscious or simply registered) consequences of these variables was not explicitly examined until much later; for example, although Katz (1911) established the dependence of perceived whiteness on the conditions of illumination, he never measured perceived illumination. Therefore he could not present direct evidence of the correlation between perceived whiteness and perceived illumination. Similarly, although Thouless showed that perceived shape depended on the conditions for observing orientation, he never measured perceived orientation. A systematic investigation of the correlation between perceived shape and perceived orientation was not undertaken until Stavrianos' (1945) study, although fragmentary data were made available by Eissler (1933). Even in the case of size constancy, Koffka, as late as 1935, observed that direct evidence bearing on the exact relationship between perceived size and perceived distance was lacking.* Nor did the otherwise admirable study by Holway and Boring (1941) redress the situation, for perceived distance was not assessed. Judging from Kunnapas' (1968) results, had Holway and Boring secured reports of apparent distance, they would have found that apparent size and apparent distance covaried lawfully. It is not possible to say with confidence why the relationships between estimates of the intrinsic or focal attribute and the extrinsic property were not examined during the early period. One possibility is that formulations of the algorithms for constancy in terms of percept-percept relations seemed to imply strongly that constancy depended on cognitive operations. This implication was not likely to be received happily by the Gestalt-oriented psychologists who were in the forefront of research on the constancies. Another reason may have been a tendency to dismiss as scientifically unrespectable an account of one percept which rested on its purported dependence on another percept. The implicit rule may have been that percepts. cannot cause percepts. Whatever the reasons may have been for the neglect in the earlier work, the assessment of percept-percept relationships within the framework of the constancy experiment has been one of the principal concerns of the last 25 years.

The subject variable that affected the degree of constancy was the observer's attitude of observation. Writers during the preexperimental period had noted that three different aspects of an attribute could be distinguished: the distal, the proximal, and the perceptual. These distinctions, plus one added between the

*Some readers, familiar with the older literature on the size-distance relationship that attends oculomotor adjustments, as well as the visual disorders known as micropsia and macropsia, may wonder whether Koffka's assessment is justified. The point to be remembered is that although a relationship was recognized by the early writers it was not systematically examined. In fact, it was recognized early (Donders, 1864) that these conditions pose special difficulties insofar as size-distance relationships are concerned and may have deterred detailed examination.

perceptual and judgmental modes, are explicit in the writings of Helmholtz (1890) and Hering (1879). The distinctions were vigorously emphasized by the Gestalt psychologists who insisted that the proper concern of a psychology of perception is to examine and explain perceptual reports. For this purpose the phenomenological method was advocated. The observer was urged to report how the world looked to naïve immediate observation. Careful instructions to induce the phenomenal attitude of viewing became critical to sound experimentation. The next step, due largely to Brunswik and his students, was the recognition that comparisons among judgments secured under various attitudes of viewing can provide useful information about the perceptual process. Brunswik compared size judgments secured with four instructions: "naive perceptual," "analytical perceptual," "realistic betting," and "analytical betting." These instructional variations, which direct the observer to immediate impressions, perspectival representation, and objective and projective (retinal) properties, proved to be highly potent sources of variance in the judgments secured in constancy experiments.

As the scope of the constancy experiments broadened to include an increasing number of perceptual properties it became clear that relative constancy was a pervasive feature of perception. At the same time it became clear that perfect constancy was a rare occurrence. The most common result of laboratory studies was a perceptual compromise, a reported percept that seemed to fall between two poles defined by the proximal and distal aspects of the focal property.

The most convincing demonstrations of this tendency were Thouless' (1931, 1932) studies of constancy. It was Thouless who coined the term perceptual compromise to refer to this outcome. This term, coupled with the use of the Thouless-Brunswik ratio to express the degree of constancy, leads to the view that the final percept is the result of compromising two competing initial percepts—one correlated with the distal stimulus and the other with the proximal stimulus. One obvious difficulty with this construction is the occurrence of deviations from constancy that do not fall between the proximal and distal poles. Overconstancy is one such deviation. Another way of interpreting the occurrence of the compromise (e.g., underconstancy) was to attribute it to underestimations of the extrinsic variable; for example, in Thouless' shape constancy experiment the perceived shape fell between the proximal and distal poles, not because of a compromise process but because orientation was underestimated. Koffka favored the latter account. This should not be surprising because the alternative implies that at some level there is a one-to-one correspondence between the local proximal stimulus (e.g., a projective shape) and a percept. This was a claim the Gestalt psychologists were determined to reject. In any event, one consequence of the experimental observations was that it became plain that in designing an explanation of perfect constancy a theory was need that could also account for the frequently occurring departures from constancy.

CONCLUSION

This essay has presented an overview of the history of interest in the perceptual constancies which spanned a period that saw psychological investigators leave the armchair for the laboratory. Obviously some aspects of the question have been neglected and others merit more intensive consideration, but because the primary emphasis in this volume is on contemporary work more extensive analysis would be out of place. The present treatment is sufficient to exhibit the continuity of theory and experimentation in this area. The familiar aphorism cautions us that he who is ignorant of history is destined to relive it. To this prognosis might be added that he will relive it without knowing that he does so. Perhaps this essay will serve to lessen these risks.

REFERENCES

Berkeley, G. *An essay toward a new theory of vision.* 1709.

Boring, E. G. The Gibsonian visual field. *Psychological Review,* 1952, 59, 141-148.

Boring, E. G. *Sensation and perception in the history of experimental psychology.* New York: Appleton, 1942.

Bower, T. G. R., Broughton, J. M., & Moore, M. K. The coordination of visual and tactual input in infants. *Perception & Psychophysics,* 1970, 8, 51-53.

Brunswik, E. *Perception and the representative design of psychological experiments.* Berkeley: University of California Press, 1947.

Cooper, L. A. & Shepard, R. N. Chronometric studies of the rotation of mental images. In W. G. Chase (Ed.), *Visual information processing.* New York: Academic, 1973.

Dembitz, A. *Beiträge zu experimentellen Untersuchengen der Bewegungswahrnehmungen durch das auge.* 1927.

Descartes, R. *La dioptrique,* 1637. P. J. Olscamp, trans. *Discourse on method, optics, geometry, and metereology.* New York: Bobbs Merrill, 1965.

Donders, F. C. *Accommodation and refraction.* (New Sydenham Society Translation). London, 1864.

Eissler, K. Die Gestalticonstanz der Sehdinge bei Variation der Objecte und ihre Einwirkungsweise auf den Wahrnehmenden. *Archiv für die gesamte Psychologie,* 1933, 88, 487-550.

Epstein, W. The process of "taking-into-account" in visual perception. *Perception,* 1973, 2, 267-285.

Epstein, W. & Park J. Shape constancy: Functional relationships and theoretical formulations. *Psychological Bulletin,* 1963, 60, 265-288.

Epstein, W., Park J., & Casey, A. The current status of the size-distance hypothesis. *Phychological Bulletin,* 1961, 58, 491-514.

Gibson, J. J. *The perception of the visual world.* Boston: Houghton Mifflin, 1950.

Hamlyn, D. W. *Sensation and perception: A history of the philosophy of perception.* New York: Humanities Press, 1961.

Hatfield, G. & Epstein, W. The sensory core: From Kepler to Titchener. Presented to Eighth Annual Meeting of Cheiron. Washington, 1976.

Helmholtz, H. von. *A Treatise on physiological optics, 1890 (Vol. 3). (J. P. C. Southall, Ed.* and Trans.). New York: Dover, 1962. (Originally published, 1925.)

Hering, E. *Beitrage zur Physiologie.* Leipzig, 1879.

Hering, E. Grundzuge der Lehre vom Lichtsinn. In W. Englemann, Ed. *Handbuch der gesammten Augenheilkunde.* Leipzig: Springer, 1920.

Hochberg, J. Higher-order stimuli and inter-response coupling in the perception of the visual world. In R. McLeod and H. Pick (Eds.), *Essays in Honor of James J. Gibson.* Ithaca: Cornell University Press, 1974.

Hochberg, J. Perception I: Color and shape. Perception II: Space and movement. In J. W. Kling and L. A. Riggs (Eds.), *Experimental psychology.* New York: Holt, Rinehart and Winston, 1971.

Holaday, B. E. Die Grossenkonstanz der Sehdinge bie Variation der inneren und aussern Wahrnehmungs bechingungen. *Archiv für du gesaemte Psychologie,* 1933, 88, 419-486.

Holway, H. H. & Boring, E. G. Determinants of apparent visual size with distance variant. *American Journal of Psychology,* 1941, 54, 21-37.

Ittleson, W. *Visual space perception.* New York: Springer, 1960.

James, W. *Prinicples of psychology.* New York: Holt, 1890.

Kaiser, P. K. Perceived shape and its dependency on perceived slant. *Journal of Experimental Psychology, 1967,* 75, 345-353.

Katz, D. *The world of color.* R. B. MacLeod and C. W. Fox, trans. London: Kegan Paul, 1935.

Kepler, J. *Ad Vitellionem,* 1604.

Koffka, K. *Principles of Gestalt psychology.* New York: Harcourt, Brace, 1935.

Kohler, W. *Gestalt psychology.* New York: Liveright, 1929.

Kunnapas, T. Distance perception as a function of available cues. *Journal of Experimental Psychology,* 1968, 77, 523-529.

Lawrence, W. H. Perspective. *Encyclopedia Britannica* (14th ed.), 1929.

Mach, E. *The analysis of sensations.* New York: Dover, 1959.

Martius, G. Ueber die scheinbare Grösse der Gegenstände und ihre Beziehung sur Gross der Netzhautbilder. *Philosophische Studiern,* 1889, 5, 601-617.

Piaget, J. & Inhelder, B. *The psychology of the child.* New York: Basic Books, 1969.

Porterfield, W. *A treatise on the eye.* Edinburgh, 1759.

Postman, L. & Tolman, E. C. Brunswik's probalistic functionalism. In Koch, S. (Ed.), *Psychology: A study of a science.* New York: Mcgraw-Hill, 1959.

Rock, I. *An introduction to perception.* New York: Macmillan, 1975.

Rock, I. *The nature of perceptual adaptation.* New York: Basic Books, 1966.

Shepard, R. N. Studies in the form, formation and transformation of internal representations. In R. Solso (Ed.), *Information processing and cognition. The Loyola Symposium.* Potomac, Maryland: Lawrence Erlbaum Associates, 1975.

Shepard, R. N. & Metzler, J. Mental rotation of three-dimensional objects. *Science*, 1971, 171, 701-703.

Stavrianos, B. K. The relation of shape perception to explicit judgments of inclination. *Archives of Psychology*, 1954, No. 296.

Sully, J. The question of visual perception in Germany. *Mind*, 1878, 3, 1-23, 167-195.

Taylor, J. G. *The behavioral basis of perception*. New Haven: Yale University Press, 1962.

Thouless, R. H. Phenomenal regression to the real object, I. *British Journal of Psychology*, 1931, 21, 339-359.

Thouless, R. H. Phenomenal regression to the real object, II. *British Journal of Psychology*, 1932, 22, 1-30.

Wallach, H. Perception of motion. *Scientific American*, 1959, 201, 56-60.

Wallach, H. Brightness constancy and the nature of achromatic colors. *Journal of Experimental Psychology*, 1948, 38, 310-324.

Wallach, H. On constancy of visual speed. *Psychological Review*, 1939, 46, 541-552.

CHAPTER

2

VISUOMOTOR COORDINATION IN VISUAL DIRECTION AND POSITION CONSTANCIES

WAYNE L. SHEBILSKE
University of Virginia

Visuomotor coordination is a two-sided coin. On the visual side coordination is required because visual information is dependent on and altered by self-produced movements; on the motor side coordination is required because eye, head, or body movements can be initiated by and controlled by visual information. This chapter focuses on the visual side of visuomotor coordination; we consider how self-produced movements of the retina affect the ability of the visual system to build an internal representation of the environment. It should become apparent, however, that the two sides of the coin are inextricable. Although the visual side can be the center of focus, it cannot be understood without some knowledge of motor control, especially eye movement control.

Visual contact with the environment depends on the ambient array of light that stimulates the retina. For a stationary observer the structure of the ambient array has a relatively simple relationship with the structure of the environment. Consequently it would be a simple matter for a stationary observer to build an internal representation of his environment, but when an observer moves a continual flux is created in the array of light that enters the eyes. Despite this flux perception is not usually disturbed by observer movements. Thus the visual system must be able to build an internal representation of the environment despite input perturbations caused by an observer's mobility. How does the visual system cope with self-produced optical motions? To answer this question we must first break it down into components that can be investigated in single experiments.

Because the eyes can move, a one-to-one correspondence does not exist between the visual direction of objects with respect to the eye as center and the self as center. The first is called *oculocentric* direction and is assumed to have the fovea as its origin (Rubin & Walls, 1969). The second, called *egocentric* direction, has its origin at the *egocenter* which is located behind the eyes in the proximity of the turning point of the head (Roelofs, 1959). When the eyes move, oculocentric direction of a stationary object changes but egocentric direction does not. Correspondingly, despite this change, *apparent* egocentric direction of objects tends to remain constant; for example, look at an object that is straight ahead and then direct your eyes to one side of it. No matter where you direct your gaze, the object continues to be straight ahead. This invariance of apparent egocentric direction despite changes in oculocentric direction is called *visual direction constancy* (VDC), a name chosen because apparent direction is the visual attribute that remains constant. Eye movements and VDC are one subdivision of the general question of perceptual stability.

In contrast to eye movements, which change oculocentric but not egocentric direction, head and body movements can change both oculocentric and egocentric direction; for example, look at an object that is directly in front of your head and then move your head laterally. After the head movement the object is no longer straight ahead. Its egocentric direction is changed. Correspondingly, the object *looks* like it is in a different egocentric direction. Notice an invariance of another kind. The object appears to be in the same position in space before and after head movements. This invariance of apparent object position despite changes in egocentric direction is called *visual position constancy* (VPC), a name chosen because apparent visual position is the perceptual attribute that remains constant. Head movements, body movement, and VPC are another subdivision of the general question of perceptual stability.

Experiments related to VDC during eye movements and VPC during head and body movements are discussed in separate sections, each of which is further divided into subsets of experiments that investigate specific problems. These subdivisions are based on distinctions that are useful in discussing current theories of perceptual stability, as is done in the last section of this chapter. We evaluate the reafference theory of von Holst and Mittelstaedt (1950), the taking-into-account theory of Epstein (1973), the higher order retinal cue theory of Gibson (1966), and the evaluation theory of MacKay (1973). It is argued that no one of these theories can account for all the features of perceptual stability considered.

EYE MOVEMENTS AND VISUAL DIRECTION CONSTANCY

For stationary eyes there is a simple relationship between the visual direction of an object and the retinal location of the image that it projects; for example, when an object moves, the image that it projects moves by a proportional amount. Similar image movements can occur when the eyes move and the object remains stationary. Therefore, although the perceptual system can supposedly encode retinal location (Bishop & Henry, 1971), other kinds of information besides retinal location must be processed before an unambiguous judgement of egocentric direction can be given.

Different theorists have suggested two kinds of cue system that could supplement retinal location information. According to one group of theorists (e.g., Bower, 1974; Gibson, 1957; 1966; 1968, Linksz, 1952), observers process higher order retinal cues that remain invariant with eye movements. On this account the location of retinal images is important only with respect to one another. Visual direction is not seen by means of the absolute retinal location of an image but rather by means of retinal cues that are provided by the structure of the ambient array. According to a contrasting point of view, absolute retinal location is processed along with extraretinal information that allows the observer to take his eye movements into account. When the eyes turn, images are displaced on the retina by an amount (measured in degrees of visual angle) equal in magnitude to the amount of eye turn (measured in degrees of ocular rotation). Hence a simple linear combination of these two cues could yield an invarient perception of egocentric direction in the face of eye movements (e.g. Epstein, 1973, Hill, 1972; Matin, 1972).

Experimental evidence shows that both the higher order retinal and extraretinal cue systems play a role in making judgments of static egocentric direction, but the evidence is not definitive for judgments of the apparent rest of objects with respect to the egocenter. The latter can be regarded as a dynamic component of VDC. The evidence is clear that retinal filters, which differ from higher order retinal cues, are major determinants of the dynamic component of VDC. There is conflicting evidence, however, about whether extraretinal information is taken into account to supplement the filters.

The evidence behind these generalizations is reviewed here in considerable detail. Each experiment investigates only part of the specific questions that are related to retinal and extraretinal effects or static and dynamic components of VDC. Hence it is useful to categorize them according to the specific questions they probe. The review that follows is divided into four sections which are tied together in the theoretical section at the end of the chapter: retinal-static, retinal-dynamic, extraretinal-static, extraretinal-dynamic.

Retinal Cues and Direction

Several theorists have noted that retinal information could provide unambiguous cues to visual direction regardless of eye position, but nobody has supported the actual use of these cues; for example, Linksz (1952) has proposed that static perspective information could provide cues to direction and Bower (1974) has suggested that the nose may provide an important visual direction cue. Neither of these interesting propositions has been tested, however. Therefore I decided to gather some evidence on each of them. I made informal observations to test Linksz's hypothesis and conducted an experiment to test Bower's.

Linksz (1952, p. 497) illustrates how static perspective cues are correlated with the visual direction of an object. The object he chose to illustrate was a lamp shade. At the horizon the base of a lamp shade would project a single horizontal line on the retina; above the horizon the bottom of the same shade would project an ellipse. The shape of the ellipse would indicate the direction of the shade with respect to the horizon. It seems most likely that such static perspective cues are employed in judging visual direction but evidence is lacking.

To get some idea whether static perspective could be used as a direction cue I made direction judgments of luminous objects with and without static perspective cues. Luminous targets were mounted on an optical bench. A continuous loop of string, which ran through pulleys, allowed me to move the targets horizontally in the frontal plane. The targets were a luminous ring that provided perspective cues and a luminous point source that did not. The ring was oriented perpendicular to the frontal plane. Except for the targets, the room was completely dark. There was no question that the ring could be set to the straight ahead much more accurately than the point source. This suggests that static perspective cues can be used to judge visual direction, but formal tests will be required before we can be sure.*

Another potential retinal cue to visual direction was proposed by Bower (1974) who noted that

> all objects that are straight ahead are symmetrically projected onto the retinas with respect to the projection of the nose and regardless of eye position. All objects that are not straight ahead are asymmetrically projected

*Naive observers could not set the ring better than the point. For informed subjects static perspective influenced cognitive rather than perceptual processes. Oculocentric orientation is influenced by egocentric orientation and direction. Therefore, an object's egocentric orientation would have to be remembered before static perspective could indicate direction. Static perspective does not supply a simple invariant relationship for the perceptual processes.

with respect to the nose. The relative symmetry of the projection of an object could thus serve as the stimulus to specify straight ahead (p. 54).

In other words, Bower has suggested that the retinal projection of the nose provides a reference that is used to judge visual direction regardless of eye position.

This is a fascinating hypothesis, but it seems unlikely in light of the adjacency principle according to which the effectiveness of cue relationships between objects is maximal when objects are in the same plane and reduced when they are separated in depth (see Gogel, this volume, for a general discussion). A large separation in depth usually occurs between the nose and any object that is judged with respect to direction. Therefore the nose should have little if any effect on the perception of direction according to the adjacency principle. However, the nose might be an exception because it is almost always present in the ambient array. Hence the nose hypothesis calls for an empirical test.

There are no published data to test Bower's hypothesis, but Steve Nice and I have made preliminary observations that are suggestive. In all conditions we asked subjects to set a luminous target so that it looked straight ahead of them. A continuous loop of cord ran through pulleys which allowed subjects to move the target in the frontal plane. Some of the conditions were designed to study the effects of an external reference on the straight ahead settings. The external reference was a luminous dowel which was 2 in. long and .25 in. in diameter. When the dowel was present, it was always placed in the straight-ahead direction, but the distance between it and the observer was varied. Four external reference conditions were tested on an equal number of trials: the dowel was placed directly in front of the target (at 182 cm), 68.25 cm in front of the subject, on the subject's nose, or it was absent altogether. Other conditions were designed to study the effectiveness of the nose as a reference to straight ahead. On half the trials the nose was not visible because there was complete darkness except for the luminous stimuli. On the other half of the trials the nose was illuminated by a small light that was focused from above. The light itself could not be seen. Black cloth was draped over the subject's shoulders and everything else that might have otherwise reflected light. The subject reported that he could see his nose, the luminous target, and nothing else when the nose was illuminated. Each observer made four straight-ahead settings in each of the eight conditions. The main dependent measure was the average absolute deviation of these four settings from the objective straight ahead. A bite plate containing dental impressions was mounted on a device that allowed fine adjustments in the x, y, and z axes. As a result the head was positioned precisely with respect to the objective straight ahead.

We found that subjects were slightly less accurate in setting a target to straight ahead when they could see their noses. However, this difference was negligible and not statistically significant. In contrast, there was a highly significant error reduction caused by seeing a reference near the target plane, or three-eighths of the way between the target and the observer. In agreement with the adjacency principle the effectiveness of a visual straight-ahead reference was maximal when the reference was near the target plane, less when it was three-eighths of the way between the target and the observer, and mininal for the nose which was the largest separation we tested. We call this the Pinocchio effect because the results suggest that the nose would be a critical cue to straight ahead if only the nose were longer.

It is not safe to generalize until more experiments are run, but it looks like the significance of retinal cues to direction may be determined by the adjacency principle. In accordance with that principle, informal observations support the functional significance of static perspective, or cues, which are based on relationships between contours that are adjacent or nearly so. Also, in agreement with the adjacency principle the Pinocchio Effect suggests that normal-length noses are relatively unimportant in judging the direction of objects that are separated in depth from the nose. Both the static perspective and nose hypotheses merit further tests. We shall not have a complete understanding of static judgments of visual direction until retinal cues are evaluated more completely.

Retinal Processes and Motion

The perception of motion with respect to the self can be considered a dynamic aspect of VDC in contrast to the static VDC judgments that we have been discussing. The distinction between these two components of VDC is best illustrated by an armchair experiment.

You can easily verify the observation that a luminous stimulus viewed in a dim light will streak, smear, or jump in the opposite direction of the eye rotation during voluntary saccades (Hill, 1972; Matin, Clymer, & Matin, 1972). This is a breakdown in the dynamic component of VDC. Although direct evidence is lacking, it appears that this dynamic malfunction does not affect the static component of VDC. You can judge this for yourself in a dim room by looking back and forth from an object on one side of a luminous stimulus (such as a lighted dial clock or TV screen) to another object on the other side. Dynamic VDC will probably fail; that is, you will see an illusory disturbance in the apparent stability of the luminous object during eye movements. Having experienced this, analyze static VDC. Look at the apparent direction of the lighted object and other objects before and after saccades or, better yet, have an assistant observe how well you can point to objects with your unseen hand. If you are like the people informally observed in my lab, static VCD will remain intact; you will have no trouble determining the visual direction of any objects, including the luminous one that was unstable during eye movements. Clearly, dynamic and static VDC are distinct.

A closely related experiment suggests two possible retinal processes that could determine dynamic VDC. The experiment can be carried out in the same room as before. Simply turn the lights on and observe the same luminous object. If the added lighting is sufficient, the object will appear stable during saccades. How can the extra light make such a dramatic difference? We consider two possible answers: (a) the additional lighting illuminates higher order retinal cues that are used to give dynamic VDC; (b) the additional lighting facilitates the filtering out of motion signals during saccades by reducing the contrast ratio of the luminous object and by adding contours that increased masking effects during saccades.

It has been suggested that higher order perspective cues not only provide static cues to visual direction but also dynamic ones as well; for example, Gibson (1966, p. 39) notes that

> neural input caused by self-produced action is simply different from the neural input caused by an intruding stimulus. The two kinds of inputs are different in their sequential properties, they are different kinds of transformations or changes, and the simultaneous pattern of nerve fibers might be widely dispersed.

Self-produced and externally produced optical motions are distinct (Gibson, 1957, 1968). Hence the perceptual system could distinguish them if the observers could detect the relevant higher order retinal cues. Although we shall see that these cues play an important part in the dynamic aspects of VPC (i.e., during head and body movements), retinal cues are probably not even seen during saccades.

To convince yourself of this add one further refinement to the above experiment. Fold a 3 x 5 card in half to form an approximately 40° wedge. Cut a 1-mm wide slit out of a portion of the 3-in. fold. Hold the slit close to your eyes by resting it on your eyebrow and cheekbone. This simple apparatus will allow you to repeat observations similar to those described by Woodworth (1938). You can use the same stimulus objects we had in the above experiment. Position the slit so that you see the luminous stimulus when you look through the slit but not when you look at objects on either side of the target. Then look with a single eye movement from one peripheral object to the other. When the room is dim, you will see streaking, smearing, or jumping, as before, but when the room is light you will not see the object at all! If you do, have an assistant watch your eyes. You are probably making involuntary fixations in order to look through the slit. To make it more interesting have the assistant substitute other objects or reading material in place of the luminous target. With a single eye movement from one peripheral target to the other you will be unable to identify the objects that are substituted.

Having seen this, I think you will agree that it is unlikely that we derive retinal data during saccades in an illuminated heterogeneous visual field. Hence we can dismiss the idea that higher order retinal cues are processed at that time. It seems more likely that contrast ratios and masking contours are the important factors that determine whether an object is seen as stable during a saccade. This conclusion is in agreement with a filtering hypothesis according to which retinal information is filtered out.

Several factors contribute to this filtering process. One is the critical duration over which light is integrated. During saccades the target falls on each retinal element for a period shorter than this critical duration so that the contrast ratio between target and background is reduced and the threshold consequently rises. Mitrani, Mateeff, and Yakimoff (1970, 1971) referred to this as smearing of the retinal image and provided psychophysical evidence that it is a substantial factor in filtering out retinal motion signals during saccades. Physiological support for a related critical duration filtering device comes from Wurtz (1969a, b) who found that drift detectors of monkeys have a critical image velocity of about 150° per second above which their output falls to zero or becomes inhibitory. Additional reduction in retinal sensitivity is caused by mechanical strains set up in the retinal layers by the rapid acceleration during saccades (Richards, 1968).

Probably the most significant cause of retinal filtering is masking (e.g., Kahneman, 1968). The inhibition effects of temporally backward and spatially lateral masks during saccades have been demonstrated in psychophysical studies by Matin, Clymer, and Matin (1972); the forward, simultaneous, and backward inhibition effects of laterally displaced moving contours have been shown psychophysically by MacKay (1970) and Mitrani et al. (1971) and physiologically by Brooks and Holden (1973). In all these studies the masking spots or contours were laterally displaced for the sake of control and substantial effects were obtained. In a well illuminated heterogeneous visual field there also would be additional masking in which the mask and the inhibited stimulus fall on the same retinal area before, during, and after ocular rotations. Undoubtedly masking-induced inhibition of retinal signals is a primary factor in our seeing a stable visual world during saccades.

The preceding factors that contribute to VDC have been called retinal either because their effects depended on the retinal image (e.g., static perspective, retinal smearing, and masking) or because the effect is localized in the retina (e.g., mechanical strains). For the static components of VDC (perceiving egocentric direction) retinal factors actually provide information about direction. Therefore these retinal factors were called cues to direction. For the dynamic component of VDC, perceiving stability despite potential motion signals, the important retinal factors are not cues to motion. They are retinal filtering processes. Both retinal cues to direction and retinal filtering of motion signals probably play a significant role in VDC. Unfortunately there is little evidence on the functional significance of retinal cues to direction.

In the following extraretinal signals are considered in relation to perceiving egocentric direction and stability despite potential motion signals. Extraretinal signals are neither initiated by retinal signals per se nor are they localized in the retina. It is argued that they can provide essential information about direction.

Therefore they can be thought of as cues to direction. The essential information they provide is direction of gaze. The fidelity and sources of this eye position information are considered. It is also argued that extraretinal signals probably do *not* contribute significantly to the apparent rest of objects during saccades.

Extraretinal Cues and Direction

Researchers have isolated the effects of extraretinal cues to direction by measuring VDC without retinal cues; for example, Matin and his co-workers (cf. Matin, 1972) used a method of constant stimuli to determine the location of a target flash (comparison) that appeared in the same egocentric location as a fixation target (standard). The comparison was viewed before, during, or after a saccade; the standard was viewed and extinguished before eye movements. The stimuli were low intensity and separated by "sufficient" time to minimize or eliminate afterimages that could have made the discrimination one of "simultaneously viewed" stimuli. Without this control subjects could have judged relative visual direction instead of absolute visual direction. With this control we can rest assured that subjects compared the absolute egocentric direction of the standard and comparison. The retinal image location of the stimuli determined an oculocentric direction for each. When the eye moved between the presentation of the standard and comparison, the egocentric direction of the standard's location was represented by a different oculocentric direction. This provided an opportunity to measure VDC. If VDC were lacking, the comparison would have to stimulate the same oculocentric direction (i.e., the same retinal location) as the standard in order to be seen as the same egocentric direction. In contrast, if VDC were complete, the same physical position as the standard would represent the same egocentric direction despite changes in oculocentric direction. The results show a high degree of constancy.

The constancy was apparently caused by extraretinal signals because extraneous retinal information was excluded. Static perspective and nose cues were eliminated because the visual field was completely dark except for the experimental stimuli. Also, the potential influence of dynamic perspective cues was virtually eliminated because the only stimulus that was ever present during saccades was a single 1-msec test flash. Accordingly, the results suggest that extraretinal signals can supplement the ambiguous oculocentric information to yield a high degree of VDC.

Although we know that there is substantial VDC under reduced conditions, the exact amount has not been established. A few experiments provide relevant data but are not in complete agreement. Some of them (Monahan, 1972; Matin, 1972; Pola, 1973) use Matin's procedure, which is described above. Others (Hill, 1972; Morgan-Paap, 1974) use a procedure that was introduced by Hill. Results from the two procedures do not converge. Specifically, experiments using Matin's procedure show no systematic trend toward under- or overconstancy, whereas experiments based on Hill's procedure show a small but definite trend toward underconstancy. There are procedural differences that could account for the discrepancies, but until they are explored we shall not know the exact fidelity of extraretinal signals in VDC.

Experiments using Matin's procedure have been concerned with the temporal development of VDC and have not focused directly on the question of consistent under- or overconstancy. We can, however, pick out some data that are relevant, albiet not so complete as we would like. Matin (1972) plotted the extent of compensation for three subjects at 450 msec after the beginning of saccades to targets located 2° 11 min to the side. Estimating from Figure 10 in Matin's report, the results showed underconstancy for two subjects of about .17 and .67° and overconstancy for one subject of about .33°. Averaging over the three subjects yields a negligible underconstancy of .17°. Monahan (1972) used Matin's procedure to measure VDC about 63 msec after the beginning of saccades to targets that were displaced by 2° 11 min. The pattern of results was identical to those of Matin. Estimating from Figure 1 in Monahan's report, two subjects showed slight underconstancy of about .25 and .40° and one subject showed an overconstancy of about .25°. The average yields a negligible .20° underconstancy.

Pola (1973) reported the extent of compensation after saccades to targets located at 5 and 8°, the largest eye turns measured in Matin's paradigm. One subject was tested 200 msec after the saccade and showed an overconstancy of 1.0 and 1.57°, respectively. The other subject was measured 500 msec after the beginning of saccades and showed an underconstancy of .50 and 2.13°, respectively. Averaged over both subjects, the results show a .25° overconstancy for a 5° eye turn and a .23° underconstancy for an 8° saccade.

In general, there is no sign of a constant error associated with VDC when it is measured by Matin's technique. Admittedly the data come from few subjects. In fact, they would not seem to be worth mentioning if it were not the case that extremely consistent trends toward underconstancy have been observed by Hill (1972) and Morgan-Paap (1974) who use a different procedure.

Although Matin specified the standard visually, Hill (1972) indicated the standard by name in his measure of VDC. He had observers adjust a target so that it appeared "straight ahead." The subject's eyes were directed straight ahead or toward a laterally displaced fixation point. The retinal cue reduction was not so complete as Matin's because two stimuli were on at once during the eye movement to the laterally displaced point. However, extraneous retinal information was reduced by keeping the visual field completely dark except for the stimuli. If VDC were lacking in this reduced cue condition, any target that stimulated the fovea (oculocentric straight ahead) would appear to have a straight-ahead egocentric direction. Thus, when the eyes fixated a laterally displaced point, the test target would have to overlap the fixation point to appear straight ahead. In contrast, if there were complete constancy, the test target would be set to the objective straight ahead even when the eyes were displaced.

Hill asked his subjects to direct their eyes straight ahead in one condition to provide a base line and had others fixate targets that were displaced 30° left or right in the constancy conditions. He found a reliable departure from constancy in the direction of underconstancy. On the average fixation left or right caused settings to be displaced 2.59° from the base line in the direction of the fixation point.

Underconstancy could be caused either by underestimating the degree of eye turn or by overestimating the distance between the fovea and the stimulated retinal area. These two possibilities were tested by having observers turn their heads in the direction of the 30° fixation targets after they were turned off. In one condition head turns were made from a straight-ahead starting postition after observers fixated the 30° targets. Subjects attempted to maintain the same fixation point and turn their heads in that direction after the target disappeared. This condition minimized retinal displacement information because the targets fell on the fovea and it measured the accuracy of eye position information. If observers underestimated eye position, their head turns should have revealed an underestimation of the target position. In another condition head turns were made from a straight-ahead starting position after observers looked straight ahead and viewed the displaced targets "out of the corners of their eyes." After the target disappeared subjects attempted to remember where it had been and to turn their heads in that direction. This condition minimized eye-displacement information because the eyes were straight ahead and it measured retinal displacement information. If observers overestimated retinal displacement, their head turns should have revealed overestimation of the target position. Measures were made with respect to a base-line condition in which head turns were made with targets on. The results showed that observers underestimated the position of their eyes by about 4° and underestimated retinal displacement by about 2°. The former result was statistically significant, the latter was not. Following the above logic, these results would predict a 6° underconstancy. Because the observed underconstancy was only 2.59°, the two experiments do not jibe quantitatively. Nonetheless, the results suggest that the observed underconstancy is due to underestimation of eye position rather than overestimation of retinal position.

Hill's conclusion about the locus of underconstancy is only in partial agreement with the results of Morgan-Paap (1974). In one of her studies Morgan-Paap had observers point with their unseen hands at peripheral targets either while looking (a) at a straight-ahead fixation point or (b) at the fixation target itself. The results, which were corrected for pointing bias, showed that registered retinal position measured by (a) was consistently *overestimated* by about 2.5° for all magnitudes of eye turn and that registered eye position measured by (b) was consistently *underestimated* by an amount that increased monotonically with eye turn; there was .19, 1.01, 1.71, and 2.47° of underestimation for 12, 22, 32, and 42° of eye turn. Morgan-Paap concluded that the underconstancy of VDC is due to a combination of retinal and eye-position errors. It will take future experiments to determine the cause of the incongruity in the retinal error findings of Hill and Morgan-Paap.

There is little relevant data on this question from studies that used Matin's procedure. The little there is, however, suggests that there is no retinal error; for example, Shebilske (1976) had subjects move their eyes from a straight-ahead position to a target displaced about 20° to the right. Subjects were able to move their eyes very close to the refixation target even when it went off before the main saccade. Sometimes the main saccade landed off-target, but then the subjects were able to correct the error even though there was no visual feedback. The average final error was −.32° for one subject and +.58° for another. This suggests that the subjects knew where the refixation target was located on the basis of the original peripheral stimulation. No significant error was apparent in the retinal signal.

Even more pressing than the question of the locus of undercompensation is the question whether there is a consistent trend toward underconstancy when retinal cues are reduced. Hill found a significant tendency toward undercompensation but Matin did not.

It would be impossible to evaluate this discrepancy if it were not for a subsequent study by Morgan-Paap (1974). Hill's results may not be comparable to Matin's because a much larger eye turn was used and the measures were taken much longer than 63 to 500 msec after the eye turn. Morgan-Paap's results suggest that the discrepant results could not be attributed to these factors. Morgan-Paap also indicated a "straight-ahead" standard by name and followed the same basic rationale as Hill. Crucial changes in her methodology, however, make it possible to compare her results with those obtained with Matin's procedure.

Morgan-Paap's experimental trials were as follows: observers opened their eyes and looked straight ahead for a short time until a target appeared at 2, 12, 22, 32, or 42° left or right of the observers' subjective straight ahead. Observers moved their eyes to fixate this displaced target which remained on for 700 msec. After 600 msec a test flash appeared for 100 msec at 0, 2, 4, 6, 8, or 10° left or right of the predetermined subjective straight ahead. By pressing a button subjects indicated whether the second light appeared to the right or left of straight ahead which was always defined with respect to the nose. Although Morgan-Paap did not measure eye movements, which is always done in Matin's procedure, she did add a subjective test that required subjects to look within 1° of the peripheral target. Another factor that makes Morgan-Paap's results comparable to Matin's is the timing of the test flash. Assuming that there was an approximate 200 msec latency in the saccade to the displaced target, the comparison stimulus was given about 400 msec after the beginning of the saccade in Morgan-Paap's experiment. This is within the 63 to 500 msec range used with Matin's procedure. One other common feature in Morgan-Paap's and Matin's procedures is that both used the method of constant stumuli to secure their data.

Morgan-Paap's results are particularly interesting because they include degrees of eye turn comparable to those used in Matin's procedure and in Hill's experiment. Morgan-Paap found that the straight-ahead point of subjective equality (PSE) showed a consistent undercompensation that monotonically increased in magnitude as the degree of eye turn increased in either direction. Averaged over both directions the PSEs were 1.03, 1.89, 2.66, 3.80 and 5.30° of arc for 2, 12, 22, 32, and 42° of eye turn. The 2 and 12° eye turns are comparable to those for which no consistent constant error is found with Matin's procedure. Recall that in the eight individual subject results that are reported above for the data obtained by Matin's technique five showed underconstancy and three showed overconstancy; the average was negligibly different than zero constant error. In Morgan-Paap's experiment six out of six subjects showed underconstancy in both directions for 2 and 12° eye turns. The average underconstancy was significantly different from zero. Although they vary from Matin's results, Morgan-Paap's data are in general agreement with Hill's for 30° eye turns. The two agree in suggesting that there is significant undercompensation when subjects attempt to set a target straight ahead while their eyes are turned to the side.

The above suggests that systematic differences may appear in the results obtained by the Hill and Morgan-Paap procedures and by Matin's procedure. A number of methodological differences may contribute to the discrepancies:

1. The stimuli were presented on a curve such that all points were equidistant from the cyclopean eye in Hill's and Morgan-Paap's experiments. In contrast they were presented in the frontal plane in Matin's, Pola's, and Monahan's experiments. Observers may be more accurate with the latter arrangement because of their experience with looking at objects on a wall, but this advantage is gained at the expense of letting convergence angle and target distance covary with direction. It is unlikely that subjects ever have the opportunity outside the laboratory to judge the direction of objects that are an equal distance away from the cyclopean eye, but at least this procedure holds distance constant. It does not, however, hold convergence angle constant; this would require a horopter surface. In fact, with stimuli on an equidistant curve convergence angle would covary in a way that would be unfamiliar with respect to experience obtained. At any rate, it would be helpful to test the same subjects with stimuli on an equidistant curve, in the frontal plane, and perhaps on a horopter surface.

2. Another difference between the procedures was the number of stimuli present during the judgment. At least one other stimulus was always visible be-

sides the one to be judged with the eyes turned in Hill's and Morgan-Paap's experiments. In contrast no two stimuli were ever present simultaneously in Matin's procedure. Perhaps Matin's control is important in eliminating all static perspective cues. It seems unlikely, however, that the minimal perspective cues in Hill's procedure could account for the underconstancy.

3. A final difference is that instructional set may not have been the same in the two procedures. It may make considerable difference whether subjects focus on how the stimulus appears to them ("apparent" instructions) or whether they attempt to judge how things really are ("objective" instructions). The objective mode of viewing tends to increase the degree of compensation in size and shape constancy, and in the constancy of object orientation (cf. Carlson, this volume; Ebenholtz & Shebilske, 1973). Therefore we can expect that it might do the same for VDC. This is supported by preliminary results collected by Steve Nice and myself. We told one group of subjects to set a target so that it *looked* straight ahead and not to worry about whether it was *in fact* straight ahead (apparent instructions). We told another group the opposite (objective instructions). Hill's original procedure was replicated in all other details. We found that although underconstancy is observed with apparent instructions overconstancy is obtained with objective instructions. Earlier researchers have not reported their instructions about direction judgments clearly enough to be sure if they were objective or apparent. In all probability they induced a set somewhere between the extremes created by our instructions.

This suggests a possible explanation for the inconsistency between Hill's and Matin's procedures. Perhaps subjects tend to be more objective when they judge a target in relation to a visual standard, as in Matin's procedure, than when they set the comparison in relation to a standard that they remember by name, as in Hill's procedure. Admittedly, Matin did not obtain overconstancy as we did, but this may be because his objective set was not so explicit as ours. Ours were designed to be as direct as possible which is unlikely for experiments that are not directly interested in the effects of instruction.

The inconsistency between Hill's and Matin's procedures concerns the exact amount of VDC that can be obtained with extraretinal signals. The evidence reviewed above suggests that we may not know the precision of extraretinal VDC signals until the fidelity of VDC is measured under a greater variety of conditions. At present it is not safe to say that there is a general tendency for undercompensation when retinal cues are reduced. There may be under some conditions and not others.

The focus on inconsistencies should not cause us to lose sight of the fact that five separate experiments agree that a high degree of VDC occurs without retinal cues. There is agreement that extraretinal signals can cause substantial VDC. In the next section we consider the source of these extraretinal signals.

Source of Extraretinal Signals

Two rival theories concerning the source of extraretinal eye position information emerged in the late nineteenth century. According to one, outflow theory, eye position is registered by outgoing signals from cerebral motor centers to the eye muscles; according to the other, inflow theory, the perceptual system monitors eye position by feedback from the extraocular muscles. The history of research on these two hypotheses has been reviewed recently in several places (e.g., Howard & Templeton, 1966; MacKay, 1973; Matin, 1972). The conclusions of Howard and Templeton and MacKay were in agreement with what could be called the "widely accepted" view that outflow is overwhelmingly favored by the available evidence. Matin is the only recent reviewer to suggest that the evidence supporting outflow theory is questionable.

The unique feature of Matin's analysis is his evaluation of the role of the gamma efferent system. The basis for his reasoning is the well-known observation that spindle activity can be modified not only by extrafusal muscle stretch or contraction but also by direct stimulation of the intrafusal muscles by gamma efferent signals. Because gamma efferents, a component of outflow, can modify inflow from muscle spindles, Matin called this feedback system a "hybrid mechanism." He noted that because of it inflow could be present without an actual change in extrafusal muscle length. He also noted that according to the results of Whitteridge (1959) inflow could be greatly reduced during extrafusal muscle stretch that is not accompanied by gamma efferent signals. It follows that inflow may be greatly reduced during eye movements caused by an external force. As we shall see, these deductions undercut most of the evidence that is traditionally cited in favor of outflow. Matin also noted that because of the influence of gamma efference spindles could not reliably indicate the position of limbs that are continually subjected to varying loads. Muscle load and eye position, however, are in an essentially invariant relationship under normal conditions. Therefore spindles may be able to monitor eye positon reliably. The following extends Matin's analysis and argues that the functional significance of inflow has been underestimated.

OCULOMOTOR CONTROL AND OUTFLOW

The acceptance of outflow theory was due in part to studies of oculomotor control that suggested that inflow information is unavailable; for example, Robinson (1964) and Stark (1971) fixed one eye of human subjects so that attempted saccades were transformed into an isometric contraction. Because there was no modulation of tension, it was concluded that muscular feedback was not available. Similarly, Keller and Robinson (1971), recording from units in the abducens believed to be motoneutons, failed to find a stretch reflex in extraoculor muscles of rhesus monkeys. Shebilske (1976) noted that the first result does not refute the existence of muscular feedback. Spindle feedback may not distinguish between actual contraction and isometric contraction because gamma efferent signals can cause spindle feedback in the absence of extrafusal muscle stretch (e.g., Granit & Kaada, 1952; Hunt & Kuffler, 1951; Leksell, 1945). In addition, it was noted that instruction-induced mental set is crucial for the presence of feedback signals in humans (e.g., Skavenski & Steinman, 1970). Therefore the results of Keller and Robinson may have been due to a failure to achieve the proper mental set in their monkeys. Accordingly, the procedure used to induce eye movements in monkeys may be critical.

The importance of feedback to oculomotor control is now supported by mounting evidence in (a) studies of refixation errors and corrective movements that are made without visual feedback (Barnes & Gresty, 1973; Becker, 1972; Becker & Fuchs, 1969; Shebilske, in press; Weber & Daroff, 1972), (b) studies of eye position maintenance (Matin, Matin, & Pearce, 1970; Skavenski, 1971, 1972; Skavenski & Steinman, 1970; Timberlake, Wyman, Skavenski, & Steinman, 1972), (c) electromyographic studies of stretch effects in human extraocular muscles (Breinin, 1957; Maruo, 1964), and (d) eye movement recordings of a patient suffering from Macro square wave jerks, which is an instability of the saccadic eye movement system (Dell'Osso, Troost, & Daroff, in press).

Gamma contingent inflow from muscle spindles may not be the only source of extraretinal feedback. The findings of corrective movements without visual feedback is also consistent with an internal monitor source (Weber & Daroff, 1972; Shebilske, 1976). According to this hypothesis, the output of motoneurons is fed into a simulator model of the dynamic characteristics of the oculomotor system; the output of this model provides feedback to the central saccadic system in the form of an estimated eye position. If this deviates from the intended eye position, the central saccadic system is capable of issuing a corrective command. Because an internal monitor would provide inflow if, and only if, a central efferent command were issued, the internal monitor can be thought of as a hybrid mechanism.

Gamma contingent inflow from muscle spindles and inflow from an internal monitor are not mutually exclusive alternatives. In fact, the evidence suggests that both may provide feedback under different circumstances; for example, Dell'Osso et al. (in press) argue that an internal monitor can account for their observations of Macro square wave jerks but feedback from muscle spindles cannot. On the other hand, an internal monitor could not explain the maintenance of eye position in the dark against an external force (Skavenski & Steinman, 1970) nor could it explain electromyographic results that support inflow (Breinin, 1957; Maruo, 1964). Thus it may be that some evidence of feedback cannot be explained by gamma contingent inflow from muscle spindles, whereas other evidence cannot be explained by an internal monitor. It is too early to see a pattern in these findings, but it is interesting that the observations of Dell'Osso et al. (in press) involve the vergence system and evidence in favor of gamma contingent inflow involves the version system.

CONSCIOUS EYE POSITION

Another fact often cited in favor of outflow theory is the observation that humans have a poor conscious sense of eye position (Brindley & Merton, 1960; Irvine & Ludvigh, 1936; Skavenski, 1972). The early observations can be considered inconclusive in light of the gamma contingent inflow hypothesis. This is not true, however, of Skavenski's results which show that awareness of eye position is much less accurate than would be expected if it were based on an inflow signal.

Brindley and Merton (1960) and Irvine and Ludvigh (1936) demonstrated that subjects had no conscious awareness of eye position when their eyes were pulled by hand-held forceps. These gross uncontrolled external forces may have distracted subjects and interfered with their deliberate attempts to maintain an eye position. Without deliberate central innervation, gamma signals may have been absent, thus causing reduced spindle sensitivity. Therefore the fact that subjects were unaware of eye position does not necessarily mean that conscious eye position is not given by gamma contingent inflow. Skavenski (1972) used controlled, gradually applied external force which had been shown to be accompanied by precise feedback signals. He found that although subjects could reliably detect the direction in which their eyes had been pulled their conscious sensitivity was much lower than their ability to use feedback to control eye position in the dark.

The fact that conscious eye position may not be based on inflow does not mean that eye position information in VDC is not. The finding of VDC with reduced retinal cues implies that extraretinal signals supply a *registered*, not a *conscious* eye position. The importance of this distinction is not unique to VDC (see Epstein, 1973 for general discussion). Most perceptual dimensions are indicated by many sources of information, and the value of perceptual dimensions are often determined for more than one purpose; for example, body-tilt information is provided by the vestibular apparatus and transducers in the skin, muscles, joints, and viscera; and one may need to know body orientation for different purposes such as which way to seim to reach the surface of for judging the orientation of an object with respect to gravity. There is no *a priori* reason to assume that all the sources of information will be given the same weight in evaluating different dimensions. In fact, evidence suggests that they are not for body orientation because there is a general lack of relation between errors in apparent body orientation and setting a target to vertical (e.g. Ebenholtz, 1970; Ebenholtz & Shebilske, 1973). One explanation is that one source of information is used to determine *registered* body orientation associated with judgments of object orientation and other sources are processed to determine *felt* body orientation. Similarly, because VDC is so much more precise than judgments of felt eye position, it seems unlikely that they are based on the same source of information.

ABNORMAL EYE MOVEMENTS

The direct measures of VDC, which traditionally have been interpreted in favor of outflow, involve abnormal eye movements in that they are mechanically restrained, externally caused, or inhibited by paralysis. Each of these conditions is designed to manipulate outflow and inflow independently. Because of the hybrid nature of inflow mechanisms, abnormal eye movement conditions do not achieve this goal. Consequently the results are equally compatible with outflow or inflow models.

An experiment by Skavenski, Haddad, and Steinman (1972) describes an attempt to manipulate inflow and outflow by mechanically restrained and externally caused eye movements. In one condition subjects adopted a corrective set which enabled them to maintain a defined eye position even though a load was applied that would otherwise have moved the eye. They found that a target viewed monocularly by the loaded eye appeared to shift in the opposite direction as the external force. The amount of apparent movement was proportional to the load. This is analogous to the results of mechanically restrained eye-movement studies (e.g., Brindley & Merton, 1960) in which targets appear to shift in the direction of the intended eye movement or, in other words, in the opposite direction from the external force restraining the eye. Skavenski et al. (1972) assumed, as others had before them, that inflow does not change when the eye remains in one place. They concluded that the observed shift in visual direction could not have been caused by inflow. As noted above, this assumption is unwarranted because inflow is likely to be present at high levels during isometric contraction. Therefore inflow could explain the shift as well as outflow.

In a second condition Skavenski et al. (1972) had subjects adopt a passive mode of viewing to enable the load to pull the eye out of position. Subjects viewed a movable target with the passively moved eye and a stationary fixation point with the other. In agreement with other studies that used external force it was found that direction constancy was absent during passive rotations. It was assumed that because muscles changed their lengths an inflow signal should have registered the passive movement, but because the passive rotation was not registered it was concluded that inflow does not provide the VDC mechanism with eye-position information.

This result is not so damaging to inflow theory as it might seem. It is exactly what would be expected if muscle spindles were the source of extraretinal feedback. As noted by Matin (1972) and Shebilske (in press), it is likely that spindle feedback is reduced below the tolerance range of the compensatory mechanism during passive stretch because spindle afference is substantially reduced in the absence of gamma efferents (Bach-Y-Rita, 1972; Witteridge, 1959). Thus, although the breakdown of VDC during passive rotations supports Helmholtz's (1866) conclusion that "effort of will" is necessary for direction constancy, the further inference that outflow is a direct source of registered eye position is unjustified. It may be that "effort of will" is a prerequisite for gamma efferent signals which are necessary to inflow availability.

One other observation, often cited in favor of outflow, is the apparent lack of visual stability that occurs during eye-muscle paralysis. If eye muscles are paralyzed by a clinical pathology or by drugs, any attempt to move the eyes results in an apparent movement of stationary surroundings in the same direction as the attempted movement (e.g., Helmholtz, 1866; Jackson & Paton, 1909; Kornmuller, 1930. Outflow is certainly present in this situation and could account for the apparent shift, but the illusory shift is not definitive proof of outflow theory unless we can be sure that all inflow is eliminated. As Matin (1972) reported, this assumption had been challenged by the finding that there is no illusion of spatial displacement when subjects are *totally paralyzed* with high doses of a neuromuscular blocking agent (Siebeck, cited by Matin, 1972). Surely outflow is present to the same extent in total paralysis, so why is the illusion absent? The answer may be that outflow is not the relevant extraretinal signal in seeing a stable world. Then why should there be an illusion during partial paralysis? The answer that Matin (1972) suggested is that all inflow may not be eliminated unless total paralysis is produced.

Because of the importance of this observation, Stevens, Emerson, Gerstein, Neufeld, & Rosenquist (1976), replicated experiments on partial and total paralysis. For the partial paralysis studies three observers, including Stevens himself, participated; for the total paralysis observations he was the only subject. The results of partial paralysis are reported as follows:

> When an attempt was made to make a fast saccade upward for example, the visual world would disappear or "jerk" and reappear above its original spatial locus. This was described as a sensation of displacement rather than actual movement. "The world did not move . . . it was not as if you had taken the stimulus and moved it across the screen . . . when I moved my eyes up, the whole screen was displaced up . . . (the stimulus) disappeared and then opened up again in another place
>
> The *displacement* was preceded either by a very rapid jump or a blanking out of the visual input during the saccades. ACR and RCE felt that it was a jerk or jump and JKS felt that it was sometimes a jerk and sometimes a blanking out.

The results of the total paralysis experiments are also best described by the observer's firsthand reports.

> In the first experiment JKS reported no *movement* or *displacement* during attempted saccades. "I tried to move my eyes as hard as I possibly could and nothing happened, the world was just there . . . I simply could not move my eyes." The earlier curare experiments described above made it clear that effort was a critical factor. Therefore, the study was repeated, and JKS was reminded at frequent intervals to exert great effort. JKS moved a finger on the tourniquet-protected arm to indicate an attempted eye movement. Careful study of the video tape showed no actual eye movements during these periods. Again JKS reported that he was very much aware that his eyes were paralyzed. "I know I did not move my eyes, I was trying very hard." However, unlike the first total paralysis experiment, "when I looked to the right I felt that if I had to touch anything . . . I would have had to reach over to the right." JKS felt that his perceptions were much the same as seen during the low dose experiments, but this *displacement* was not punctuated by *jumping*. That is, no jerk, jump, or blanking out of the visual input was perceived during the attempted saccade during total paralysis. The jumping had been very striking in the low dose experiments and had made the displacement illusion quite apparent. JKS emphasized that the displacement perception was not necessarily visual in nature, and found it very difficult to describe.
>
> Later, a third total paralysis experiment was carried out, and again JKS reported the same perceptions described in the second total paralysis experiment. When he attempted to move his eyes to the right, they felt paralyzed, yet the visual world was spatially relocated to the right. As before, he emphasized the perception was not visual (Stevens et al., 1976).

The first total paralysis experiment replicated Siebeck in detail and thus supported inflow theory. The other total paralysis experiments are rather difficult to interpret. Stevens et al. concluded that *displacement* rather than *motion* must be the perception associated with outflow. They speculated that Siebeck probably expected motion and therefore did notice the displacement. Other interpretations are possible, however.

Steven's claimed that it was difficult to describe his experience of displacement. His most descriptive statement was that "when I looked to the right I felt that if I had to touch anything. . . I would have had to reach over to the right." He assumed that this feeling occurred because outflow had produced a change in his registered direction of gaze. This experience could have had other causes; for example, if a person's head were turned to the right, he would have to reach to the right to touch things that he saw. This would be true even if the eyes were in the primary position in the head. Could it be that registered head position changes with extreme effort to look to one side? Perhaps Siebeck exerted a lower level of effort, as Stevens did in his first experiment. The extreme effort that Stevens used in his second and third total paralysis experiments may have caused changes in registered head position.

If you would like to try a related introspective experiment, turn your head to the straight-ahead position and pick out some object that is about 10° to your right. Close your eyes and imagine that you are looking at the object that you picked. Where do you feel your head is pointing? Now, with your eyes still closed, make an extreme effort to look at something way off to the right. Where do you feel your head is pointing? My head feels straight when I try to look a little to the right, but it feels turned to the right when I make extreme efforts to look way to the right. It feels turned even when I know it is not. It is possible that Stevens overlooked a change in his registered head position.

The problem of interpreting the discrepancy between the results of Siebeck and Stevens is indicative of the inherent deficiency of the introspective method. Perhpas Siebeck was looking for motion and failed to see displacement, or perhaps Stevens was looking for a change in registered eye position and failed to notice a change in registered head position.

At any rate, both Siebeck and Stevens agree that there is an important difference in apparent visual stability during partial and total paralysis. This difference is difficult to explain from an outflow point of view. Therefore paralysis experiments at present are no more definitive than the other abnormal eye-movement experiments, but further exploration of the differences between partial and total paralysis is bound to be fruitful, especially if the dependence on introspective reports can be overcome (cf. Matin, in press).

UNRESTRAINED VOLUNTARY SACCADES

Few experiments have investigated the source of extraretinal eye-position information during unrestrained voluntary saccades. The fact that extraretinal signals begin to grow before the beginning of unrestrained saccades (Matin, Matin, & Pearce, 1970) was taken as support for outflow under the assumption that an inflow signal would be too late. However, it was noted subsequently (Matin,1972; Shebilske, 1976) that inflow could account for this result because muscle spindle contraction may precede extrafusal muscle contraction (e.g., Elder, Granit & Merton, 1953).

The only other evidence of extraretinal sources that has been obtained during unrestrained voluntary saccades was presented by Shebilske (1976). The first experiment showed that when a main saccade terminates off target corrective movements (CMs) can be made without visual feedback. This agreed with the findings of other investigators (Barnes & Gresty, 1973; Becker, 1972; Becker & Fuchs, 1969; Weber & Daroff, 1972). The CMs without fisual feedback were truly corrective in that they were highly correlated with the magnitude of error after the main saccade and they significantly reduced the discrepancy between eye and target position. In addition, corrections were almost as good in the dark as they were with visual feedback.

Shebilske (1976) considered four possible explanations of these results: preprogramming, persistence, inadvertent visual feedback, and extraretinal servocontrol. The first three attempted to explain the results within the general framework of traditional oculomotor control models. The last one assumed a different source of dysmetric error and a different source of corrective information than traditional models.

The preprogramming hypothesis was originally proposed by Becker and Fuchs (1969) who suggested that CMs might be part of a preprogrammed package of two movements. This was an attractive explanation for the CMs they observed for 40° eye movements. For these large saccades there is little diversification in the excursion couplets. For saccades of less than 30°, however, there is high intra- and intersubject variability in frequency, direction, and magnitude of CMs. Therefore preprogramming is an unsatisfactory general account of CMs because it does not explain a salient feature of the phenomena.

Another possibility that can be rejected is the persistence hypothesis. On this account subjects have no information about the actual degree of dysmetria error without visual feedback. Instead, they become aware of small CMs on visual feedback trials and simply persist in the dark to make small saccades after the main one. This hypothesis can be rejected because of the significant correlation between errors and subsequent corrections in the dark.

Another argument that attempts to explain the results within the tradtional framework is the suggestion that visual feedback was inadvertently introduced by having some visible background. This possibility has been ruled out by careful controls. Shebilske (in press) even controlled for the possibility that an external auditory reference could guide CMs.

The last explanation considered by Shebilske was suggested by Weber and Daroff (1972) who proposed an extraretinal servocontrol model. According to this model, CMs in the dark are based on a comparison of two high fidelity extraretinal signals, a reference input from the cerebral motor centers to the oculomotor nuclei (outflow), and a feedback signal (inflow). The outflow signal represents the intended eye position and the inflow signal encodes the actual eye position within some margin of error. If these two signals differ, the system is capable of making a CM to correct the discrepancy.

The extraretinal servocontrol model is supported by the following reasoning: when there was no visual feedback after the beginning of a saccade, subjects were able to make truly corrective CMs. Hence they must have known the correct goal before the beginning of the main saccade. The preprogramming hypothesis suggests that even after dysmetric saccades the intended eye position is in correspondence with the actual eye postion. It assumes further that the eyes are intended to be on target only after the second saccade. This hypothesis is rejected because it does not explain the extreme intra- and intersubject variability in frequency, magnitude, and direction of CMs. Alternatively, it is assumed that the initial main saccade is intended to bring the eye to the correct position but that some noise in the oculomotor pathways or in the musculature causes the eye to land off target. This explains the variability in dysmetria. Because discrepancies between actual and intended angle of gaze after main saccades are detected and corrected without visual feedback, some extraretinal feedback signal (inflow) must provide information about the actual eye position. In other words, after dysmetric saccades outflow and inflow are compared and any discrepancy is corrected by a CM.

The extraretinal servocontrol model suggests a different source of dysmetria and a different source of corrective information than assumed by traditional models. An important assumption for traditional outflow models was that outflow faithfully represents eye postion. Thus, although outflow theorists were aware of saccadic dysmetria, they assumed that it was caused by an error in visual localization of the refixation target (e. g., Ludvigh, 1952). This meant that although the intended eye position may be wrong the actual and intended eye position would always be in agreement even after dysmetric saccades. Of course, it was also assumed that visual feedback was necessary for CMs (e.g., Ludvigh, 1952). This would be required if the source of error were visual. Therefore the finding of CMs without visual feedback is inconsistent with the traditional visual error-visual correction hypothesis and suggests an extraretinal error-extraretinal correction hypothesis.

The traditional oculomotor control models made it difficult to test VDC during normal voluntary saccades. Indeed, no such tests were developed, but if the extraretinal servocontrol model is correct then testing VDC during unrestrained voluntary saccades would be relatively easy. According to the extraretinal servocontrol model, outflow and inflow differ after main saccades that land off-target. It is assumed that outflow corresponds to the intended eye position, whereas inflow corresponds to the actual eye position. Shebilske (1976) noted therefore that outflow and inflow theories of VDC could be tested by measuring apparent visual direction at this critical moment.

Shebilske (1976) ran two test conditions. Because most large dysmetric saccades were undershoots, the critical trials for both tests were those in which main saccades landed to the left of the refixation target. (This target was always to the right of a straight-ahead starting position.) In both test conditions the target went off at the beginning of the main saccade and a brief test flash was given 70 msec after the end of the main saccade. Because the latency of CMs is longer than 70 msec, the flash always occurred before CMs. In one condition it appeared in the same place in which the refixation target had been. In the other it was delivered to the fovea.

The results on critical trials for both test conditions were contrary to outflow and in support of inflow theory. In the first the oculocentric position of the test flash indicated that it was to the right of the point at which the fovea was directed. If eye position were registered by outflow which encodes the intended angle of gaze, the fovea would have been registered as being directed toward the refixation target. Therefore the test flash should have appeared to the right of the target. This outflow prediction was not supported; the test flash appeared in the same place as the refixation target, a result that agrees with inflow theory. The direction of the test flash would be seen correctly even though the eye landed off-target if the actual eye postiion were registered by means of inflow. It is also possible however, that the small errors predicted by outflow are within a tolerance range (e. g., Mack, 1970; Matin, Matin, & Pola, 1970) and thus not detected. Hence the second test condition was crucial.

In the second test condition the oculocentric direction indicated that the test flash was in the same position in which the fovea was directed. Therefore, if the direction of gaze was registered to be at the refixation target in accordance with the outflow signal, the test flash should have been seen in the same place as the refixation target. If the null result in the first condition was caused by a tolerance range, the test flash should have been seen in the same place as the refixation target in the second condition as well. It was not. In agreement with inflow theory, the test flash was seen to the left of the refixation target.

In summary, according to the extraretinal servocontrol model only inflow encodes the actual eye position after dysmetric saccades. Outflow encodes the intended eye position that would cause subjects to be unaware of the actual direction of objects for a brief moment (the latency of CMs) after dysmetric saccades. Because subjects were aware of the actual direction of objects after dysmetric saccades, Shebilske (in press) concluded that the VDC processes register eye position by inflow signals during unrestrained voluntary saccades.

SOURCES OF INTERNAL DISTURBANCES

The inflow-outflow question takes on a new look once it is admitted that noise in the oculomotor pathways or musculature may cause a discrepancy between intended and actual eye position. A mutually exclusive test based on this assumption favors inflow over the "widely accepted" outflow theory. Furthermore, the possibility of internal noise suggests that outflow may be an unreliable source of eye-position information. This is especially true to the extent that large discrepancies appear between the point to which the cerebral motor centers command the eyes to go and the point at which the eyes actually fixate after a main saccade. Only small amounts of saccadic dysmetria have been observed to date; but an examination of the possible causes of internal noise shows that most are eliminated by controls in laboratory conditions. Therefore the data we have may underestimate the amount of saccadic dysmetria that occurs under everyday viewing conditions.

A number of possible sources of internal noise are beginning to surface. Variability in the relationship between the discharge rate of motoneurons and eye position may be one source of internal noise; for example, Keller and Robinson (1971) found that the standard deviation of these firing rates averages 4.5% of the mean with a range of 2.5 to7.2%. Error may be averaged out over a number of motoneurons that influence a single contraction. This remains to be shown, however. It is also possible that temporary metabolic fluctuations affect the muscles and limit the precision of the system to the observed variance. This source of variability alone may account for the dysmetria that has been observed under laboratory conditions. Other sources of noise may increase dysmetria in everyday viewing conditions.

To date laboratory conditions have been ideal for maximum accuracy of the oculomotor system. The head has been held in one place, and eye movements are usually made along one plane at eye level and often distributed evenly over the left and right visual fields. Under these conditions the saccadic dysmetria measured may represent minimal error. Let us examine what happens in everyday viewing conditions when the head is continually moving and the distribution of eye fixations is not so carefully controlled.

Head movements are accompanied by postural reflex eye movements that are not registered centrally (Duke-Elder & Wybar, 1973); for example, when the head is tilted back, the doll reflex rotates the eyes downward, and these reflexive eye movements are not registered centrally (Ebenholtz & Shebilske, 1975). Unregistered postural reflexes could increase dysmetria because starting position must be accurately registered before normometric saccades can be executed. The reason is that, depending on the position of the eye, extraocular muscle planes change in relation to the eyeball. Hence the effect that a particular muscle contraction has on the direction of eye rotation is altered for each eye position (Duke-Elder & Wybar, 1973). Therefore during head movements postural reflex eye movements could cause erroneously registered eye position which in turn could increase dysmetria.

Postural reflex eye movements change from moment to moment as the head moves about. Shebilske and Fogelgren (in preparation) wondered what would happen if the head were tilted back for an extended period and then returned to upright. They noted that the doll reflex simulates the effects of wedge prisms in that it causes an illusory perception of visual direction. Specifically, when the head is tilted back, objects appear elevated (at least in reduced cue conditions) because registered eye position is *higher* than actual eye position (Ebenholtz & Shebilske, 1975). Because adaptation to wedge prisms can cause a change in registered eye position (e.g., Kalil & Freedman, 1966; McLaughlin & Webster, 1967), Shebilske and Fogelgren wondered whether adaptation to the doll reflex could do the same. Therefore they had subjects read while their heads were tilted back 20° from upright. A short passage of text was centered at eye level perpendicular to the line of sight when the eyes were in their normal straight-ahead position with respect to the head. A control group read the same passage with their head upright and the passage mounted so that eye position was the same for both groups. The results were exactly analogous to wedge prism studies. Pretest and posttest measures of negative aftereffects showed changes in registered eye position after 3, 6, and 9 minutes. In two separate experiments the shift in the experimental group with respect to the control group was highly significant in the predicted direction. When the head was returned to upright, registered eye position was *lower* than actual eye position.

It is commonplace for us to move our heads as we move our eyes. It is also routine to move our eyes with our heads tilted back for more than 3 minutes and then to return to upright. Both conditions can cause erroneously registered eye position which in turn might increase dysmetria. Therefore the fact that head position is controlled in the laboratory means that saccadic dysmetria may be underestimated in comparison with everyday viewing conditions.

Next, let us look at the potential effects of asymmetries in the distribution of eye fixations, which are often avoided in laboratory conditions but are common in everyday situations. Asymmetries in the overall time spent looking in one direction occur whenever the object of regard is not exactly centered with respect to the head. Asymmetries in the direction that the eyes move (e.g., left to right or right to left) occur whenever ther is a preferred direction of scanning, as in reading.

Slight asymmetries in the direction of gaze cause large shifts in registered eye position (Ebenholtz & Wolfson, 1975; Paap, 1975); for example, Paap studied magnitude of inducing eye turns of 12, 22, 32, and 42° from subjective straight ahead for durations of 30, 60, and 120 seconds. Posttest minus pretest differences in subjective straight ahead were taken as measures of shift in registered eye position. The mean shift in the direction of the inducing eye turns was 1.28, 2.86, 2.53, nd 3.33°, respectively, for the four magnitudes and 1.88, 2.55, and 3.99 for the three fixation times. Even the weakest inducing condition (12° for 30 sec) produced slightly more than a 1° shift. These shifts represent a discrepancy between actual and registered eye position. As noted above, this discrepancy could, in turn, cause dysmetria.

The mechanism behind these shifts in registered eye position is assumed to be posttetanic potentiation, which is the tendency for muscle tension to remain higher than the normal resting state after continued contraction. When the eye is held in an asymmetric position, the muscles on one side are contracted more than the muscles on the other. Hence one set of muscles becomes potentiated in relation to the other. Therefore, when both sets reçeive equal voluntary innervation, the eye will deviate in the direction of the potentiated muscles. The deviation is apparently unregistered.

There is also a bias for contracting one set of muscles more frequently than another contralateral set whenever there is a preferred scanning direction; for example, most college readers move their eyes from left to right along each line of print, making few right to left movements within a line and only one right to left movement at the end of each line to the beginning of the next (e.g., Shebilske, 1975). This means that although the overall direction of gaze may be balanced there is a much higher frequency of rightward saccades than leftward. During each rightward saccade the muscles on the right side of the eye are contracted, whereas those on the left are relaxed. Therefore posttetanic potentiation may cause the muscles that pull the eye rightward to become potentiated in relation to the contralateral muscles. If the effects are anlogous to those obtained with a biased direction of gaze, there should be an unregistered rightward shift in the direction of gaze.

I tested this hypothesis by having subjects scan targets left to right or right to left for 1 minute. Five targets, each of which was separated by 2°, were centered at each subject's subjective straight ahead, which was determined by a pretest. Subjects moved their eyes from target to target at the beat of a metranome which was set at one beat per second. To make sure that subjects looked at the targets, they were required to indicate whether the target line was horizontal or vertical. When subjects came to the last target, they made eye movements in the opposite direction back to the first target. Thus 48 of the 60 saccades made in each exposure period were in one direction and 12 in the opposite direction. A control group fixated the center dot for 1 minute. The experiment is still in progress and only partial results can be reported. There was no pretest-posttest shift for the control group. The group that made 48 right to left movements and 12 left to right shifted 4° to the left on the posttest. This difference was significant and in the predicted direction. Hence the preliminary results suggest that biases in the direction of eye movements have effects that are analogous to biases in the direction of gaze.

It is concluded that asymmetries in the direction of gaze and eye movements can cause shifts in registered eye position which could increase dysmetria. Because the mechanism underlying these shifts appears to be posttetanic potentiation, it should be noted that fixation asymmetries could also directly cause dysmetria. They could cause overshoots or undershoots, depending on whether the agonist or antagonist were potentiated. Therefore the fact that laboratory conditions usually control fixation symmetry but everyday viewing conditions do not mean that laboratory conditions probably underestimate the frequency and magnitude of dysmetric saccades that normally occur.

In summary, head movements and asymmetric distributions of fixations could cause rather large discrepancies between the point to which the cerebral motor centers command the eyes to go and at which the eyes actually fixate after a main saccade. Because these factors have been controlled in laboratory measures, saccadic dysmetria may have been underestimated by the available experiments. On this account outflow may be a labile source of eye-position information. Although experimental evidence is lacking, it may be that in the case of everyday viewing conitions gamma contingent inflow or inflow from an internal monitor may be a much more reliable source for the VDC processes. Thus it remains true, as Matin (1972) concluded, that we have no clear picture of the sources of extra-retinal eye-position information used by the VDC processes. However, it is beginning to appear that the role of inflow may have been underestimated.

Extraretinal Signals and Motion

The preceding conclusions regarding the fidelity and source of extraretinal signals must be limited to the static component of VDC. We saw that retinal mechanisms operated in different ways for judgments of static visual direction and the apparent rest of objects with respect to the egocenter. It may be that extraretinal signals also have different effects on static and dynamic judgments. In fact, there is evidence to suggest that although extraretinal signals provide eye-position information for static judgments of visual direction development of these signals is not synchronized well enough with eye movements to account for the dynamic component of VDC (cf. Matin, 1972).

This evidence was obtained by Matin and his colleagues in a series of experiments in which the comparison stimulus for static judgments was presented before, during, and after voluntary saccades. The development of VDC was determined by plotting the magnitude of compensation as a function of the delay of the compariason stimulus. The conclusions of these experiments were summarized by Matin (1972, p. 346):

> The growth of extraretinal signals is very much slower and more prolonged than the saccade itself, beginning to grow considerably before the saccade begins, changing only slightly when measured with flashes presented during the saccade, and showing considerable growth both before the saccade and for some time after the saccade is completed.

This conclusion rested on Matin's ability to rule out the possibility that the observed temporal development was due to retinal rather than extraretinal effects. Matin (1972) argues that afterimages of the standard reduced compensation for eye movements. Hence experiments were required to show that the increasing amount of compensation over time was not caused by a diminishing influence of retinal afterimages. Such experiments were possible because of a fortunate trial-to-trial variability that caused the eyes to be in different positions 75 msec after the beginning of 2° 11 minute saccades. An afterimage should diminish as a function of time and not eye position. Therefore, if the amount of compensation was a function of an afterimage, there should have been no increase in compensation as a function of eye position with time constant. In contrast, with time held constant at 75 msec, compensation was a linear function of eye position with a slope of .50 minute of compensation per 1 minute of eye rotation. This same function was plotted by Pola (in press) at 15, 25, and 200 msec after the beginning of an 8° saccade. At each of these times the amount of compensation was a linear function of eye position, but the slope of the function depended on the delay, .67, .40, .62, and .75 respectively. The size of the time effect is small in relation to the eye-position effects and the shape of the time function suggests that it is not simply due to a fading influence of retinal persistence. Taken as a whole, these experiments buttress the conclusion that VDC depends on an extraretinal signal that grows considerably before saccades and continues to grow during and for sometime after.

It is difficult to see how retinal motion signals could be canceled out or compensated for by extraretinal signals that are out of phase with eye movements. This asynchrony is inferred from judgments of static visual direction. Perhaps judgments of apparent rest are determined by totally different processes. Therefore the role of extraretinal signals should be directly assessed for judgments of apparent rest or motion of objects during eye movements.

Unfortunately direct tests have yielded conflicting results. Wallach and Lewis (1965) designed an ingenious direct test to determine whether extraretinal signals cancel out or compensate for retinal motion signals. Their rationale was simple. If extraretinal signals cancel retinal motion information, a mismatch between the amount of image movement and eye movement should cause apparent motion. Abnormal image displacements were created with an optical device that projected an image of the pupil onto the retina. The target disk, the projection of the pupil, could be made to move at a slower or faster rate than the eye itself. They found that abnormal image movement did not reliably cause reports of perceived target movements. Conflicting results, however, were obtained by Mack (1970) who ran a modified version of the Wallach and Lewis test and who found that image movements 20 or 40% as large as concurrent eye movements caused apparent target motion during saccades, whereas normal image displacements did not. Wallach and Lewis concluded that the perceived stability of objects during eye movements is not caused by compensating for retinal motion signals but by suppressing them. Mack, on the other hand, concluded that motion-signal suppression is not complete and that apparent stability depends in part on a compensation mechanism that compares image-movement information with extraretinal eye-movement information.

Wallach and Lewis' conclusion may be regarded as a precursor of the filtering hypothesis that was presented earlier. We saw that there is compelling psychophysical and physiological evidence to show that retinal filtering is a major determinant of the perceived rest of objects during eye movements, but the filtering mechanisms described above cannot explain why Mack's subjects saw no motion with normal image movement and motion with abnormal. The critical duration filter cannot account for the apparent motion because motion was seen when the image movement was with the eye movement, which would decrease retinal smearing, and in the opposite direction to the eye movement, which would increase retina smearing. Similarly, mechanical strains and masking cannot explain the apparent motion because both factors were the same as in normal image displacement trials in which no motion was seen. Hence Mack's results suggest that retinal filtering is supplemented by a compensatory process that takes eye position into account. Before we accept this conclusion, however, we should resolve the discrepancy between Mack's findings and those of Wallach and Lewis.

It is difficult to see how extraretinal signals could be synchronized with eye movements well enough to account for apparent rest of normal image displacement. Therefore it seems desirable to look for artifacts in Mack's experiment. One possible artifact, which has not been considered, is stroboscopic motion. Perhaps Mack's subjects thought they saw real motion, but they actually saw stroboscopic motion, which can appear to be similar or identical to real movement. A necessary condition for stroboscopic motion is that subjects detect an apparent displacement of an object in egocentric space (Rock & Ebenholtz, 1962). Mack (1970) used stimulus conditions much like those used by Matin in the VDC experiments reviewed above. On the basis of the high degree of constancy observed in Matin's experiments it seems likely that in Mack's experiments abnormal displacements of the retinal image were registered as changes in egocentric direction. Thus, if the brightness contrast ratio happened to be just right for the time interval and distance used in Mack's experiment, we would expect her subjects to see stroboscopic movement. The reasons why the subjects in the Wallach and Lewis (1965) experiment did not see movement during saccades may be because these critical factors were not right for stroboscopic movement or because a change in egocentric direction was not registered. The latter is possible because their stimuli were foveally centered disks, the contours of which fell in an area that has a low sensitivity to displacements (Timberlake, Wyman, Skavenski, & Steinman, 1972).

The presence or absence of the conditions for stroboscopic movement may also account for the fact that when an attempted eye movement is prevented by mechanical restraints or paralysis the visual world sometimes apparently disappears and then reappears in another place and other times appears to jump to a new place (Stevens et al., 1976). The jumping appearance may be stroboscopic movement. The same could be true for the apparent movement of afterimages during saccades. Presently, the alleged role of stroboscopic movement is an untested hypothesis. It could be tested by studying whether the reports of motion with abnormal image displacements, restrained eye rotations, and afterimages depend on the same brightness, time, and distance parameters as stroboscopic movement.

In summary, although extraretinal signals provide eye-position information for static judgments of visual direction, the signals may not be synchronized with eye movements well enough to cancel out retinal motion signals. Retinal filtering of motion signals is a major determinant of the perceived stability of objects during eye movements but it cannot explain the apparent motion caused by abnormal image displacements, restrained eye movements, or afterimages. It was suggested that these situations may introduce apparent motion as an artifact of stroboscopic movement. Presently, however this is an untested hypothesis.

HEAD AND BODY MOVEMENTS AND VPC

In the beginning of this chapter we distinguished between two parts of an observers internal representation of his environment: (a) the egocentric direction of objects and (b) the absolute position of objects in space. The separation of these two components was the basis for the distinction between VDC (the invariance of apparent egocentric direction despite changes in oculocentric direction) and VPC (the invariance of apparent object position, despite changes in egocentric direction). A change in egocentric direction is a relative change between the self and an object. Hence it could be caused by object movement, self-movement, or both. Accordingly, the presence of VPC implies that an observer is able to see egocentric direction correctly and to attribute the proper amount of egocentric change to himself. In other words, VDC is a subroutine of the VPC process which requires an additional attribution stage. In the experiments that have been reviewed to this point head and body position have been fixed. Therefore the results were interpreted in terms of VDC. In effect, we made the tacit assumption that with head and body fixed the additional attribution stage did not have significant effects on the results. Now we focus on the attribution stage to consider VPC in experiments that manipulate head and body position.

Having encoded a change in egocentric direction, the observer must attribute the proper amount of change to himself in order to perceive object position correctly. The attribution stage could be serviced by either or both of two kinds of cue system, a higher order retinal cue system and/or an extraretinal cue system; for example, Gibson (1966) notes that perspective transformations caused by self-produced changes in egocentric direction are different from perspective transformations caused by object movement. Hence the difference in this higher order retinal cue could be used to attribute the proper amount of egocentric direction change to self-produced action. On the other hand, the amount of self-movement could be assessed by extraretinal information such as vestibular information (e.g., Hay & Goldsmith, 1973; Melvill Jones & Gonshor, in press). Although a simple linear combination of retinal and extraretinal information is adequate for VDC, a more complex algorithm would be required for VDC because the geometry of the transformation is more complex (Gogel & Tietz, 1973; Graham 1958; Wallach, Yablick, & Smith, 1972). For our purpose it is adequate to note that an assessment of the amount of self-movement could cause an expected amount of egocentric direction change that could then be compared with the actual amount of change.

There is evidence to show that extraretinal information is used by the VPC process during head movements. Higher order retinal cues are also used. During body movements evidence is lacking for extraretinal effects on VPC, but there are extraretinal effects on a closely related constancy. Finally, there is no question that the VPC processes rely heavily on higher order retinal cues during body movements.

Extraretinal Information and Head Movements

To isolate the effects of extraretinal information on VPC experiments have measured VPC in the absence of extraneous retinal information. Wallach and Kravitz (1965a) designed a method of measuring VPC that was analogous to the method used by Wallach and Lewis (1965) to measure extraretinal effects on the dynamic component of VDC. Wallach and Kravitz reasoned that if extraretinal signals are taken into account a mismatch between the amount of egocentric direction change and head movements should cause apparent motion. Their method is described in detail because it is employed in many of the experiments reviewed below. The experimenter controls the amount and direction of target motion in relation to head movement with an apparatus that transmits an observer's head rotation to a target. The proportion of target to head displacement is called the displacement ratio (DR) and is usually transformed into percent (% DR). Two dependent variables are established by a modified method of limits: (a) a range of DRs that leads to the perception of a stationary target, the no-motion range (NMR), and (b) the midpoint of the NMR, the no-motion point (NMP).

In additon, a faster one-trial test procedure was developed by Wallach and Frey (1969, p. 250):

> Instead of presenting *S* with a stationary target, the test was so arranged that *S*'s head movements caused the target spot to become vertically displaced; when *S* turned his head to the right the target moved objectively upward, and left turning made the target move down. Thus, the same head movement that brought forth the apparent horizontal target displacement caused by the adaptation effect would simultaneously elicit an objective

vertical displacement for the target. We had hoped, and indeed found, that the two displacements, although one was objective and the other apparent, add vectorially to produce an apparent oblique target motion: after the adaptation training Ss reported target motion at a slant, and they were able to reproduce the slant of the apparent motion path by setting a rod that could be turned in the frontal plane.

VPC WITHOUT HIGHER ORDER RETINAL CUES

In their germinal study Wallach and Kravitz (1965a) measured VPC without extraneous retinal information. In an otherwise dark room the target was a 7-cm diameter light spot viewed at a distance of 2 m. The median NMR was 5.4%DR for 22 randomly selected observers. The midpoint of each person's NMR was averaged over subjects who revealed a small but reliable constant error in the NMP of 1.5% DR *with* the head movements. In a subsequent study Wallach and Kravitz (1965b) showed no difference in the NMR when a 7-cm spot was used as the target and when a patterned visual field subtending 16° of arc was used. In the latter case the whole textured ambient array moved by some fraction of the observers head movement; thus the objective movement produced the same kind of perspective transformation as self-produced motion. Yet the objective component of the movement could be precisely discriminated from the identical change caused by the head movements. These studies indicate that the visual system is equipped with a VPC process that evaluates extraretinal information.

VPC ADAPTATION WITHOUT HIGHER ORDER RETINAL CUES

This conclusion was also supported by adaptation experiments. It was noted earlier that if the VPC mechanism integrated retinal and extraretinal information the process would be as follows: extraretinal information would cause an expected amount of egocentric direction change that would then be compared with the actual amount of change. This process is similar to other sensorimotor processes that can adapt to altered sensorimotor contingencies (e.g., Rock, 1966). Therefore, if the VPC mechanism integrates retinal and extraretinal information it should be able to adapt in order to be better suited to altered head movement-optical motion contingencies. After experience with altered contingencies new expected egocentric changes should be associated with head movements. On the other hand, if high-order retinal information is the only source used by the VPC mechanism, then, in the absence of higher order retinal cues, there should be no adaptation when head movements cause abnormal optical motion. Wallach and Kravitz (1965a, 1965b) found significant adaptation to altered head movement-optical motion contingencies in the absence of higher order retinal information. Thus the importance of extraretinal information was supported.

In related experiments Wallach and his associates showed that adaptation to VPC is similar to other sensorimotor adaptation processes in that adaptation increases with exposure time and the magnitude of the optical transformation (Wallach & Floor, 1970; Wallach & Frey, 1969). In addition, they argued that experimentally produced VPC adaptation is a modification of an antecedent adaptation to normal environmental conditions. This was supported by evidence along three lines:

1. After adaptation is experimentally produced it spontaneously dissipates (Wallach & Floor, 1970; Wallach & Frey, 1969; Wallach & Frey, 1972), which suggests that long-term memory retains the normal relationship only temporarily superseded because of a recency effect and a high concentration of specific conditions that dominated short-term memory (cf. Ebenholtz, 1969).
2. Pretraining which provides temporal concentration of the specific conditions that produce adaptation diminishes the effect of immediately succeeding adaptation (Wallach & Floor, 1970). It is as though pretraining strengthens the normal compensation process that reduces the short-term effects of experimental manipulations.
3. Adaptation that takes a direction different from the usual has distinct properties. In normal viewing conditions head movements to the right cause retinal displacement to the left and vice versa. Hence the natural conditions of adaptation outside the laboratory is optical motion *against* the head movement. When experimentally produced adaptation is *with* head movements, it is greater than adaptation *against* head movements and the latter decays more rapidly (Wallach & Frey, 1969; Wallach, Frey, & Romney, 1969). In addition, adaptation perpendicular to the normal direction of adaptation is less than adaptation parallel to the normal direction and does not dissipate over a decay period (Wallach, Frey, & Romney, 1969).

Wallach and Kravitz (1968) were among the first to study the locus of VPC adaptation. They observed that there were three possible sites; (a) in the retinal system, (b) in the extraretinal system, and (c) in the process that integrates these two sources. They found that little if any VPC adaptation can be attributed to a modification of extraretinal information about head position because VPC adaptation has virtually no affect on measures of auditory direction that depend also on registered head position.

Hay (1968) produced evidence that agreed with the conclusion of Wallach and Kravitz (1968) and went a step further in pinpointing the locus of VPC adaptation. He found that head movements alone in the presence of a covarying stimulus will not produce adaptation unless the eyes move with the covarying target and eye movements alone over the same path taken when they follow the covarying target will not produce adaptation. Apparently VPC adaptation will occur if, and only if, a new correlation between eye and head movements is required during exposure. In a subsequent study Hay (1971) showed that in addition to this training requirement one must also engage the eye-head movement system after exposure during the test. When the head is held stationary and an objective movement path is adjusted to appear vertical, no aftereffect is observed. The implication of these studies is that VPC adaptation is localized in the compensatory eye-movement system.

This hypothesis is supported by studies that directly measured eye movements during head rotations before and after adaptation (Melvill Jones & Gonshor, in press). The main findings were that the vestibuloocular reflex measured with the eyes open in complete darkness could not only be attenuated by adaptation to optical reversal but could actually be reversed. In other words, when the head turned to the right, the induced slow phase of the vestibuloocular reflex was to the right and vice versa. These results are dramatic proof of the placticity of the compensatory eye-movement system and as such add credence to the hypothesis that VPC adaptation is localized in that system. Melvill Jones and Gonshor (in press) also review neurophysiological evidence that supports the idea that the vestibuloocular reflex can be modulated centrally, probably through the vestibular cerebellum.

SOURCE OF EXTRARETINAL VPC INFORMATION

The above results are compatible with a model of VPC that was proposed by Hay and Goldsmith (1973), according to which stationary objects are perceived as stationary despite changes in egocentric direction if, and only if, a VPC comparator receives an optical motion that matches the expected optical motion. In their model the comparator is the retina itself and the expected optical motion elicited by a head movement is the reflexive compensatory eye movements. In other words, the vestibuloocular reflex itself is the hypothesized source of extraretinal VPC information.

Although this model is compatible with the evidence reviewed above, it is incomplete. For one thing, it can not account for the possible effects of distance. Compensatory eye movements can stabilize images only of objects at the same distance from the observer as the fixated object because of the optics of motion parallax described by Graham (1958, p. 870).

> When a subject's eyes move with respect to the environment or when the environment moves with respect to a subject's eyes, a differential angular velocity exists between the line of sight to a fixated object and the line of sight to any other object in the visual field. This condition of differential angular velocity leads to such discriminations as are concerned with the response that *near* objects move *against* the direction of movement and *far* objects move *with* the direction of movement (of the head or environment).

There is evidence that distance can be taken into account by the VPC mechanism (Gogel & Tietz, 1973; Gogel & Tietz, 1974; Hay & Sawyer, 1969; Wallach, Yablick, & Smith, 1972). Therefore Hay and Goldsmith's model is incomplete.

These distance results could be accounted for by a slightly modified model. If it is assumed that gain in the vestibuloocular reflex covaries with convergence, the distance affects could be explained under the assumption that subjects converge to a distance that corresponds to the apparent rather than the actual target distance. Neither premise of this hypothesis has been tested. The hypothesis is consistent with our everyday experience, however. If there is no means of taking distance into account other than convergence-yoked eye movements, then in full cue situations, objects at different distances from the fixated object should appear in motion. To some extent they do, as noted in a statement by Gibson, Gibson, Smith, and Flock (1959, p. 40). "Motion parallax is the optical change of the visual field of an observer which results from a change of his viewing position. It is often defined as the set of 'apparent motions' of stationary objects which arise during locomotion." To see motion parallax simply fixate a finger at arms length and rotate your head from side to side. Notice the motion of distant objects. Now fixate one of the distant objects and repeat the head movement. This time notice the apparent motion of your finger. If distance is taken into account simply by linking the reflexive compensatory eye movements to convergence angle, then apparent motion parallax should equal optical motion parallax. This prediction remains to be tested.

Whether or not the distance effects can be accounted for, the Hay and Goldsmith model is incomplete in other ways; for example, Skavenski, Winterson, and Steinman (1975) found that "the slow compensatory system was capable of almost completely nulling movements of the visual scene that would be produced by normal body movements when such movements were passively induced and the subject was provided with both visual and vestibular inputs to his slow compensatory system (p. 2)." *But* the slow system compensated for few of the natural head movements that occcurred when subjects attempted to fixate a target placed at optical infinity while trying to hold their heads still while free of artificial support. It was concluded that the emphasis on the role of eye movements in maintaining visibility of the visual world seems unwarranted. It could be that other sources of extraretinal information are used to supplement the vestibuloocular reflex and/or it could be that higher order retinal cues make a contribution.

Higher Order Retinal Cues and Head Movements

Some evidence shows that higher order retinal cues contribute to the VPC process during head movements. Wallach and Kravitz (1965a) attempted to measure a no-motion range in the context of barely visible stationary vertical lines. A gap formed a wide channel for the path of the shifting target. All other conditions were identical to the experiment in which they found a NMR of 5.4% DR with a homogeneous dark background. The addition of the strips provided retinal information that transformed when the head moved but not when the object moved. The subjects were able to use this information to reduce their no-motion range dramatically. In fact, the smallest DR (.28%) that the apparatus could produce always caused a noticeable target motion whether it was moved *with* or *against* the head direction. This sharp reduction in the NMR supports Gibson's (1968) hypothesis that object-caused optical motion can be descriminated from the self-produced because motion perspective associated with the latter "entails change in the *whole* of the textured ambient array whereas the alteration of perspective caused by an objective motion entails only change in *part* of the ambient array, the remainder being frozen" (Gibson, 1968, p. 341).

Extraretinal and Retinal Information and Body Movements

No clear evidence that extraretinal information contributes to VPC during body movements has been produced. However, it has become apparent that it does for a closely related constancy. Wallach, Stanton, and Becker (1974) noted that when an observer moves he views objects to the side of the movement path from different directions. This causes the objects to change their orientation as well as their direction with respect to the observer; in relation to the observer objects undergo a partial rotation when they are passed by the observer. Yet the apparent orientation of the object in space tends to appear stable. Wallach et al. (1974) measured the fidelity of this constancy with and without the availability of higher order retinal cues. Their procedure was based on the same rationale employed by Wallach and Lewis (1965) for eye movements and Wallach and Kravitz (1965a) for head movements. Instead of measuring a no-motion range, Wallach et al. (1974) measured a no-rotation range. They found a substantial amount of constancy but it was much less precise than VPC during head movements. Also, the presence of a stationary visual context had no effect on the size of the no-rotation range. This contrasts to the dramatic effects of a visual context on the no-motion range during head movements. The results suggest that extraretinal signals contribute to this spatial rotation constancy and that high-order retinal information does not. Conclusions in regard to retinal cues should be held in abeyance, however, until a wider variety of visual contexts has been tried.

It would be interesting to see the Wallach et al. (1974) experimental procedure applied to VPC during active body movements. Perhaps extraretinal effects are there as well, but, be that as it may, there is no question that higher order retinal cues are of major importance for VPC during active walking (Lishman & Lee,

1973) and passive movement in a vehicle (e.g., Gibson, 1950; Rock, 1968). When optical motions normally caused by self movement are simulated while a subject is stationary, the subject will attribute the optical motion to himself, thus causing a compelling illusion of self-movement (e.g., Brandt, Wist, & Dichgans, 1975; Dichgans, Brandt, & Held, 1975; Young, Oman, & Dichgans, 1975). This shows that the attribution stage of VPC relies heavily on higher order retinal cues during body movements.

THEORIES OF PERCEPTUAL STABILITY

If there is one thing that should be clear in this review, it is that the visual side of the visuomotor coin has many facets. The question of how self-produced movements of the retina affect the ability to build an internal representation of the environment is too general to be answered experimentally. Experiments necessarily probe only part of the general problem; for example, some examine VDC, whereas others examine VPC, some investigate static components, whereas others investigate dynamic, and some assess retinal effects, whereas others assess extraretinal effects. Similarly, theories often apply only to certain aspects of perceptual stability. We should not be surprised to learn that no one theory can explain all facets. To explain phenomena as distinct as the static and dynamic components of VDC is probably too much to ask of one theory. In addition, it seems unlikely that one theory could explain the integration of higher order retinal cues and the integration of retinal and extraretinal information. Each theory considered below explains some facet but no one theory can explain all aspects at a process level.

The reafference model of von Holst and Mittelstaedt (1950) explains in one way how it is that we see a stable world. The reafference model assumes that the sensation of motion, which is usually associated with retinal displacement, is somehow canceled by outgoing signals that command the eyes to move. Specifically, the model assumes that the observer unconsciously deduces whether an input is self-produced or externally caused. The deduction depends on a series of processes that copies and stores the outgoing command to the muscles and subsequently compares the retinal input with the command copy. If they match, it is decided that the input was self-produced; if the retinal input and command copy do not match, it is decided that the retinal input was externally caused. The reafference model has stimulated more physiological and psychophysical research (cf. Evarts, 1971; Matin, 1972) than any other model of perceptual stability. Despite its heuristic value, however, it is not adequate as a general model. It cannot explain the effects of higher order retinal cues nor can it explain accurate perception of visual direction when outflow does not correspond to actual eye position.

MacKay and Mittelstaedt (1974) separate the specific reafference model from the reafference principle according to which "reafference and exafference are distinguished by comparison of the total afference with the state of the system—the 'command'" (Mittelstaedt cited by MacKay & Mittelstaedt, 1974, p. 78). This increases the generality of the theory but it is still too specific to account for the possibility that the functionally significant extraretinal signal is inflow instead of outflow command. Therefore it may be preferable to consider both the reafference model and the reafference principle as exemplars of the more general taking-into-account theory of perceptual stability (e.g., Epstein, 1973).

According to the taking-into-account theory, "the perceptual system must combine information present in the retinal counterpart of the to-be-discriminated distal variable with information about other variables that affect the state of the retinal counterpart" (Epstein, 1973, p. 267). When applied to VDC, the taking-into-account hypothesis assumes that the retinal counterpart of egocentric direction is retinal image location which can also be affected by eye position. Like the foregoing theories, the taking-into-account theory assumes that retinal information and eye position information are compared, but the taking-into-account theory is more general with respect to the possible sources of eye-position information. Hence it is compatible with the possibility that eye-position information is given by gamma contingent inflow or an internal monitor instead of the outflow command.

The evidence reviewed above shows that extraretinal eye-position information contributes to the static component of VDC and VPC during head movements. Hence taking-into-account processes cannot be disregarded by a complete theory of perceptual stability. We must, however, guard against overextensions of the hypothesis, as Epstein (1973) has warned. In particular, the taking-into-account theory should not be applied at a process level to higher order retinal cue effects. At a descriptive level we could think of higher order retinal cues as providing eye-position information which is then compared with retinal-image location, but at a process level this model seems unlikely. Certainly it does not describe what Gibson (1966) meant by higher order retinal cues.

Gibson's higher order retinal cue theory falls in the same category as Wallach's relational theories of brightness constancy (Wallach, 1948) and speed constancy (Wallach, 1959). As Epstein has noted, these relational theories assume no taking-into-account process. Apparent whitness is given directly by intensity relationships within the retinal image, and apparent velocity is given directly by the rate of retinal displacement relative to the retinal size of the movement field. Similarly, in Gibson's higher order retinal cue theory apparent egocentric direction and apparent object position are given directly by relational retinal cues; that is, cues that are based on a relationship within the ambient array that remains invariant in the face or eye, head, or body movements. At a process level this account of higher order retinal cue effects is more appealing.

Higher order retinal cues affect the static component of VDC and VPC during head and body movements. Hence the higher order retinal cue theory cannot be ignored by a complete theory of perceptual stability. Like the taking-into-account theory, however, it should not be overextended. Gibson (1966, 1968) implied that extraretinal signals have no functional significance in viewing a stable world, but that was not his main point, for he notes himself that "action sensitivity or movement sensitivity does not depend on specialized receptors. . . . Proprioception considered as the obtaining of information about ones own action does not necessarily depend on the proprioceptors, and exteroception considered as the obtaining of information about extrinsic events does not necessarily depend on exteroceptors" (Gibson, 1966, p. 34). Gibson's main point was that it is not parsomonious to postulate a process that copies, stores, compares, matches, and decides when a much simpler explanation is available. Presumably Gibson would have no objection, in principle, to the present proposal that VDC and VPC depend in part on the extraction of combined sensory evidence, including extraretinal feedback signals.

Because both extraretinal information and higher order retinal cues play a role, it is proposed that both taking-into-account processes and higher order retinal cue processes are utilized in VDC and VPC. However, all the support for extraretinal effects comes from reduced cue experiments. Thus we cannot be sure that taking-into-account processes play a role in full cue situations. We must ask how taking-into-account processes and higher order retinal cue processes combine their information when both are available. One possible answer was proposed by Epstein (1973) for speed constancy. He related an experiment by Robertson (1973, cited by Epstein, 1973) that was designed to compare the taking-into-account hypothesis for velocity and Wallach's relational hypothesis. The results suggested that relational cues almost completely dominate the taking-into-account process when the two are put in conflict. Consequently Epstein suggested that the application of the velocity taking-into-account processes should be narrowed to targets that are presented under reduced cues.

I tried a similar manipulation with informal observations on the effects of perspective cues to directions. The cues were provided by a ring oriented perpendicular to the frontal plane. Earlier it was reported that I had found the perspective information useful in setting the ring to straight ahead. To get some idea of how information would be combined by the taking-into-account processes and higher order retinal cue processes I induced large changes in registered eye position by means of posttetanic potentiation. I could still easily set the ring to straight ahead; therefore higher order retinal cues appeared to dominate as they did in the velocity experiment. It was difficult, however, to assess whether the ring actually *looked* straight ahead. In other words, I knew when it was straight ahead as my ability to set it there showed, but I was not sure that it looked straight ahead when my setting was made. Perhaps cognitive factors induced by an "objective" mental set affect how information from higher order retinal processes and taking-into-account processes are weighted by the perceptual system.

There is a need for more research on the question how information from the higher order retinal processes and the taking-into-account processes is combined in a variety of contexts. The most interesting context, of course, is the full cue situation. It is not only important to know how cue systems combine their information with other cue systems but we must also understand how each cue system operates. Many unanswered questions remain; for example, we have more to learn about which of the potential higher order cues are, in fact, used, and we have more to learn about which source of information is used in taking-into-account eye position. Moreover, we have not yet established the fidelity of either cue system. Hence reduced cue conditions will remain important for developing process models of each of them.

The taking-into-account theory and the higher order retinal cue theory both answer the question of perceptual stability by assuming that an unchanging combination of information is extracted from the sensory input. MacKay and Mittelstaedt (1974) propose an alternative answer to this question which we have called the visual side of the visuomotor coordination question.

> The answer proposed by MacKay on which the present authors are agreed, is that in a relatively stable environment it would be informationally efficient to make stability of the representation the norm, and use the sensory stystem to explore and sample the environment, sensitive to any evidence of change. This is in contradistinction to the presupposition that seems to have been made by von Holst (and by most theorists of perceptual stability), according to which the sensory afference albeit modulated by signals fed in, fed foreward or fed back, would itself finally constitute the internal representation of the world, so that any changes in sensory input would have to be prevented from including changes in that representation by eliminating them from the afferent signals at a lower level. On this view perceptual stability depends on extracting an unchanging residue from the sensory input by a process of continuous and elaborate "modification" under the guidance of information from motor centers (MacKay & Mittelstaedt, 1974, pp. 75-76).

MacKay's evaluation theory assumed a principle that he called "informational inertia," according to which the perceptual system adopts the "null hypothesis" of stability and maintains it until sufficient evidence is received to the contrary.

Like the other theories that we have considered, the evaluation theory is not satisfactory as a general account of perceptual stability. Its inadequacy as a general model is illustrated best in the context of reduced retinal cue experiments in which the only retinal information is retinal image location. Suppose we try to explain the results of these experiments (e.g., Matin, 1972) by the informational inertia principle. In this case the null hypothesis would be that egocentric direction of objects is unchanged despite changes in oculocentric direction. What would constitute "sufficient evidence" that the null hypothesis could be rejected?

To answer this question let's analyze the available sensory evidence, retinal image location, and extraretinal eye-position information. A change in retinal image location per se is not sufficient for discrimination. Retinal image location is different for targets that are in different egocentric directions than the original fixation point. Retinal image location is also different for targets that are in the same egocentric direction as the original fixation point. Extraretinal signals are not sufficient either. It would be absurd, for example, to think that the discrimination could be made with the eyes closed. Thus, because neither source is sufficient, the evidence for rejecting the null hypothesis would have to be a combination of retinal image location and extraretinal information. This cominational use of sensory evidence is no longer a simple rejection of a null hypothesis; as Matin (1972) has shown it is the establishment of a precise alternative hypothesis by means of a simple linear combination.

To say that combined sensory evidence is used to establish a precise percept of egocentric direction is tantamount to stating that the sensory evidence (including extraretinal signals) itself finally determines the internal representation of egocentric direction. It follows that it would *not* be informationally efficient to make stability of egocentric direction the norm. It would be an unnecessary step to form the null hypothesis of stability when the sensory information required to reject that hypothesis could in itself establish a precise internal representation of egocentric direction.

Like the other theories that we have considered the evaluation theory cannot be ignored by a complete explanation of perceptual stability. It explains phenomena that the other theories do not; for example, I wonder how many people have seen their lighted dial alarm clocks jump around during eye movements before the phenomena was pointed out. I suspect that some did not because they ignored the instability just as most of us ignore double images that fall outside Panum's fusional area. Similarly, Wallach et al. (1974) suggested that the constancy mechanisms have low fidelity for the spatial rotation dimension of perceptual stability during walking. Yet, we do not normally see objects as rotating in space as we pass them by. Apparently, there can be processing beyond the extraction of invariant combinations of sensory information. The evaluation theory, with its principle of informational inertia, may account for these secondary processes.

Thus our theme is supported. The visual side of the visuomotor coordination question has many facets and requires a many faceted theory to explain it. Our decision to move away from the reafference principle to the more general taking-into-account theory was motivated by recent developments in our understanding of eye-movement control. This demonstrates the inextricable relationship between the visual and motor sides of visuomotor coordination. The taking-into-account theory, the higher order retinal cue theory, and the evaluation theory each explain aspects of perceptual stability that the others cannot. Hence we need to be eclectic in our choice of theories until a more general theory can satisfactorily account for these phenomena.

REFERENCES

Bach-Y-Rita, P. Extraocular proprioception and muscle function. *Bibliotheca Ophthalmoologica*, 1972, 82, 56-60.

Barnes, G. R. & Gresty, M. A. Characteristics of eye movements to targets of short duration. *Aerospace Medicine*, 1973, 44, 1236-1240.

Becker, W. The control of eye movements in the saccadic system. *Bibliotheca Ophthalmologica*, 1972, 82, 233-243.

Becker, W. & Fuchs, A. F. Further properties of the human saccadic system. Eye movements and corrective saccades with and without visual fixation points. *Vision Research*, 1969, 9, 1247-1259.

Bishop, P. O. & Henry, G. H. Spatial Vision. *Annual Review of Psychology*, 1971, 22, 119-160.

Bower, T. G. R. *Development in infancy*. San Francisco: Freeman, 1974.

Brandt, T., Wist, E. R., & Dichgans, J. Foreground and background in dynamic spatial orientation. *Perception & Psychophysics*, 1975, 17, 497-503.

Breinin, G. M. Electromyographic evidence for ocular muscle proprioception in man. *Archives of Ophthalmology*, 1957, 57, 176-180.

Brindley, G. S. & Merton, P. A. The absence of position sense in the human eye. *The Journal of Physiology*, 1960, 153, 127-130.

Brooks, B. & Holden, A. L. Suppression of visual signals by rapid image displacement in the pigeon retina: A possible mechanism for saccadic suppression. *Vision Research*, 1973, 13, 1387-1390.

Dell'Osso, L. F., Troost, B. T., & Daroff, R. B. Macro square wave jerks. *Neurology*, in press.

Dichgans, J., Brandt, T., & Held, R. The role of vision in gravitational orientation. Sonderdruck aus "Fortschritte der Zoologie," 1975, 23, 255-263.

Duke-Elder, S. & Wybar, K. Ocular motility and strabismus. In S. Duke-Elder (Ed.), *System of ophthalmology* (Vol. 6). St. Louis: Mosby, 1973.

Ebenholtz, S. M. The possible role of eye-muscle potentiation in several forms of prism adaptation. *Perception*, 1974, 3, 477-485.

Ebenholtz, S. M. Perception of the vertical with body tilt in the median plane. *Journal of Experimental Psychology*, 1970, 83, 1-6.

Ebenholtz, S. M. Transfer and decay functions in adaptation to optical tilt. *Journal of Experimental Psychology*, 1969, 81, 170-173.

Ebenholtz, S. M. Adaptation to a rotated visual field as a function of degree of optical tilt and exposure time. *Journal of Experimental Psychology*, 1966, 72, 629-634.

Ebenholtz, S. M. & Paap, K. R. The constancy of object orientation: Compensation for ocular rotation. *Perception & Psychophysics*, 1973, 14, 458-470.

Ebenholtz, S. M. & Shebilske, W. L. The doll reflex: ocular counterrolling with head-body tilt in the median plane. *Vision Research*, 1975, 15, 713-717.

Ebenholtz, S. M. & Shebilske, W. L. Instructions and the A and E effects in judgments of the vertical. *American Journal of Psychology*, 1973, 86, 601-612.

Ebenholtz, S. M. & Wolfson, D. M. Perceptual aftereffects of sustained convergence. *Perception & Psychophysics*, 1975, 17, 485-491.

Elder, E., Granit, R., & Merton, P. Supraspinal control of the muscle spindles and its significance. *The Journal of Physiology*, 1953, 122, 498-523.

Epstein, W. The process of 'taking-into-account' in visual perception. *Perception*, 1973, 2, 267-285.

Evarts, E. V. Feedback and corollary discharge: A merging of the concepts. *Neurosciences Research Progress Bulletin*, 1971, 9, 86-112.

Gibson, E. J., Gibson, J. J., Smith, O. W., & Flock, H. Motion parallax as a determinant of perceived depth. *Journal of Experimental Psychology*, 1959, 58, 40-51.

Gibson, J. J. What gives rise to the perception of motion? *Psychological Review*, 1968, 75, 335-346.

Gibson, J. J. The senses considered as perceptual systems. Boston: Houghton Mifflin, 1966.

Gibson, J. J. Optical motions and transformation as stimuli for visual perception. *Psychological Review*, 1957, 64, 288-295.

Gibson, J. J. *The perception of the visual world.* Boston: Houghton Mifflin, 1950.

Gogel, W. C., & Tietz, J. The effect of perceived distance on perceived movement. *Perception & Psychophysics*, 1974, 16, 70-78.

Gogel, W. C., & Tietz, J. D. Absolute motion parallax and the specific distance tendency. *Perception & Psychophysics*, 1973, 13, 284-292.

Graham, C. H. Visual Perception. In S. S. Stevens (Ed.), *Handbook of experimental psychology.* New York: Wiley, 1958.

Granit, R. & Kaada, B. R. Influence of stimulation of central nervous structures on muscle spindles in cat. *Acta Physiologica, Scandinavia*, 1952, 27, 130-160.

Hay, J. C. Does head-movement feedback calibrate the perceived direction of optical motion. *Perception & Psychophysics*, 1971, 10, 286-288.

Hay, J. C. Visual adaptation to an altered correlation between eye movement and head movement. *Science*, 1968, 160, 429-430.

Hay, J. C. and Goldsmith, W. M. Space-time adaptation of visual position constancy. *Journal of Experimental Psychology*, 1973, 99, 1-9.

Hay, J. C. & Sawyer, S. Position constancy and binocular convergence. *Perception & Psychophysics*, 1969, 5, 310-312.

Helmholtz, H. von. *A treatise on physiological optics* (Vol. 3). (J. P. C. Southall, Ed. and Trans.). New York: Dover, 1963. (Originally published, 1866.)

Hill, A. L. Direction constancy. *Perception & Psychophysics*, 1972, 11, 175-178.

Holst, E. von & Mittelstaedt, H. The principle of reafference: interactions between the central nervous system and the peripheral organs. In P. C. Dodwell (Ed. and trans.), *Perceptual processing: Stimulus equivalence and pattern recognition.* New York: Appleton, 1971. (Reprinted from *Die Naturwissenschaften*, 1950.)

Howard, I. P., & Templeton, W. B. Harman spatial orientation. New York: Wiley, 1966.

Hunt, C. C. & Kuffler, S. W. Further study of efferent small-nerve fibers to mammalian muscle spindles. Multiple spindle innervation and activity during contraction. *The Journal of Physiology*, 1951, **113**, 283-297.

Irvine, S. & Ludvigh, E. Is ocular proprioceptive sense concerned in vision? *Archives of Opththalmology*, 1936, **15**, 1037-1049.

Jackson, J. H. & Paton, L. On some abnormalities of ocular movements. *Lancet*, 1909, **176**, 900-905.

Kahneman, D. Methods, findings, and theory in studies of visual masking. *Psychological Bulletin*, 1968, **70**, 404-425.

Kalil, R. E. & Freedman, S. J. Persistence of ocular rotation following compensation for displaced vision. *Perceptual & Motor Skills*, 1966, **22**, 135-139.

Keller, E. L. & Robinson, D. A. Absence of a stretch reflex in extraocular muscle of the monkey. *Journal of Neurophysiology*, 1971, **34**, 908-919.

Kornmuller, A. E. Eine experiementelle anaesthesic der auberen augenmuskeln am menschen und ihre auswirkungen. *Journal of Physiological Neurology*, 1930, **41**, 354-366.

Leksell, L. The action potential and excitatory effects of the small ventral root fibers to skeletal muscles. *Acta Physiologica, Scandinavia*, 1945, **10**, Supplement 31.

Linksz, A. *Physiology of the eye*, (Vol. 2), *Vision*. New York: Grune & Stratton, 1952.

Lishman, J. R. & Lee, D. N. The autonomy of visual kinaesthesis. *Perception*, 1973, **2**, 287-294.

Ludvigh, E. Possible role of proprioception in the extraocular muscles. *Archives of Ophthalmology*, 1952, **48**, 436-441.

Mack, A. An investigation of the relationship between eye and retinal image movement in the perception of movement. *Perception & Psychophysics*, 1970, **8**, 291-298.

MacKay, D. M. Elevation of visual threshold by displacements of retinal image. *Nature*, 1970, **225**, 90-92.

MacKay, D. M. Visual stability and voluntary eye movements. In R. Jung (Ed.), *Central processing of visual information*. New York: Springer-Verlag, 1973.

MacKay, D. M. & Mittelstaedt, H. Visual stability and motor control (reafference revisited). In W. D. Keidel (Ed.), *Cybernetics and Bionics*. Munich: Oldenbourg, 1974.

Maruo, T. Electromyographical studies on stretch reflex in human extraocular muscle. *Japanese Journal of Ophthalmology*, 1964, **8**, 96-111.

Matin, E., Clymer, A. B., & Matin, L. Metacontrast and saccadic suppression. *Science*, 1972, **178**, 179-182.

Matin, L. Eye movements and perceived visual direction. In D. Jameson & L. M. Hurvich (Eds.), *Handbook of sensory physiology* (Vol. 7). Heidelberg: Springer-Verlag, 1972.

Matin, L. A possible hybrid mechanism for modifications of visual direction associated with eye movement—the paralyzed-eye experiment reconsidered. *Perception*, in press.

Matin, L., Matin, E., & Pearce, D. G. Eye movements in the dark during the attempt to maintain a prior fixation position. *Vision Research*, **10**, 1970, 837-857.

Matin, L., Matin, E., and Pola, J. Visual perception of direction when voluntary saccades occur: II. Relation of visual direction of a fixation target extinguished before a saccade to a subsequent test flash presented before the saccade. *Perception & Psychophysics*, 1970, **8**, 9-14.

McLaughlin, S. C. & Webster, R. G. Changes in straight ahead eye position during adaptation to wedge prisms. *Perception & Psychophysics*, 1967, **2**, 36-44.

Melvill Jones, G. & Gonshor, A. Goal-directed flexibility in the vestibuloocular reflex arc. In *Basic mechanisms of ocular motility and their clinical implications* (Vol. 24), Wenner-Gren Center Symposium Series. Oxford: Pergamon, in press.

Mitrani, L., Mateeff, St., & Yakimoff, N. Smearing of the retinal image during voluntary saccadic eye movements. *Vision Research,* 1970, 10, 405-409.

Mitrani, L., Mateeff, St., & Yakimoff, N. Is saccadic suppression really saccadic? *Vision Research,* 1971, 11, 1157-1161.

Monahan, J. S. Extraretinal feedback and visual localization. *Perception & Psychophysics,* 1972, 12, 349-353.

Morgan-Paap, C. L. *The constancy of egocentric visual direction.* Unpublished doctoral dissertation, University of Wisconsin, 1975.

Paap, K. R. *Perceptual consequences of post-tetanic-potentiation: An alternative explanation for adaptation to wedge prisms.* Unpublished doctoral dissertation, University of Wisconsin, 1975.

Pola, J. The relation of the perception of visual direction to eye position during and following a voluntary saccade. Unpublished doctoral dissertation, Columbia University, 1972.

Pola, J. Voluntary saccades, eye position, and perceived visual direction. In R. A. Mantz & J. W. Senders (Eds.), *Eye movements and psychological processes.* New Jersey: Lawrence Erlbaum Associates, in press.

Richards, W. A. Visual suppression during passive eye movement. *Journal of the Optical Society of America,* 1968, 58, 1159-1160.

Robinson, D. A. The mechanics of human saccadic eye movements. *The Journal of Physiology,* 1964, 174, 245-264.

Rock, I. The basis of position constancy during passive movement of the observer. *American Journal of Psychology,* 1968, 81, 262-265.

Rock, I. *The nature of perceptual adaptation.* New York: Basic Books, 1966.

Rock, I., & Ebenholtz, S. Stroboscopic movement based on change of phenomenal rather than retinal location. *American Journal of Psychology,* 1962, 75, 193-207.

Roelofs, C. O. Consideration on the visual egocentre. *Acta Psychologia,* 1959, 16, 226-234.

Rubin, M. L. & Walls, G. L. *Fundamentals of visual science,* Illinois: Thomas, 1969.

Shebilske, W. L. Extraretinal information in corrective saccades and inflow vs. outflow theories of visual direction constancy. *Vision Research,* 1976, 16, 621-628.

Shebilske, W. L. Reading eye movements from an information-processing point of view. In D. W. Massaro (Ed.), *Understanding language: An information-processing analysis of speech perception, reading, and psycholinguistics.* New York: Academic, 1975.

Shebilske, W. L. & Fogelgren, L. Changes in straight ahead eye position during reading with head tilted back in median plane: adaptation to the doll reflex (in preparation).

Skavenski, A. A. Extraretinal correction and memory for target position. *Vision Research,* 1971, 11, 743-746.

Skavenski, A. A. Inflow as a source of extraretinal eye position information. *Vision Research,* 1972, 12, 221-229.

Skavenski, A. A., Haddad, G., & Steinman, R. M. The extraretinal signal for the visual perception of direction. *Perception & Psychophysics,* 1972, 11, 287-290.

Skavenski, A. A. & Steinman, R. M. Control of eye position in the dark. *Vision Research,* 1970, 10, 193-203.

Skavenski, A. A., Winterson, B. J., & Steinman, R. M. *The Minivor—a natural means of stabilizing retinal image motion—how good is it?* Paper presented at the Symposium of the Center for Visual Science, Rochester, May 1975.

Stark, L. The control system for versional eye movements. In P. Bach-Y-Rita, C. C. Collins, & J. E. Hyde (Eds.), *The control of eye movements.* New York: Academic, 1971.

Stevens, J. K., Emerson, R. C., Gerstein, T. K., Neufeld, G. R., Nichols, C. W., & Rosenquist, A. C. Paralysis of the awake human: visual perceptions. *Vision Research*, 1976, **16**, 93-98.

Timberlake, G. T., Wyman, D., Skavenski, A. A., & Steinman, R. M. The oculomotor error signal in the fovea. *Vision Research*, 1972, **12**, 1059-1064.

Wallach, H. Perception of motion. *Scientific American*, 1959, **201**, 56-60.

Wallach, H. Brightness constancy and the nature of achromatic colors. *Journal of Experimental Psychology*, 1948, **38**, 310-324.

Wallach, H. & Floor, L. On the relation of adaptation to field displacements during head movements to the constancy of visual direction. *Perception & Psychophysics*, 1970, **8**, 95-98.

Wallach, H. & Frey, K. J. Adaptation in distance perception based on oculomotor cues. *Perception & Psychophysics.*, 1972, **11**, 31-34.

Wallach, H. & Frey, K. J. Adaptation in the constancy of visual direction measured by a one-trial method. *Perception & Psychophysics*, 1969, **5**, 249-252.

Wallach, H., Frey, K. J., & Romney, G. Adaptation to field displacement during head movement unrelated to the constancy of visual direction. *Perception & Psychophysics*, 1969, **5**, 253-256.

Wallach, H. & Kravitz, J. H. Adaptation in the constancy of visual direction tested by measuring the constancy of auditory direction. *Perception & Psychophysics*, 1968, **4**, 299-303.

Wallach, H. & Kravitz, J. H. The measurement of the constancy of visual direction and of its adaptation. *Psychonomic Science*, 1965a, **2**, 217-218.

Wallach, H. & Kravitz, J. H. Rapid adaptation in the constancy of visual direction with active and passive rotation. *Psychonomic Science*, 1965b, **3**, 165-166.

Wallach, H. & Lewis, C. The effects of abnormal displacement of the retinal image during eye movements. *Perception & Psychophysics*, 1966, **1**, 25-29.

Wallach, H., Stanton, L., & Becker, D. The compensation for movement-produced changes of object orientation. *Perception & Psychophysics*, 1974, **15**, 339-343.

Wallach, H., Yablick, G. S. & Smith, A. Target distance and adaptation in distance perception in the constancy of visual direction. *Perception & Psychophysics*, 1972, **11**, 3-34.

Weber, R. B. & Daroff, R. B. Corrective movements following refixation saccades: type and control system analysis. *Vision Research*, 1972, **12**, 467-475.

Weber, R. B. & Daroff, R. B. The metrics of horizontal saccadic eye movements in normal humans. *Vision Research*, 1971, **11**, 921-928.

Whitteridge, D. The effect of stimulation of intrafusal muscle fibers on sensitivity to stretch of extraocular muscle spindles. *Quarterly Journal of Experimental Physiology*, 1959, **44**, 385-393.

Woodworth, R. S. *Experimental Psychology*, New York: Holt, 1938.

Wurtz, R. H. Response of stirate cortex neurons to stimuli during rapid eye movements in the monkey. *Journal of Neurophysiology*, 1969a, **32**, 975-986.

Wurtz, R. H. Comparison of effects of eye movements and stimulus movements on striate cortex neurons of the monkey. *Journal of Neurophysiology*, 1969b, **32**, 987-994.

Young, L. R., Oman, C. M., & Dichgans, J. M. Influence of head orientation on visually induced pitch and roll sensation. *Aviation, Space and Environmental Medicine*, 1975, **46**, 264-268.

CHAPTER

3

THE CONSTANCIES IN OBJECT ORIENTATION: AN ALGORITHM PROCESSING APPROACH

SHELDON M. EBENHOLTZ
University of Wisconsin

Unlike object properties such as size, depth, shape, and color, an object's orientation need not exhibit invariance over changes in spacial position. Thus a cube may be made to translate along any of the three main axes of space or to rotate around any of these axes, yet its object qualities of size, shape, depth, and color will remain unaffected. This is not so in the case of an attribute such as orientation, for orientation represents a relational property whose specification therefore requires the identification of a frame of reference. There are three broad reference systems in relation to which object orientation may be specified: the gravitational, egocentric, and object relative systems, respectively.

Several of the studies reported were supported in part by Grant MH13006, NIMH. The aid of Judith Callan in the study of the aftereffects of maintained head turn on egocentric orientation is gratefully acknowledged.

Orientation may be defined in relation to the direction of gravity in the sense that for any plane suface that constitutes a portion of an object a line on that surface may intersect another line representing the direction of gravity or be parallel with it. The angle of intersection* then represents the orientation of the line in relation to the direction of gravity. This is represented in *Figure 3-1* where line 1 is oriented at angle ϕ in relation to the gravitational direction. The tilt of line 1 is, of course, constant, regardless of the vantage point of the observer 0, represented to the left in Figure 3-1. Yet from the observer's point of veiw, the relative position of 0 and 1 is quite critical; for example, when viewed from the left, as in Figure 3-1, the line is tilted in 0's median plane and it is likely that the perception of the direction of the line relative to 0 would hinge largely on the presence of depth cues signaling the relative distance of the top and bottom of the line from 0. On the other hand, when viewed from the front, the line is tilted in 0's frontal plane, clockwise or counterclockwise, with no depth cues at all required to perceive its relative direction; for example, head-toe, left-right.

When variation in the relative positions of 1 and 0 causes the line to present different aspects to 0, it may be said to have an orientation in relation to 0, that is, an egocentric orientation. It is necessary, however, to identify the egocentric coordinate system more precisely if egocentric orientation is to have an unambiguous meaning, for the self or ego contains many subsystems that can be useful in specifying orientation. Among them are the apparent direction of gaze, in which case, for example, a line may be judged as perpendicular to the direction of gaze or at some angle with respect to it, the apparent median plane, the apparent frontal plane defined in relation to the head, a similar plane defined in relation to the trunk, and any orientation of a line relating to the felt position of a limb. In fact, any body part for which there is proprioception, including the whole body, may be used as a basis for judgment of the *relative* orientation of a visual target and the body part in question. Thus egocentric orientation may represent a large array of coordinate systems with respect to which objects may be encoded in terms of their relative orientation.

The third reference system is entirely object-relative in that orientation is defined in terms of the angular position of one visual target with respect to another.* A most frequently encountered example of this mode of specifying orientation stems from the rod and frame effect (Witkin & Asch, 1948). Here,

*The angle is measured in the plane that contains the direction of gravity and the line in question.

*Other reference frames for orientation include the geographical (e.g., orientation with respect to the compass points) and in fact any conceptualizable spacial system will do. These are probably derivable from the gravitational, egocentric, or relative angular orientational systems already described.

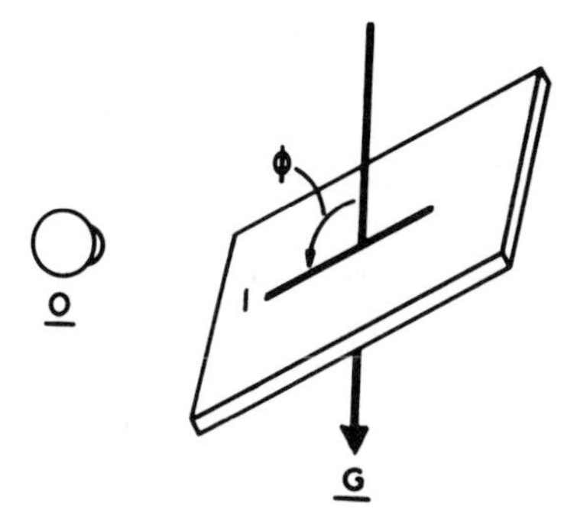

Figure 3-1 The gravitational frame of reference.

when confronted with a large luminous square frame that is tilted in the frontal plane, there is a tendency to perceive an enclosed line as upright when actually it is oriented in the same direction as the frame. Of course, only rarely is the line seen as upright when it is made fully parallel with the sides of the frame. This extreme condition would suggest complete reliance on visual determinants of orientation by obeying a rule by which a line that was parallel with another would appear not to be tilted (i.e., upright) and a line would appear tilted only if it was at some angle in relation to a reference line. In the absence of gravitational and egocentric modes of perceiving orientation a set of vertical parallel lines would appear to have the same orientation as an array of horizontal parallel lines, provided the two arrays were not simultaneously present.

THE CONSTANCY OF ORIENTATION

For each of the three reference systems it is possible to demonstrate a corresponding type or orientation constancy. For this purpose it is necessary to explicate the orientation constancy paradigm. Loosely defined, the constancy process refers to a set of covarying conditions over which a given perceptual attribute remains invariant. In the prototypical case of size constancy retinal size and registered distance represent the covarying conditions over which perceived size is invariant. There are, however, many different examples of orientation constancy and no single specific set of covarying factors is appropriate to all. Nevertheless, in general, some aspect of the visual input or proximal pattern will usually be found to covary with some aspect of the observer's posture or position in space. When perceptual orientation remains relatively invariant despite these changes in proximal stimulation, orientation constancy can be said to have been demonstrated.

In constancy for objects oriented with respect to gravity head tilt or whole body tilt in the frontal plane (i.e., left or right) will cause a rotation in the retinal projection of the line target. Nevertheless, relatively high levels of orientation constancy have been obtained for frontal plane tilts of head and trunk (e.g., Wade, 1968). In analogous fashion, when 0 is tilted backward or forward in his median plane, a gravitationally upright line target will project a pattern of binocular disparities that increases with body tilt and the monocular projections will become increasingly foreshortened. Despite these transformations in the proximal pattern, and even though systematic errors occur in judgments of the upright, relatively high degrees of constancy have prevailed (Ebenholtz, 1970; 1972). The relative positions of 0 and target for the two conditions of gravitationally based orientation are represented in *Figure 3-2.*

Orientation constancy based on egocentric frames of reference has not received much attention. One example of this type of constancy is presented in *Figure 3-3.* The standard line is at some angle θ in relation to 0's frontal plane and is also displaced to 0's left. With 0's head fixed straight ahead, the task is to match the orientation of the standard with that of a comparison line capable of

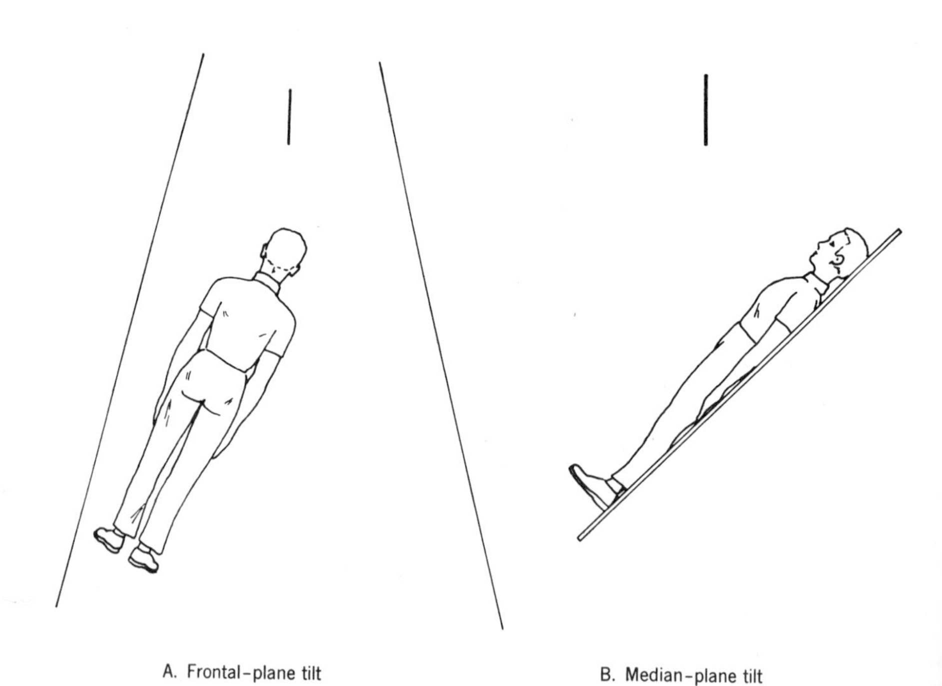

Figure 3-2 Relative positions of observer and target in frontal *(a)* and median plane *(b)* tilts.

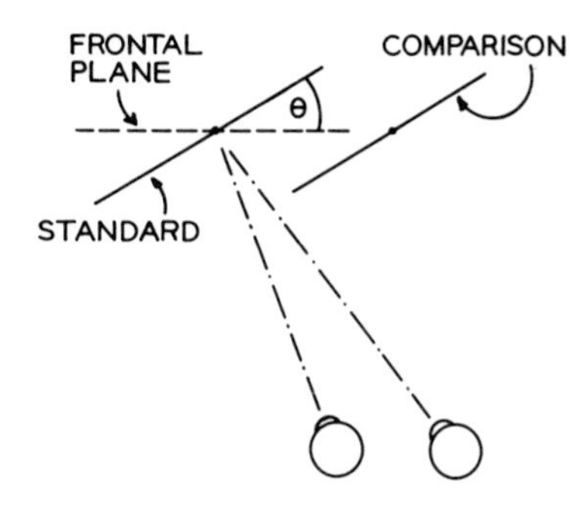

Figure 3-3 Egocentric orientation constancy for tilt in relation to the frontal plane.

rotating around a vertical axis. This type of task, in which orientation is defined in relation to 0's subjective frontal plane, is quite independent of 0's perception of the gravitational direction. It qualifies as a constancy because the retinal projection of the line target is systematically transformed as it is displaced to 0's left or right, and this change in proximal stimulation covaries with 0's direction of gaze. Studies have shown a substantial degree of constancy with a slight tendency toward undercompensation for lines displaced 25° left and right of straight ahead (Ebenholtz & Paap, 1973). Furthermore, evidence is available that the direction of gaze does indeed play a critical role in the compensation process (Ebenholtz & Paap, 1976). In these studies the apparent direction of gaze was independently manipulated to produce an error in apparent displacement of the target. Errors in apparent orientation were obtained in accordance with prediction made with the orientation constancy process. These and an additional study of egocentric orientation constancy are described below.

Constancy for the relative angular orientation of objects seems to lack direct investigation. This is true provided that simple rotation of angles, for example, in the frontal plane, is not counted as a case of orientation constancy. This kind of transformation in the proximal pattern may indeed lead to only slight variations in the apparent magnitude of the angle (e.g., Weene & Held, 1966) as a function of degree of rotation, but because such rotation does not alter the projected angle size it does not qualify as entailing a constancy process. On the other hand, rotation of the object in depth provides the grounds for a constancy of relative angular orientation. The right-angled object represented in Figure 3-4 will provide a changing pattern of binocular disparity as well as a transformation in its monocular projections when the apex angle is rotated toward or away from 0. A constancy of relative angular orientation would obtain if 0 were to judge the angle as constant when seen at various positions in space. It is important to point out that constancy for the relative angular orientations of lines and planes is implied in the definition of shape constancy which in addition refers to a con-

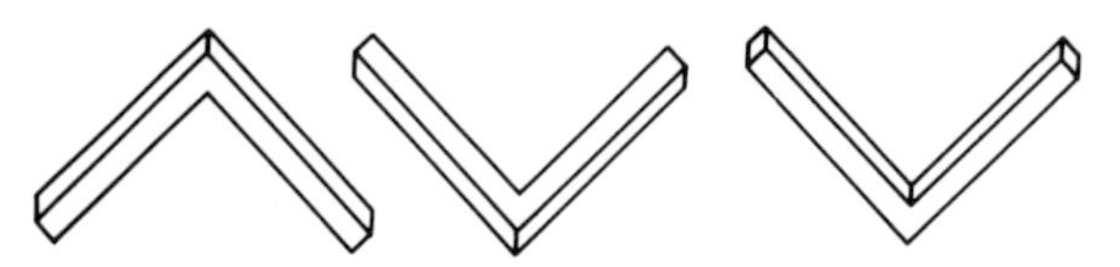

Figure 3-4 Three spacial positions, hence three different retinal projections, of two mutually perpendicular lines.

stancy for the distances separating the lines and planes of a given object. Studies of the constancy of relative angular orientation may therefore represent a more analytic approach to the study of shape constancy than is now the case.

The Perceptual System as an Algorithm Processor

The maintenance of an invariant perceptual quality throughout changes in the visual input on which the quality is thought to depend is a central characteristic of perception and one in need of explication. One solution to this problem is, first to view the perceptual system as though it had access to the proximal pattern and the conditional events on which the changes in visual inputs were dependent. Second, it must be assumed that the perceptual attribute in question is some joint function of the proximal pattern and the conditional events and furthermore that this function, or some related algorithm, is used in analog form by the perceptual system for computation, the outcome of which results in a stable perception (Ebenholtz, 1970; Ebenholtz & Paap, 1973; Ebenholtz & Shebilske, 1973; Epstein, 1973). In constancies for gravitationally oriented targets the visual input must be processed jointly with body tilt information, for in the absence of the latter visual input is totally insufficient to convey information about gravitational orientation. Based on an earlier formulation (Ebenholtz, 1970), an algorithm for lateral body tilt that permits the derivation of apparent object orientation (AOO) may be suggested in terms of registered body orientation (RBO) and the angle α, the latter representing the angle on the retina between the vertical retinal meridian and the image of the line target, namely,

$$\text{AOO} = \alpha + \text{RBO}$$

According to this formulation, when RBO signals a 30° clockwise tilt and α equals 30° counterclockwise, the resultant must indicate an object at zero degrees of tilt, that is, upright. Of course, RBO may not signal head or body tilt accurately over the entire range of body tilt (Schöne & Udo de Haes, 1971) and although α may be assumed to be registered correctly the eye in fact does *not* tilt as much as the head but undergoes counterrolling in the opposite direction

(Miller, 1962). Thus orientation constancy is not perfect, but the departures from full constancy are explicable, at least in principle. Consider, for example, the typical error function generated in studies in which 0 adjusts a line target to appear gravitationally upright from various body tilts, as represented in *Figure 3-5*. The two components of error labeled the A-and E-effects (Muller, 1916) represent departures from constancy, although the overall level of constancy is high nonetheless. The error function can be derived from two separate sources. First, on the premise that the ocular counterrolling is not registered in the perceptual system compensation for the counterrolling cannot take place. The algorithm will·identify a target as vertical when the angle α is exactly equal but opposite in sign to the magnitude of registered body tilt. Because of the countertorsion, however, the vertical retinal veridian will be tilted somewhat less than the head (or body) and therefore overcompensation must occur. This is represented in Figure 3-4 as the E-effect. The second error component, or A-effect, represents undercompensation which is probably due to the failure of the gravity receptors (i.e., the utricle and saccule) to register tilt accurately except near the upright and in fact to decrease in effectiveness with increases in body tilt. Independent evidence for a decreasing role of the otolith organs with increased body tilt is available (e.g., Udo de Haes & Schöne, 1970; Wade, 1970a) to support the present analysis.

Studies of the apparent vertical with 0 tilted in the median plane yield error functions highly similar to those represented in Figure 3-5. Because there is

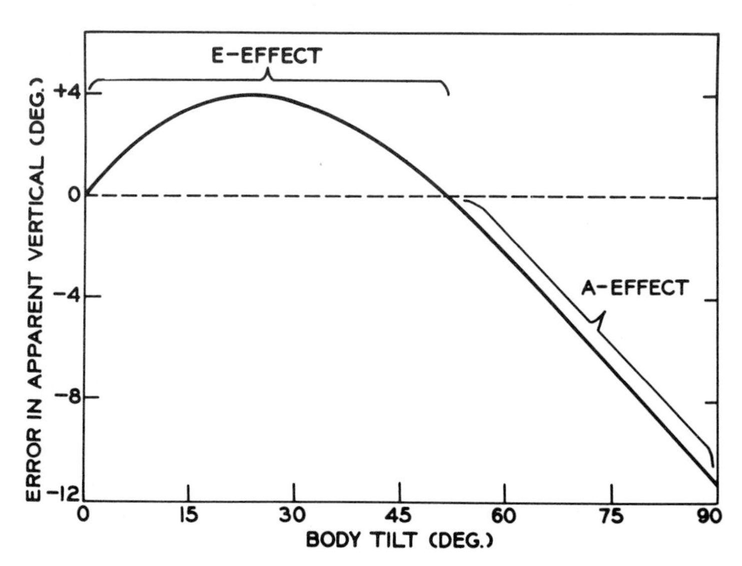

Figure 3-5 Errors in setting a line to the apparent vertical as a function of body tilt (idealized function).

evidence for a downward acting ocular reflex with backward body tilt (Ebenholtz & Shebilske, 1975) that is quite analogous with the counterrolling response to lateral tilt, analogous explanations of the error function for median plane tilts have been developed (e.g., Ebenholtz & Shebilske, 1973). In this case, represented in Figure 3-2b, it is again assumed that the perceptual system solves an algorithm for apparent orientation based on a function relating visual input and registered body orientation. The specific form of the algorithm will differ somewhat from the preceding case, however, for binocular disparity probably plays a major role in defining the input with median plane tilts. Nevertheless, at least in principle, orientation constancy can be deduced from the algorithm and departures from constancy can be accounted for in terms of special conditions associated with the terms expressed in the algorithm.

Some Evidence for Algorithm Processing

Deductions based on an algorithm-processing approach serve the purpose of formal explanation in that the algorithm, together with a few additional premises concerning limiting conditions, permits the logical deduction of the constancy in question. Independent evidence, however, is required for the hypothesis that the perceptual system is, in fact, an algorithm processor. To this end Ebenholtz and Paap (1976) sought to probe the perceptual system by specifically biasing one of the terms of the algorithm. A predictable change in perception should result. The constancy under investigation is derived from the observation that a horizontal line displaced laterally (or a vertical line displaced vertically) in the frontal plane projects a pattern of binocular disparities, when fixated that is vastly different from the same horizontal (or vertical) line when viewed from a position of symmetrical convergence (and at eye level). Yet, as noted earlier, there is a relatively high level of orientation constancy under these conditions (Ebenholtz & Paap, 1973). *In Figure 3-6* the geometrical basis for the change in disparity with displacement is represented. When viewing a line target at a displacement angle ϕ from the straight-ahead position, it must be rotated by θ from the frontal plane in order to project the identical pattern of binocular disparities as the frontal target viewed straight ahead. This follows from the fact that only lines tangent to the Vieth-Muller circle at the point of fixation will produce equivalent disparities. Accordingly, a line that is displaced but *not* rotated must produce disparities at the displaced position that are different from those produced in the straight-ahead position. Because a constancy has been demonstrated for these targets, it can be assumed that some compensation process exists such that the apparent orientation (AO) of the target in relation to the frontal plane is some joint function of the binocular disparity (η) and the displacement angle (ϕ), namely, AO $= f(\eta,\phi)$. It follows that if the visual input

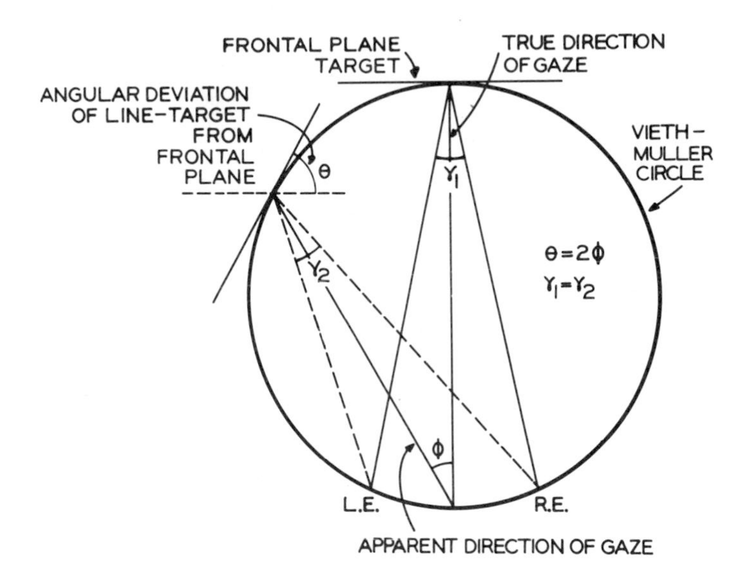

Figure 3-6 The relation between apparent displacement angle ϕ and apparent orientation angle θ.

(η) were to be held constant AO would then vary with the displacement angle ϕ. Furthermore, if the perceptual system reflects the geometry of the Vieth-Muller horopter, the orientation angle (θ) of the target should increase by twice the angle ϕ. This is represented in Figure 3-6, in which, the line is seen to the left but actually is straight ahead.

In order to vary ϕ but keep η constant two procedures were used which resulted in a change in registered direction of gaze (e.g., Ebenholtz, 1976; McLaughlin & Webster, 1967; Park, 1969). In one case Os were adapted to vertically or laterally displacing wedge prisms; in another Os maintained an asymmetrical direction of gaze (to the right or left) for about 1 minute. After exposure to these conditons they felt that the direction of gaze had been deviated either in the direction of the prism base or opposite the direction in which the eyes were maintained, while actually fixating straight ahead. The task was simply to adjust a horizontal or vertical luminous line to the apparent frontal plane while viewing the target with the eyes and head straight ahead. There were four separate experiments and in each the effect of inducing an error in the direction of gaze in one of two opposed directions, that is, up-down, or right-left, was examined. In eight independent tests all but two failed to show significant orientation effects in the predicted direction.

As a further test of this approach the effects of errors in apparent *head* position on the apparent orientation of a horizontal line were examined. The O's task was to adjust a luminous horizontal line pivoting around a vertical axis to a position in which it appeared to be parallel with his shoulders. After making an initial pair of orientation settings O rotated his head 30° right or left and maintained that position while fixating a dim red light for about 10 minutes. After O returned his head to a straight-ahead position a second pair of orientation settings was made. In order to check on the efficacy of this procedure as a method of inducing a lateral shift in the position of a visual target Os were asked to place their unseen index finger in a position directly below a small black dot representing the center of the luminous line. The finger moved in a thimblelike apparatus in a track cut with a radius of curvature of 30 cm. The track was placed on a table surface at 30 cm from a plane tangent to the front surface of O's corneas. The luminous line was placed 43.5 cm above the track at eye level and also at a distance of 30 cm from O. Measures of apparent displacement were also taken before and after the 10-minute head-rotation period and counterbalanced with orientation judgments. One group of eight Os was exposed to a leftward head turn and a second group of eight turned their heads to the right.

Aftereffects of maintained head position on the apparent location of visual targets have been demonstrated (Ebenholtz, 1976). They probably arise from the inertial tendency of muscle-innervation patterns to maintain themselves beyond the point at which relaxation should occur (Ebenholtz & Wolfson, 1975). Thus, to maintain the head straight ahead after a sustained leftward posture, voluntary innervation of the rightward-acting muscles would be required to balance the continued involuntary action of the leftward-acting muscles. It follows that O would feel that his head had been turned to the right when in fact it was straight ahead, and when it felt straight ahead it would actually be turned toward the left in the direction of the inducing posture (Howard & Anstis, 1974). Furthermore, a target fixated with head and eyes straight ahead would appear displaced to the right (Ebenholtz, 1976).

If we assume that, with eyes straight ahead, apparent orientation is some joint function of visual input and registered head position, then errors in the latter should be manifest in orientation errors, given that visual input is constant. Accordingly, if O feels that his head has turned 5° right when actually he is looking straight ahead, a horizontal line truly in O's frontal plane should appear, rotated with its right side toward O. To perceive the line as paralled with the shoulders O would have to rotate the line 5°, right side away.

The results are represented in Table 1 in which, for both displacement and orientation, the direction of the aftereffect is precisely as predicted. In addition, each mean differed significantly from zero at $p = .05$ or better. These results, together with those of Ebenholtz & Paap (1976) clearly show that orientation perception is determined by apparent displacement which in turn was mediated by changes in apparent head position and apparent direction of gaze. Similar effects on the apparent vertical have been reported by Wade (1970a) and others

Table 1

AFTEREFFECTS OF MAINTAINED RIGHT OR LEFT HEAD TURN ON APPARENT DISPLACEMENT AND ORIENTATION (DEGREES)

		Displacement		Orientation	
		Mean	σ_m	Mean	σ_m
Sustained	30° Left	3.25	.65	3.82	1.35
Posture	30° Right	-1.78	.94	-1.31	.71

Note: The minus sign denotes a leftward displacement and a rotation left side away.

as a result of adaptation of the entire body to rightward tilts of 30, 60, and 90°. After a 3-minute adaptation period a slight decrease in the E-effect (i.e., in overcompensation) of about 1° occurred at 30° of body tilt. At 90° of tilt, however, the A-effect showed an increase of about 8° which indicated a large increase in undercompensation. This is precisely the trend to be expected on the grounds that adaptation to whole body tilt causes O to register a tilt that is less than his actual angle. Unfortunately, however, Wade did not provide an independent measure of changes in registered or apparent body tilt.

A MODEL OF THE COMPENSATION PROCESS

Figure 3-7 is a flow-chart model of the algorithm processing approach in which each rectangle represents an important locus within the total compensation process.

The "visual input" represents familiar cue systems such as motion parallax, stereopsis, and monocular or pictorial cues such as linear perspective, and any other relevant aspects of the retinal image. We might also include certain eye movements that serve to convey information about orientation. These inputs fit well into the information extraction schema advanced by J. J. Gibson, but it would appear that they lead primarily to egocentric and/or relative judgments; for example, they constitute perceptions of the relative slant angle between a surface and the observer's line of sight (Rosinski, 1974) or the relative slant between intersecting surfaces. Information about the observer's position in relation to the gravitational direction or about the head in relation to the trunk are, of course, nonoptical but are required for constancy-type judgments and for some forms of egocentric orientation described earlier. It is well to point out,

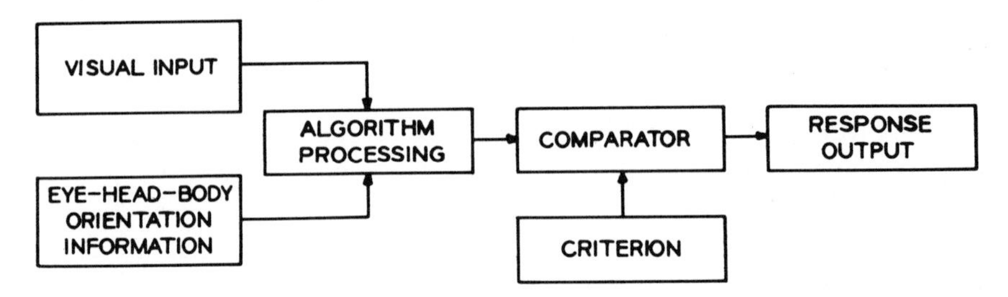

Figure 3-7 A flow chart representing the important loci in an algorithm-processing model of perceptual orientation.

therefore, that the algorithm-processing approach is not meant as a substitute for information extraction but is necessary to complement it. Insofar as the products of algorithm processing lead to the perception of attributes not signaled uniquely by information in the optic array, we might contrast the two approaches in terms of *information extraction* and *information production*. On this account the perceptual constancies as a class represent clear instances of *information production*.

The locus that represents "eye-head-body" tilt information refers to the various systems that are capable of signaling the position of the head in relation to gravity, the angular position of the eyes in relation to the head, and the relative position of all other relevant body parts. To this end the otolith organs (i.e., saccule and utricle) of the inner ear probably play a major role even though their effectiveness may diminish as head tilt departs from the vertical (Udo do Haes & Schöne, 1970). Other sources of tilt information are the joint receptors such as the Golgi-tendon organs (Skoglund, 1973) and possibly the pressure transducers, stretch receptors, and a variety of mechanoreceptors that might subserve the function of signaling a shift in pressure along the body surface or a change in the direction of gravitational force acting on the viscera.

One of the critical problems encountered in any evaluation of the algorithm-processing model stems from the absence of knowledge concerning precisely which systems actually have an input into the algorithm. Clearly, not all sources of tilt information need take part. Furthermore, not all sources of tilt information need contribute to the conscious sense of position, that is proprioception, and there is no necessity for them to be identical. Yet tests of the algorithm model frequently entail a correlation between the predicted output (i.e., apparent object orientation) and the event on which visual input is conditional; for example, body tilt (Ebenholtz, 1970) or direction of gaze (Ebenholtz & Paap, 1976). Significant correlations have not been found in several studies (Ebenholtz, 1972; Ebenholtz & Paap, 1973; 1976), even though the algorithm approach was supported in other respects. Some part of this problem may be statistical, related to the degree of reliability of the individual scores. More important, however, is the possibility that the absence of significant correlation arises from the partial overlap between the inputs to the algorithm processor and those underlying the conscious states measured by the dependent variable. This holds for other constancies as well.

"Algorithm processing" refers to the joint processing of the proximal stimulation with the events on which it is conditional according to some function, the solution to which represents some unique perceptual state. This is the locus for what has been referred to above as "information production," for it seems to share many of the creative aspects of language learning and represents a clear example of what may be termed a perceptual grammar. The boundary conditions over which the algorithm operates and beyond which it breaks down (i.e., constancy fails) and the reasons for the failure of compensation remain to be explored. Also awaiting answers are questions concerning the development of the algorithms, their innate status, and the possibility of the development of synthetic algorithms. It remains to be noted that the concept of an algorithm processor shares a direct kinship with Helmholtz's empiricism and his conception of perception as the result of a process of induction or unconscious inference. Although we need not subscribe to the idea that all perception is so determined, it is quite possible that the particular function that represents the algorithm is itself a product of experience; for example, the change in retinal size with distance and retinal orientation with body tilt. This is an empirical question.

One final point in this connection is based on the observation that the algorithm-processing approach entails nonvisual functions in the use of eye-head-body-tilt information* and in the course of solving the function-rule itself. Clearly, processing outside the visual areas is implied in the perceptual constancies. This conclusion has been supported for size (Blakemore, Garner, & Sweet, 1972) and orientation constancy (Mitchell & Blakemore, 1972).

In many tasks O must judge whether a standard appears in a certain way; for example, gravitationally upright or tilted at 30°. The output of the processor must then be compared with a criterion perceptual state. This may be accomplished directly with a comparison target or indirectly by reference to a criterion stored in memory. The outcome of this comparison will then determine O's response.

It is of some interest to note that it is at the locus of the "comparator" that the role of "objective" and "apparent" instructions may be represented. "Apparent" instructions are presumed to represent the typical output of the processor represented in the comparator, but when objective instructions are

*It is important to acknowledge the historical antecedants of the present approach in the sensory-tonic field theory of preception. Although predicated on concepts different from those developed in this chapter sensory-tonic theory emphasizes the role of postural factors in orientation perception (e.g., McFarland, Wapner, & Werner, 1962).

issued a special criterion is implied. In the case of constancy for the gravitational direction "objective" instructions may well imply that the target should be made to appear the way objects that are *known* to be upright appear—at any given body or head tilt. For large tilts vertical surfaces such as walls appear to recede from the O as they increase in height; hence the criterion would likewise require the target to appear to recede in order to fulfill the objective instruction mandate; namely, "set the target so that it *is* vertical, regardless of how it looks." Because setting the vertically oriented target to appear to recede represents a case of overcompensation for backward body tilts, a reduction in the A-effect is to be expected and has been obtained for large body tilts (Ebenholtz & Shebilske, 1973). For small body or head tilts, for example, about 25° only a small criterion shift, if any at all, should occur, for experience with vertical surfaces does not yield an obvious distortion at these tilt angles. Data bear out this possibility also because no significant effects of instruction have been obtained at small body tilts (Ebenholtz & Shebilske, 1973; Wade, 1970b). The reason for this pattern of results seems to reside in the fact that in the range of body tilts for which constancy is reasonably good experience does not indicate other than vertical objects that also *look* vertical. At extreme body tilts, however, at which constancy falls off, vertical objects would be seen as tilted. Hence an altered criterion would be brought to bear but primarily for the range over which constancy normally breaks down. A similar analysis would seem to be appropriate for an explanation of instructional effects in other constancies as well, for example, Carlson (this volume) on size constancy and Landauer (1964) on shape constancy, provided that O is capable of formulating the rule that describes the look of objects when constancy fails. An interaction of age with instructions is to be expected on this account.

FURTHER IMPLICATIONS

That apparent object orientation is a function of visual input and nonvisual factors (see Figure 3-7) permits the examination of two areas of research in terms of explanations that have not yet been thoroughly explored.

The first concerns the phenomenon of perceptual adaptation to optical tilt (Ebenholtz, 1966; Mack, 1967; Mikaelian & Held, 1964). The immediate effect of viewing through Dove prisms or a similar optical system is to cause the scene to appear rotated or tilted, the more so the stronger the optical tilt. Within about 30 min. however, adaptation may be discerned in the sense that judgments of the apparent vertical of a thin luminous line, taken in a darkened environment without prisms, will shift in the direction of the optical distortion. Therefore, although a line would appear upright for the upright head when within 1° of the gravitational vertical, after adapting to a 20° clockwise tilt for 30 minutes, it would appear upright when actually tilted clockwise by about 4°. Furthermore, by selecting rapid adaptors Mikaelian and Held (1964) have been able to demonstrate a complete adaptation aftereffect of 20° to a 20° tilt in 2 hours. What might the locus of this effect be? The algorithm-processing model shown in Figure 3-7 suggests several sources for the change in perception caused by prism adaptation:

1. The eye could be induced to undergo torsion during adaptation and the torsion might outlast the exposure period to yield aftereffects during the test procedure. This possibility can be ruled out, for there is no evidence that inspection of a tilted line can induce torsion (Howard & Templeton, 1964). Furthermore, in a context directly related to prism adaptation Russotti (1968) measured ocular torsion photographically in Os who adapted to a 20° optical tilt. Although significant levels of adaptation were obtained, no significant ocular torsion was present. To these studies may be added the observations made by Judith Callan and me on Os who matched the position of an afterimage of a vertical line with a thin luminous line. The matching was accomplished before and again after a 10-minute exposure period to a 30° optical rotation. Measures of the apparent vertical also were made. No O who showed an adaptation aftereffect also showed a comparable shift in matching to the after image.
2. The effects of optical tilt may be to induce an apparent tilt in the head in relation to the trunk or even in the whole body in relation to gravity. If such an effect were opposite the direction of optical tilt, the adaptation aftereffect could be explained. This possibility has not received support. Unpublished observations by the author indicate that the adapted O does *not* set his head at an angle when requested to position it so that it feels upright while in a darkened environment. Furthermore, Mack & Rock (1968) have obtained significant adaptation aftereffects when these were measured by having O look down at a horizontal plane to position a luminous target so that it appeared to be at a 12 o'clock position. As these authors have noted, shifts in apparent orientation of the 12 o'clock position cannot reflect changes in apparent head position, for the task entails an egocentric frame of reference that moves *with* any change in felt head position. Adaptation to optical tilt therefore cannot be accounted for in terms of induced changes in head or body orientation.
3. We are left with the alternative that prisms cause a change in the registration of the proximal pattern itself. Indeed, this is the view proposed by Mack and Rock (1968) and generally by Rock (1966) with respect to other forms of prism adaptation. The implication is that an egocentric change occurs in which a retinal image of a particular orientation represents a changed orientation after prism exposure, all other factors being constant. This change in retinal-cortical significance presumably underlies tilt adaptation. A remaining possibility, however, stems from the role of scanning eye movements in the perception of orientation.

Little is known about the role of eye movements as a source of information equivalent to visual input into the perceptual system, but it is clear, at least potentially, that these eye movements do convey *egocentric* information; that is, they are capable of encoding orientation only in relation to the head. Thus it is the interpretation of the relative direction of scanning eye movements that may change during exposure to optical tilt and may underlie the adaptation aftereffects. No direct tests of this hypothesis are presently available, although changes in the compensatory eye-head movement system (i.e., the vestibulo-ocular reflex) have been induced by prism exposure in humans and lower organisms (Robinson, 1975; Gonshor & Jones, 1973).

The second research area concerns the interpretation of the rod and frame effect (RFE), described earlier in the discussion of reference systems for orientation. The explanation, in terms of accepting the main lines of the visual field as representing the apparent vertical (Koffka, 1935), is a "relational" explanation and may operate quite independently of the algorithm processing approach. There are problems with Koffka's suggestion, however, in that direct estimates of the degree of frame tilt indicate that the frame is not perceived as less tilted than it actually is (Gogel, 1975). Although relational factors may play some role in orientation perception in the presence of visual frames of reference, another set of determinants may also be operating. It is conceivable, for example, that the frame induces an apparent tilt in the O opposite in direction to the frame tilt.* Setting the line opposite the apparent body tilt would then yield a setting *in* the direction of frame tilt and the RFE could be deduced. The algorithm-processing approach itself would then underlie this explanation, for the RFE would depend on "faulty" head or body-tilt information entering the algorithm processor.

There is evidence that surrounding frames of reference do induce an apparent body tilt (Passey, 1950; Witkin, 1949, 1952), but the method of measurement typically has been to have O adjust his tilt to the postural vertical rather than make direct estimates without body movements. This method may have contributed to the relatively small effects obtained by Passey. Furthermore, no published data directly compare the effects under identical conditions, of a reference frame on apparent line orientation with its effects on apparent body tilt. If the RFE is determined, at least in part, by the suitability of the frame as an object capable of inducing an apparent body (or head) tilt, the differential effects of a tilted mirror, tilted frame, tilted room, or tilted field of parallel lines (Singer, Purcell, & Austin, 1970. Witkin & Asch, 1948) on the RFE may not be a consequence of framework articulation but would follow from the differential

*Similar suggestions can be found in Brosgole & Cristal (1967) and Templeton (1973), although the possibility was dismissed by the first-mentioned authors.

effectiveness of these configurations as induction conditions for apparent body and/or head tilt. Likewise, the relatively low RFE obtained with a small frame (Beh, Wenderoth, & Purcell, 1971) may be attributable to the likely decrease in effectiveness of induction objects with size.

REFERENCES

Beh, H. C., Wenderoth, P. M., & Purcell, A. T. The angular function of a rod-and-frame illusion. *Perception & Psychophysics*, 1971, 9, 353-355.

Blakemore, C., Garner, E. T., & Sweet, J. A. The site of size constancy. *Perception*, 1972, 1, 111-119.

Brosgole, L. & Cristal, R. M. The role of phenomenal displacement on the perception of the visual upright. *Perception & Psychophysics*, 1967, 2, 179-188.

Ebenholtz, S. M. Additivity of aftereffects of maintained head and eye rotations: An alternative to recalibration. *Perception & Psychophysics,* 1976, 19, 113-116.

Ebenholtz, S. M. The constancy of object orientation: Effects of target inclination. *Psychologische Forschung*, 1972, 35, 178-186.

Ebenholtz, S. M. Perception of the vertical with body tilt in the median plane. *Journal of Experimental Psychology*, 1970, 83, 1-6.

Ebenholtz, S. M. Adaptation to a rotated visual field as a function of degree of optical tilt and exposure time. *Journal of Experimental Psychology*, 1966, 72, 629-634.

Ebenholtz, S. M. & Paap, K. R. Further evidence for an orientation constancy based upon registration of ocular position. *Psychological Research,* 1976, 38, 395-409.

Ebenholtz, S. M. & Paap, K. R. The constancy of object orientation: Compensation for ocular rotation. *Perception & Psychophysics*, 1973, 14, 458-470.

Ebenholtz, S. M. & Shebilske, W. The doll reflex: ocular counterrolling with head-body tilt in the median plane. *Vision Research*, 1975, 15, 713-717.

Ebenholtz, S. M. & Shebilske, W. Instructions and the A and E effects in judgments of the vertical. *American Journal of Psychology*, 1973, 86, 601-612.

Ebenholtz, S. M. & Wolfson, D. M., Perceptual aftereffects of sustained convergence. *Perception & Psychophysics,* 1975, 17, 485-491.

Epstein, W. The process of taking-into-account in visual perception. *Perception*, 1973, 2, 267-285.

Gogel, W. C. & Newton, R. E. Depth adjacency and the rod-and-frame illusion. *Perception & Psychophysics*, 1975, 18, 163-171.

Gonshor, A. & Jones, G. M. Changes in human vestibulo-ocular response induced by vision-reversal during head rotation. *The Journal of Physiology*, 1973, 234, 102 p.

Howard I. P. & Anstis, T. Muscular and joint-receptor components in postural persistence. *Journal of Experimental Psychology*, 1974, 103, 167-170.

Howard, I. P. & Templeton, W. B. Visually-induced eye torsion and tilt adaptation. *Vision Research,* 1964, 4, 433-437.

Koffka, K. *Principles of gestalt psychology*. New York: Harcourt, Brace; 1935.

Landauer, A. A. The effect of viewing conditions and instructions on shape judgments. *British Journal of Psychology*, 1964, 55, 49-57.

Mack, A. The role of movement in perceptual adaptation to a tilted retinal image. *Perception & Psychophysics*, 1967, 2, 65-68.

Mack, A. & Rock, I. A re-examination of the Stratton Effect: Egocentric adaptation to a rotated visual image. *Perception & Psychophysics*, 1968, 4, 57-62.

McFarland, J. H., Wapner, S., & Werner, H. Relation between perceived location of objects and perceived location of ones own body. *Perceptual & Motor Skills*, 1962, 15, 331-341.

McLaughlin, S. C. & Webster, R. G. Changes in straight-ahead eye position during adaptation to wedge prisms. *Perception & Psychophysics*, 1967, 2, 37-44.

Mikaelian, H. & Held, R. Two types of adaptation to an optically rotated visual field. *American Journal of Psychology*, 1964, 77, 257-263.

Miller, E. F. II. Counterrolling of the human eyes produced by head tilt with respect to gravity. *Acta Otolaryngology*, 1962, 54, 479-501.

Mitchell, D. E. & Blakemore, C. The cite of orientational constancy. *Perception*, 1972, 1, 315-320.

Muller, G. E. Uber das Aubertsche Phänomen. *Zeitschrift für Psychologie und Physiologie der Sinnesorgane*, 1916, 49, 109-244.

Park, J. N. Displacement of apparent straight ahead as an aftereffect of deviation of the eyes from normal position. *Perceptual & Motor Skills*, 1969, 28, 591-597.

Passey, G. E. The perception of the vertical. IV. Adjustment to the vertical with normal and tilted visual frames of reference. *Journal of Experimental Psychology*, 1950, 40, 738-745.

Robinson, D. A. How the oculomotor system repairs itself. *Investigative Ophthalmology*, 1975, 14, 413-415.

Rock, I. *The nature of perceptual adaptation.* New York: Basic Books, 1966.

Rosinski, R. R. On the ambiguity of visual stimulation: A reply to Eriksson. *Perception & Psychophysics*, 1974, 16, 259-263.

Russotti, J. S. The measurement of eye torsion during adaptation to a tilted visual field. Unpublished M. A. thesis, Psychology Department, Connecticut College, New London, 1968.

Shöne, H. & Udo de Haes, H. A. Space orientation in humans with special reference to the interaction of vestibular, somaesthetic and visual inputs. In H. Drischel and N. Tiedt (Eds.), *Biokybernetik,* (Vol. 3.,). Jena: Gustav Fischer, 1971, pp. 172-191.

Singer, G., Purcell, A. T., & Austin, M. The effect of structure and degree of tilt on the tilted room illusion. *Perception & Psychophysics*, 1970, 7, 250-252.

Skoglund, S. Joint receptors and kinaesthesis. In A. Iggo (Ed.), *Somatosensory system*, Chapter 4, pp. 111-136. *Handbook of sensory physiology* (Vol. 2). New York: Springer-Verlag, 1973.

Templeton, W. B. The role of gravitational cues in the judgment of visual orientation. *Perception & Psychophysics*, 1973, 14, 451-457.

Udo de Haes, H. A. & Schöne, H. Interaction between statolith organs and semicircular canals on apparent vertical and nystagmus. *Acta Otolaryngology*, 1970, 69, 25-31.

Wade, N. J. Effect of prolonged tilt on visual orientation. *Quarterly Journal of Experimental Psychology*, 1970a, 22, 423-439.

Wade, N. J. Effect of instructions on visual orientation. *Journal of Experimental Psychology*, 1970b, 83, 331-332.

Wade, N. J. Visual orientation during and after lateral head, body, and trunk tilt. *Perception & Psychophysics*, 1968, 3, 215-219.

Weene, P. & Held, R. Changes in perceived size of angle as a function of orientation in the frontal plane. *Journal of Experimental Psychology*, 1966, **71**, 55-59.

Witkin, H. A. Further studies of perception of the upright when the direction of the force acting on the body is changed. *Journal of Experimental Psychology*, 1952, **43**, 9-20.

Witkin, H. A. Perception of body position and of the position of the visual field. *Psychological Monographs*, 1949, **63**, (No. 7), Whole No. 302.

Witkin, H. A. & Asch, S. E. Studies in space oreintation. IV. Further experiments on perception of the upright with displaced visual fields. *Journal of Experimental Psychology*, 1948, **38**, 762-782.

CHAPTER

4

STEREOSCOPIC DEPTH CONSTANCY

HIROSHI ONO AND JAMES COMERFORD
York University

In many ways depth constancy is similar to size constancy, the most studied of the perceptual constancies. In both, we are dealing with an observer's ability to judge that a given linear extent remains the same when the viewing distance varies. In depth constancy, however, the judgment is of the linear extent of a stimulus in a saggital plane, and in size constancy the judgment is of the linear extent in a fronto-parallel plane; for example, imagine that you are holding a pencil 50 cm away and pointed at your nose. If the pencil appears to be the same length when seen 25 cm away, then depth constancy has occurred. If the pencil is viewed broadside and appears to be the same length at the two distances, size constancy has occurred. These constant judgments emerge in spite of

We wish to thank R. Angus, T. Anstis, J. Foley, M. Komoda, I. P. Howard, and W. B. Templeton for comments on an earlier version of this chapter, W. Gogel and W. Richards for checking the accuracy of our discussion of their work, and G. Gonda for her help in all phases of its preparation.

changes in the relevant proximal stimuli when the viewing distance is changed. Although the judgment required is similar, the information necessary for the two constancies is different, and they are usually treated as different topics.

Depth constancy is discussed in this chapter, but the discussion is restricted to the perception of depth that develops from binocular or retinal disparity—the difference in the location of images in the two eyes. The perception of depth, of course, can derive from other sources; for example, cues from the convergence or accomodation of the two eyes may provide information about the distance between a point in space and an observer. The depth from these cues is given only indirectly; it would be implied by the difference between two perceptions of distance. In contrast, several cues, namely, interposition, relative size, and disparity, can produce depth information directly, but they provide only ordinal information about the position of points in spece, and by themselves are not sufficient for depth constancy to occur. The disparity cue, if coupled with information concerning veiwing distance, can produce a metric of depth. (This fact is elaborated on later.) Stereoscopic depth constancy refers to the perception of a constant linear extent from binocular disparity at different distances. The discussion in this chapter centers around this topic.

In the next section we define disparity and related concepts. This section does not deal specifically with depth constancy. Its primary purpose is to introduce some of the terminology of stereoscopic vision and to discuss a caveat about computing disparity when stimuli are near an observer. In the following section consideration is given to the fact that the visual system must also process distance information for depth constancy to occur. Just as the retinal image size alone is insufficient information for size constancy, disparity alone cannot provide sufficient information for depth constancy because the disparity produced by a constant physical separation decreased in proportion to the square of the viewing distance from an observer (the inverse square law). For depth constancy to occur the visual system must somehow take into account the viewing distance as well.

The subsequent discussion is an examination of the existing data and hypotheses relevant to depth constancy. Experiments that deal explicitly with depth constancy are few, and many theories of stereopsis do not address themselves directly to the problem of determining how the visual system processes viewing distance and disparity together. This is surprising because the inverse square law has been known for a long time and stereopsis has been studied extensively. Kaufman (1974) states that "all theories are really inconsistent with the geometry of stereopsis," that is, the inverse square law. Perhaps he overstated his case, and it may be better to say that most theories are incomplete rather than inconsistent because many theories do not make explicit how information about viewing distance interacts with disparity information. Although they are not comprehensive theories of stereopsis, four hypotheses are relevant to depth constancy; in the last section of this chapter we examine them in detail with reference to the empirical evidence.

BINOCULAR DISPARITY AND RELATED CONCEPTS

If the left and right-eye images of an object are located at the same distance and in the same direction from their respective foveae, the two images are said to fall on corresponding retinal points. Alternatively, corresponding retinal points may by defined as those points that, when stimulated, give rise to identical visual directions. Although this definition is often presented, in our view it is an empirical question whether identical retinal loci produce identical visual directions. Hering (1942) illustrates corresponding points by referring to the observation of stars. Because stars are located at optical infinity, the image of a given star would be located at the same position on the two retinae if the two eyes were pointed straight ahead. The two images of a star are said to lie on corresponding points.

The two images of an object fall on corresponding points in the two eyes when points of the object lie on a Vieth-Mueller (V-M) circle. For a given fixation point the V-M circle (sometimes called a theoretical horopter) passes through the nodal points of the eyes (where all visual lines intersect) and the fixation point. When points do not lie on the V-M circle, their images fall on noncorresponding retinal points in the two eyes. The extent of noncorrespondence is generally specified by a difference in the angles formed by the lines of sight to the stimuli ($\alpha - \beta$), as in Figure 4-1, for two stimuli, A and B. Disparity is specified in this way because there is an approximate one-to-one relationship between the difference in the two angles, $\alpha - \beta$, and the difference in the retinal loci.

Disparity as we have defined it is also equivalent to the difference between the two convergence angles for fixation on A and B; that is, if an observer is fixating on A, the angle α specifies the convergence angle of the fixation point; if he is fixating on B, the angle β specifies the convergence angle of this fixation point. The disparity δ will correspond to the difference between the two convergence angles, that is, $\alpha - \beta$. We might therefore define disparity as the difference between two convergence angles. This definition, however, has the disadvantage that it might lead us to think that a change in fixation from one point to another is necessary for disparity to be an effective cue to depth, whereas it is empirically well established that, even without a change in convergence, horizontal disparity information can be processed to produce perception of depth. The second disadvantage in defining disparity as the difference between two convergence angles is that an incorrect value will result when one of the points is close to the observer because of a change in internodal distance as a function of fixation dis-

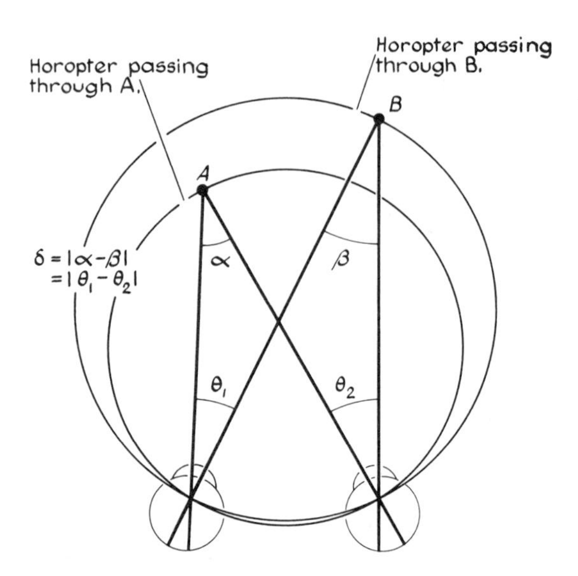

Figure 4-1 An illustration of binocular disparity. When A is fixated, the locus of corresponding points lies on the Vieth-Mueller circle drawn through A; when B is fixated, the locus of corresponding points lies on the Vieth-Mueller circle drawn through B. Binocular disparity between two points occurs when both do not lie on the same Vieth-Mueller circle. Disparity may be defined by the difference between α and β or the difference between θ_1 and θ_2.

tance. (Why this is so should become clear after further discussion of the definition of disparity.)

Another equivalent definition of disparity is the difference between the two visual angles subtended by the points A and B at each eye. Let θ_1 be equal to the visual angle subtended by A and B at the right eye and θ_2 be equal to the visual angle subtended at the left eye. (See Figure 4-1.) Because we have two triangles with equal opposite angles, the sum of the remaining two angles in each triangle is equal, that is, $\alpha + \theta_1 = \beta + \theta_2$. Therefore $\alpha - \beta$ equals $\theta_2 - \theta_1$.

The sign of the disparity specified by $\alpha - \beta$ or $\theta_2 - \theta_1$ will differ, depending on the point of fixation. If in Figure 4-1 the observer fixates on A, point B produces a disparity of δ; if he fixates on B, point A also produces the same disparity δ but of opposite sign. The convention is to label as "uncrossed" disparity whatever arises when A (the near point) is fixated and as "crossed" disparity whatever arises when B (the far point) is fixated. Uncrossed disparity arises from any point outside a V-M circle and crossed disparity from any point within the circle. The points on the circle produce no disparity and fall on corresponding points.

The difference between uncrossed and crossed disparity can be seen in a simple demonstration. Hold two fingers up in the median plane. First fixate on the near finger and close one of your eyes. If you are viewing your fingers with your right eye, the farther finger will be seen on the right side of the median plane. When you view your fingers with your left eye, the farther finger will be seen on the left side. The far finger produces an uncrossed disparity. Now fixate on the far finger and repeat the procedure. When you view with the right eye, the near stimulus will be seen on the left side of the median plane, and when you view with the left eye the near finger will be seen on the right side. The near finger produces a crossed disparity.

So far in our discussion we have not made explicit how the origin for θ or the distance between stimulus and observer are specified. At least three different pairs of reference points can be used. The amount of computed disparity will depend on the choice made, but the most practical way to specify it is to select the front surface of the cornea as the reference point. Its advantage is that it can be determined easily and its use provides a fairly good approximation to the difference between the retinal loci when the viewing distance is large. Another possible reference point, the center of rotation, has the advantage of facilitating discussion of the horopter. A third possibility is the nodal point (or the entrance pupil for blurred images). The nodal point in the human eye is about 7 mm behind the cornea and about 6 mm in front of the center of rotation. The advantage of using it is that the value of $\theta_2 - \theta_1$ then approximates the difference in the retinal loci because the nodal point is appropriate for determining the visual angle subtended by an object. A cumbersome aspect, however, is that the internodal distance gets smaller as the convergence of the two eyes increases. Nevertheless, if retinal locus is considered important, the nodal point should be used. (For an alternative view see Shipley & Rawlings, 1970.)

The differences between the calculated disparities, using the three different reference points, are very small for most experimental situations and are usually not of great concern. Recently, however, Gulick and Lawson (1976) have shown that the computed position of corresponding points differs significantly, depending on whether the nodal point or the center of rotation is used for a reference point. Following their lead, we have examined the computed values obtained by using the three reference points. The disparity values were computed for the reduced or schematic eye with an interocular distance of 6.5 cm. The nodal point and the center of rotation were placed at 7 and 13 mm behind the corneal plane, respectively. These values are identical to those used by Gulick and Lawson. The values in Table 1 are those of crossed disparities produced by a 5-cm depth on the median plane for several fixation distances. As we can see from the table, the discrepancy is small beyond 100 cm of viewing distance, but the differences become large when the viewing distance is less than 50 cm. In fact, for this amount of depth at a fixation distance of 12.5 cm the discrepancy between the corneal plane and nodal point definitions of disparity is 2°41.4′. These differences may be significant in some depth perception research.

The exact determination of disparity for all situations may become quite complicated (see Graham, 1965; Shipley & Rawlings, 1970), but in most the following approximation can be used for computing the disparity of stimuli near the median plane. For a small angle (α) specified in radians (the angle subtended by an arc of a circle equal in length to the radius of a circle) $\tan \alpha = \alpha$. Using the notation in Figure 4-1, we have

$$\theta_1 \simeq \frac{I}{D} \text{ and } \theta_2 \simeq \frac{I}{D \pm d}$$

where I is interocular distance, d is depth, and D is distance.

$$\delta \simeq \frac{Id}{D(D \pm d)} \tag{1}$$

The d in (1) is the distance between the two frontoparallel planes which include the two stimuli (A and B in Figure 4-1). Sometimes this distance is called the exocentric distance but more commonly the depth. The D in (1) is the distance between one of the stimuli and the observer or, to be exact, the distance, called the absolute or egocentric distance, between the nodal point of the eye and

Table 1

CROSSED DISPARITY FOR TWO POINTS IN DEPTH (5 cm) ON THE MEDIAN PLANE COMPUTED BY USING THREE DIFFERENT PAIRS OF REFERENCE POINTS

Fixation Distance from Corneal Plane (cm)	Disparity (in degrees and minutes of arc)		
	Corneal Plane	Center of Rotation	Nodal Point
12.5	17°42.6′	14°2.4′	15°1.2′
25	3°39′	3°15.6′	3°21′
50	49.2′	46.8′	47.4′
100	12′	11.4′	11.4′
200	3′	3′	3′

frontoparallel plane of one of the stimuli. The disparity δ in (1) is in radians. To convert radians to seconds of arc multiply the value by 206,265; to convert to minutes of arc multiply by 3,437.75; and to convert to degrees of arc multiply by 57.3.

NECESSITY OF ABSOLUTE DISTANCE INFORMATION

The inverse square law of disparity is derived from (1) and involves further approximation. If d is relatively small with respect to D, then $D \simeq D + d$. Substituting D for $D + d$, we have

$$\delta = \frac{ID}{D^2} \tag{2}$$

Equation 2 states that for a constant physical depth disparity decreases in proportion to the square of the absolute physical distance; for example, in Table 1 the disparity value for 50 cm is approximately four times larger than that for 100 cm.

Two points should be noted about the inverse square law. First, the law clearly indicates that disparity alone does not provide sufficient information concerning physical depth for depth constancy to occur. What disparity alone can provide is ordinal information about physical depth from a given fixation point. If one of two objects in the median plane is fixated, the relative distance of the two objects depends on whether the disparity of the other object is crossed or uncrossed. An uncrossed disparity would indicate that the disparate object is farther away than the fixated object. Furthermore, the greater the disparity, the greater the depth. Without absolute distance information, however, there is no way for the visual system to judge that two equal extents at two different distances are equal. Second, the law is simply a geometric statement of the relation between physical depth and disparity values and refers exclusively to the relation between two physical variables. The law by itself makes no reference to psychological function or perception. Depth constancy is a perceptual phenomenon and refers to depth being judged constant when viewing distance varies. Therefore, if depth remains constant perceptually and disparity is the cue to depth, it can be said that the visual system behaves as though it knew the inverse square law and that somehow it has processed absolute distance information.

The mechanism for processing information concerning viewing distance and disparity together is sometimes described as calibrating or mapping disparity values differently for different absolute distances. The necessity for different calibrations for depth constancy can be illustrated by a demonstration we have used for an undergraduate perception class. This demonstration is not of depth constancy, but it shows what depth should be seen with a constant disparity at different viewing distances. Its basic idea is analogous to that of the classic experiment by Holway and Boring (1941) who presented observers with stimuli of constant visual angle at several viewing distances. They found veridicality of size perception under certain conditions. Veridicality of size perception predicts that if a constant linear extent were presented at different distances it would appear to be the same; that is size constancy would occur. Likewise, if a constant disparity is presented at different distances and the perception of depth is veridical, depth constancy may be inferred. (The veridicality of depth perception is not necessary for the occurrence of depth constancy; that is, there may be a constant error in judging depth at all distances. If, however, depth perception is veridical, it necessarily follows that depth constancy is occurring; for instance, if a 5-cm depth is presented at two different distances and is perceived to be 5 cm, it is obvious that depth constancy has occurred.)

The demonstration kit consists of two plexiglass sheets that can be rotated near the bottom, where two schematic eyes are drawn. See Figure 4-2. Two lines have been drawn through the nodal point of each eye. One line corresponds to the line of sight (optic axis); the other corresponds to the visual line from another object. Because the two lines are fixed with respect to each eye, the sum of the two angles subtended by the two lines (or difference between the two

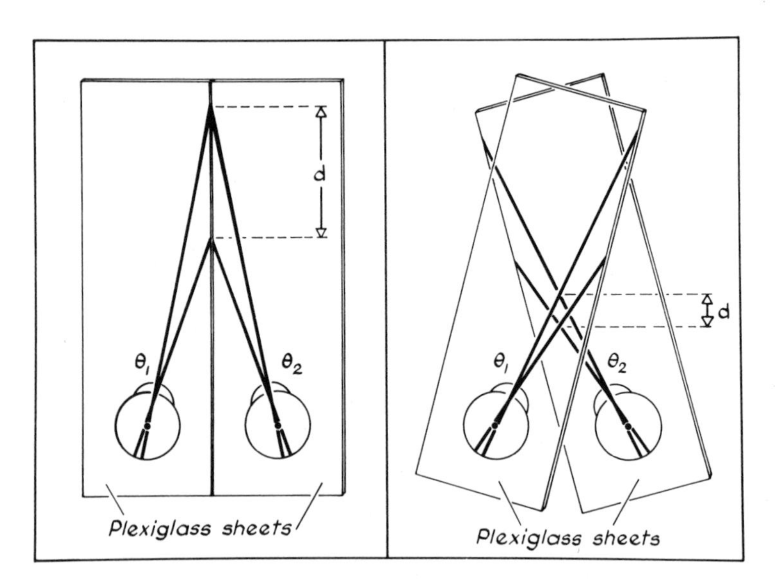

Figure 4-2 The depth *d* is associated with a given binocular disparity $\theta_2 - \theta_1$. When the fixation distance is altered and the disparity remains the same, *d* changes. This change of *d* with distance implies that for veridical depth perception to occur the visual system must process information about the fixation distance.

angles of which one has a different sign) represents the disparity that remains constant no matter where the line of sight is directed. Rotation of the two plexiglass sheets inward represents increased convergence and rotation outward represents decreased convergence. The difference in the linear extent between the intersection of two lines of sight and the intersection of the other two visual lines represents depth. As the convergence is increased by rotating the sheets (or one sheet) the depth decreases approximately in direct proportion to the square of the distance between the nodal plane and the intersection of the two lines of sight. Mathematically,

$$d \simeq \frac{D^2 \delta}{I} \tag{2}$$

This demonstration makes it clear that if we are to perceive depth correctly the visual system must interpret the same disparity produced by the two points differently for different convergence distances. To state it in another way, the visual system must calibrate disparity differently for different convergence distances in order for a veridical perceprion of depth or depth constancy to exist.

THE EXPERIMENTAL EVIDENCE FOR DEPTH CONSTANCY

The experiments reported in this section have generally not been conducted to test the existence of depth constancy; for example, Ogle (1953) was interested in the validity (i.e., veridicality) of depth perception, Foley and Richards (1972) were interested in the effect of eye movements on the perception of depth, and Fried (1973) was interested in determining the adequacy of convergence as a cue to distance. An exception is Wallach and Zuckerman's (1963) paper, "The constancy of stereoscopic depth," but because all in some way involve the study of the relation between binocular disparity and depth perception we can interpret the results of these studies with depth constancy in mind; for example, if depth perception is veridical for different distances, the existence of depth constancy necessarily follows. Although we are interested in its occurrence, if it is like other constancies we would not expect it to be exact. Hence we are concerned with whether the data from these studies suggest depth constancy with relatively loose criteria. An alternative approach is to examine how much the data are in agreement with depth constancy by using a quantitative index such as the Brunswik or Thouless ratio as in other research on constancy phenomena. In future work on depth constancy we would expect the development of such an index. For the present, however, much of the data we discuss are not amenable to such treatment.

Before dealing with the experimental evidence, we shall consider two methodological problems associated with research on depth perception. One concerns

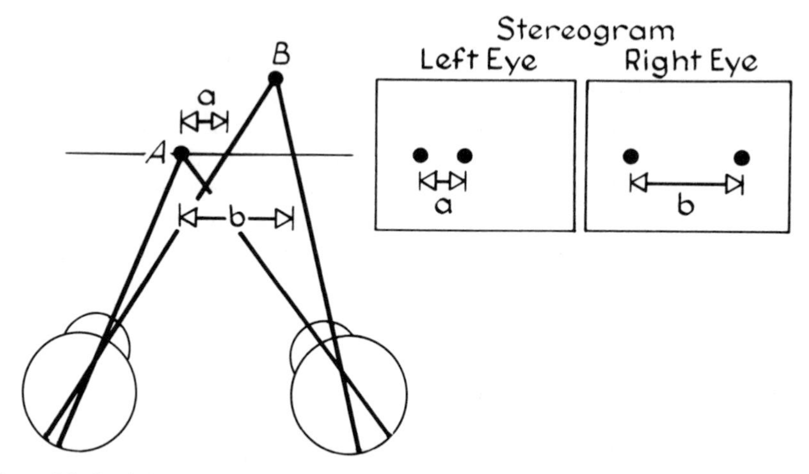

Figure 4-3 Equivalent configurations of real and stereoscopic stimuli. Two dots are placed at different distances from an observer in a configuration of real stimuli, one at A, the other at B. The stereogram provides an equivalent stimulus configuration for the real stimuli, in terms of retinal loci, when seen through appropriate lenses and prisms. The visual angle subtended by A and B by the real stimuli is the same in both eyes for both real and stereoscopic stimuli.

the stimuli to be presented to observers; the other concerns the response measures.

Stimuli can be "real" or stereoscopic. Real and stereoscopic stimuli in one sense represent equivalent configurations because the two retinal loci may be the same. A major difference, however, is that for real stimuli, no matter which is fixated, there is no mismatch between accommodation and convergence. For stereoscopic stimuli a mismatch can occur. The stereogram in Figure 4-3 is an equivalent configuration in terms of retinal loci for two dots at different convergence distances. There are two sets of stimuli to convergence provided by the stereogram; an observer can fixate A or B. The stereoscope can be made with prisms and lenses so that accommodation and convergence are matched when an observer is fixating on A in the stereogram. Although the cues from convergence (and, of course, binocular disparity) indicate that stimulus B is a particular distance behind stimulus A, cues from accommodation indicate that A and B are at the same distance. Because there is evidence that when a mismatch occurs between convergence and accommodation the perceived absolute distance is an outcome of a compromise between the two cues (Swenson, 1932; Ono, Mitson, & Seabrook, 1971), it is possible that depth seen with stereoscopic stimuli may be less than with the equivalent real configuration. Stereoscopic stimuli have two advantages that are responsible for their frequent use. One is that the cue of disparity may be studied independently of other cues to distance such as accommodation. The second is that simulation of a large range of distances is possible by varying a stimulus configuration by small amounts.

Although some of the experiments discussed below have obtained evidence consistent with depth constancy by using stimuli presented in a stereoscope, several have not. For these cases the possibility exists that discrepant information from disparity and accommodation contributed to the lack of depth constancy.

The response measures used in these studies have included kinesthetic matching measures, visually directed pointing measures, production and verbal measures. Using a kinesthetic matching measure, the observer attempts to equate the distance between two movable indicators held in his hands with the depth seen in some part of a visual array. Using a visually directed pointing measure, the observer points under a table with an unseen hand to the position where a stimulus appears to be. A difference in distance between two such positions for two stumuli corresponds to a perceived depth between the two stimuli. Using a production measure, the observer manipulates the array until his perception agrees with a particular visual criterion; for example, the task may be to set one stimulus to appear at half the absolute distance of a second stimulus. Finally, using a verbal measure, the observer reports the depth between given points in an array.

Which response measure best reflects visual perception? When discussing experimental evidence, we take different measures of the perception of depth at their face value. However, the reader should keep in mind that Foley (Reference Note 1) has found that quite different results are obtained when depth is estimated with a pointing measure compared with a verbal measure (for discussion of the problem see Foley, in press, Gogel, 1968, & Ono, 1970).

The experiments to be considered differ with respect to stimuli and response measure. They also differ in regard to whether the stimulus configuration is presented at a near or far distance from an observer. As we have emphasized in the preceding section, the visual system must take into account the viewing distance for depth constancy to occur. Thus viewing distance is a critical variable in any experiment relevant to depth constancy. As we shall see, data suggest that depth constancy occurs for stimulus configurations at near distances of stimuli, whereas no such data exists at far distances.

Perceived Depth at Near Viewing Distances

Although Ogle (1953) used only one fixation distance in his experiment, his study is reported here because he found veridical depth perception and because he stated that it was unlikely that stereoscopic depth perception for his observers was veridical for only that one distance. Ogle had observers fixate a point at 50 cm in the presence of two vertical rods, one a reference rod, the other a test rod. The reference rod could be placed by the experimenter at varying distances from the fixation point and also at several peripheral veiwing angles. The test rod was moved by the observer in the performance of three tasks: (a) setting the test rod to appear equal in depth to the reference rod, (b) setting the test rod to appear half the distance between the fixation plane and the reference rod, and (c) setting the test rod to appear as far in front of (or behind) the fixation plane as the reference rod appeared behind (or in front of) the fixation plane. The results indicated that the stimuli were localized veridically between 6 cm (approximately 50 minutes of arc uncrossed disparity) behind the fixation plane and 5 cm (approximately 50 minutes of arc crossed disparity) in front of the fixation plane. Ogle noted that as the disparities became large perceived depth tended not to increase. If veridicality of depth perception for the same range of disparity can be domonstrated for fixation distances of more than 50 cm, Ogle's results suggest that depth constancy will exist for depths up to at least 5 cm. For closer distances depth constancy would exist for smaller depth values.

Three other experiments that varied disparity in relation to a fixed reference stimulus are worthy of note. The results of two of them are consistent with Ogle's. Fry (1950) had observers use a variable visual stimulus to bisect an interval in depth defined by a stimulus on a transparent slide placed at 40 cm and a vertically displaced stimulus on a slide placed between 22 and 38 cm. The results closely approximated veridicality. Lawson and Gulick (1970) had observers report the depth seen in a stereogram by using the perceived length of the base of the stereogram as a unit. Stereograms were placed 65.4 cm from observers and disparity was varied from .24° uncrossed disparity to .24° crossed disparity. Because the perceived length of the base was not reported, it is difficult to say whether the depth perception was veridical, but a linear relation was found between depth and disparity for the range they studied, a result not inconsistent with veridical depth perception. Foley (Reference Note 1) had observers report depth by using both verbal and pointing responses. He presented two simulated points of light with a polarization stereoscope, one simulating a reference stimulus at 25 cm, the other, a test stimulus ranging in distance from 12 to 38 cm. Although veridicality of depth perception was not found, the results for the pointing response supported the von Kries hypothesis which can predict depth constancy under conditions of veridical distance perception.

Recent evidence (Richards, 1971a,b; Foley & Richards, 1972) supports Ogle's suggestion that veridicality may hold with fixation at more than one distance. Because these three studies reached similar conclusions, we discuss only one of them. Foley and Richards (1972) had observers fixate a dot on a screen which was 250 cm in distance or the same stimulus at a simulated distance of 24 cm, achieved with prisms and lenses. Two vertical bars 1.2 x .12° were presented to the observer through polarizers for 80 msec; the orientations of the polarizers were such that a bar was projected to the left of fixation in his right eye and another to the right of fixation in his left eye. Thus the two bars were equivalent to a single bar presented with a crossed disparity. Production and verbal measures of depth were obtained for each disparity. For the verbal measure the observer estimated the amount of depth of the bar in relation to the perceived distance of the fixation point. For the production measure, after viewing a stimulus presentation, he set a binocularly viewed black circular probe to the same depth as the flashed bar. The ratio of the depth (the difference between the matched position and the fixation distance) to the fixation distance was computed to represent the perceived ratio of depth to distance. For both measures the perceived ratio of depth to distance was compared with the simulated ratio.

For three of his five observers the perceived ratios agreed closely with the simulated ratios up to a simulated distance ratio of about .3. Foley and Richards found that for both distances the depth-distance ratio increased as disparity increased, and at larger disparities decreased in a smooth function. The increasing portions of the functions tended to be veridical for both distances. This suggests that the depth-distance ratio was perceived fairly correctly up to 75 cm in front of the fixation point for the 250-cm fixation distance condition and 7.2 cm for the 24-cm fixation distance condition. For larger disparities, that is, as the distance between the fixation point and flashed bar increased, the depth-distance ratio was underestimated in both conditions until for very large disparities the flashed bar appeared almost at the distance of the fixation point. The results from one of these three observers are presented in Figure 4-4.

Overall, these results lend support to Ogle's (1953) suggestion that depth perception is veridical for more than one fixation distance. The veridicality of depth perception was expressed in terms of ratios: the ratio of the perceived distances was similar to the ratio of the physical distances. A stronger case for veridicality would have been made if the perceived depth were the same as the physical depth, but this information was not reported. These results suggest that depth-distance ratios may be perceived as approximately veridical until the ratio of crossed disparity depth to the distance of the fixated point reaches .3, independent of the distance of the fixation point. In comparison, Ogle's report of veridicality for the first 5 cm of depth for a fixation distance of 50 cm corresponds to a ratio of .1. It is not entirely clear what underlies the discrepancy in these results, but it is due in part to Ogle's use of a stringent criterion for what he called validity of depth perception and our looser criterion in interpreting veridicality in the results of Foley and Richards.

Other results of Foley and Richards (1972) suggest that the range of stereoscopic depth perception may be greatly extended by allowing the observers to make eye movements while viewing the stimuli. The observer's task was to place a probe at a specified proportion of the distance to a white screen; that is, half

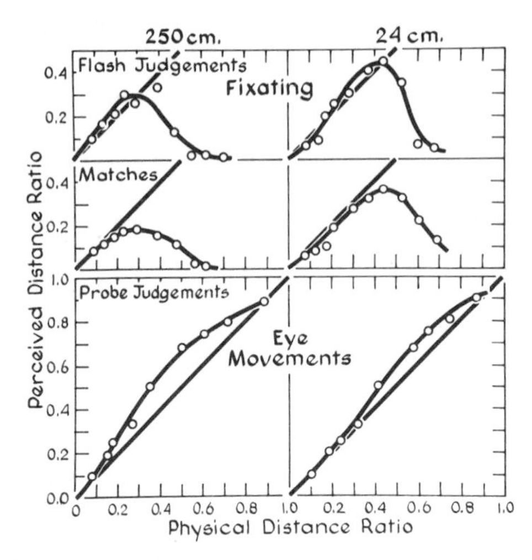

Figure 4-4 Results from one observer in a study by Foley and Richards (1972) for two viewing distances. The perceived distance ratio (the ratio of the distances of two stimuli) increases and then descreases as physical distance ratio increases when fixation is required. When eye movements are allowed, however, there is a montonic increase of perceived distance ratio with physical distance ratio. (Adapted from Foley &Richards, 1972.)

the distance. He was instructed to look back and forth between the probe and the screen while making his setting. The screen was presented at a distance of 250 cm or a simulated distance of 24 cm. They found that depth increased monotonically as a function of disparity even to very large disparities, although depth, as seen by one observer in Figure 4-4, was generally overestimated. Although the effect of eye movement on the perception of depth from large disparities is important to a general understanding of depth constancy, it is likely that more is involved than depth based on disparity alone. It is likely that the difference in absolute distance perception related to the two convergence states for the two stimuli would provide additional information about depth.

Wallach and Zuckerman (1963) attempted specifically to test for the existence of depth constancy. They noted that for depth constancy perceived depth should be dependent on the square of the distance for a given disparity, whereas for size constancy perceived size should be linearly related to absolute distance. Thus, given a change in cues to distance, perceived depth should be affected more than perceived size. Wallach and Zuckerman tried to determine whether this difference occurs between size and depth perception. Two pyramidlike objects were oriented so that their apices pointed toward the observer. The experimental pyramid was seen through a stereoscopelike mirror arrangement, and although it was placed 133 cm from the observers' eyes mirrors and lenses were adjusted to set accommodation and convergence at half this distance. The pyramid measured 12.5 cm along the base and 12.4 cm from apex to base. The second pyramid was used for a control condition. It was viewed directly 133 cm away and was half as large in all its dimensions as the experimental pyramid.

Predictions were made about depth and size perception. Depth constancy would predict that the large experimental pyramid seen with convergence at half distance would appear to have $(1/2)^2$ or one-fourth the depth it would have if it were seen without altered cues. Because the small control pyramid had one-half the depth of the larger pyramid, the perceived depth of the large and small pyramids should be in the ratio of 1:2. On the other hand, size constancy would predict that the base of the pyramid would appear to be the same size for both pyramids.

Observers estimated the depth and size of the experimental and control pyramids with a kinesthetic matching measure. The results indicated a fair degree of agreement with the predictions (see Table 2). These results, and those of a second experiment conducted by Wallach and Zuckerman (not reported here), support the idea that the visual system interprets binocular disparity according to the square of the distance.

Wallach and his associates (Wallach & Frey, 1972; Wallach, Frey, & Bode, 1972) conducted further experiments, of which only one is discussed, in which measures of perceived size and depth were compared. The experiments dealt with the effects of adaptation to altered oculomotor cues. Wallach, Frey, and Bode present evidence that this adaptation consisted in a change in the relation between oculomotor adjustments and "registered" distance, which they define as the "representation of object distance in the nervous system." They reasoned that if the adaptation involves an altering of the relationship between oculomotor adjustment and registered distance the perception of depth should also be

Table 2

WALLACH AND ZUCKERMAN'S RESULTS

	Mean Estimates and Confidence Limits at the 1% Level		Ratio (Critical/Control)		
	Critical Condition	Control Condition	Experi-Mental	Complete Constancy	No Constancy
Size	8.88 ± 0.82	7.13 ± 0.59	1.245	1.0	2.0
Depth	4.07 ± 0.40	6.43 ± 0.82	0.633	0.5	2.0

From Wallach & Zuckerman, 1963.

affected by adaptation. Because of the inverse square relation, the perception of depth should be affected more than the perception of size. To test this idea they used two pyramidlike objects with their apices pointed away from the observer. One pyramid, viewed at a 66.7-cm distance, was 11.0 cm for the base diagonal and 10 cm from apex to base. A second pyramid was viewed at 33.3 cm; it measured half the size on the base diagonal and one-fourth the depth from apex to base. Thus both visual angle and retinal disparity were almost the same for the two pyramids.

During adaptation the observer wore glasses that altered accommodation and convergence so that the optical distance was 1.5 diopters less than normal; he then walked through a college building for 20 minutes. Complete adaptation would imply that if the pyramid at 33.3 cm were viewed without glasses it would appear at 66.7 cm. Therefore perceived size would be expected to double and perceived depth, quadruple.

Sixteen observers made matches of the depth and size of the two pyramids before and after adaptation by using rods of adjustable length. Perceived size increased only 53% for one pyramid, 69% for the other. For this amount of adaptation the perceived depth increases are expected to be 134% for the first pyramid and 186% for the second because of the inverse square relation. The actual depth increase however, was 103 and 101%. Wallach, Frey, and Bode point out that this failure of the prediction to hold was related to a failure of depth constancy in the preadaptation condition. Depth estimates in the preadaptation condition for the two pyramids were in a ratio of 2.32, whereas the physical ratio was 4.0. A possible explanation is that the observers came into the experiment with an expectation of what a pyramid is shaped like (e.g., all sides should be equal). This "familiar size" was in a cue conflict with disparity during testing both before and after adaptation. In any case, there was a greater change in perceived depth than in perceived size as a consequence of adaptation, which suggests qualitatively if not quantitatively a D^2 relation between perceived depth and disparity.

Other experiments have dealt with the effect of adaptation on the perception of depth from disparity. Many of these studies are not relevant here, for they are concerned with the sensory mechanisms of stereopsis (e.g., Beverley & Regan, 1973) or with showing that the mechanism that interprets disparity as depth changes as a function of length of time in which the observer fixates on a stimulus (e.g., Howard & Templeton, 1964). Several studies have shown that there can be an adaptation of the relation between depth and disparity by altering information provided by other cues (e.g., Wallach, Frey, & Bode, 1972; Wallach, Moore, & Davidson, 1963; Epstein & Morgan, 1970). Such an adaptation may be due simply to a change in registered distance (Wallach, Frey, & Bode, 1972), but under some conditions perceived distance does not change (Epstein & Davies, 1972); this suggests that the depth-disparity relation is not fixed for a given perceived distance but may be recalibrated.

Three studies investigated explicitly the relation between depth and disparity as a function of the convergence angle to the stimulus. Below, we discuss two of these studies; one by Foley (1967b) and a second by Fried (1973). A third study (Foley, 1967a), is covered later.

Foley (1967b) found that the convergence angle to a stimulus had an effect on the relation between depth and disparity. The observers' task was to set the distance from themselves to a near variable stereoscopic point of light equal to the distance from that light to a second, more remote stereoscopic light. The second light could be placed at different convergence angles varying from slight divergence to a convergence angle corresponding to approximately 100 cm in distance. Foley found that each of his three observers used more disparity to perform the task as the convergence angle to the farther light increased. To state these results in terms of convergence distance: observers used greater disparity for smaller convergence distance. The evidence that more disparity is needed as the convergence angle increases, which Foley named "depth micropsia," is evidence of depth constancy. The extent of the increase as a function of convergence angle did not, however, follow the square of the distance. Thus the quantitative aspect of Foley's results suggests that depth constancy is incomplete and that depth perception is not veridical.

Noting that the nonveridicality of depth perception found in some experiments could be due to a mismatch between accommodation and convergence, Fried (1973) conducted an experiment with real rather than stereoscopic stimuli. In Fried's experiment the observer viewed two pinpoints of light at different distances. They were then turned off and the observer matched the remembered depth between the two points with the distance between two lights presented at 40 cm in the frontoparallel plane. The observer viewed alternately the experimental and matching stimuli; at no time were both on together. Each observer made estimates of the depth intervals produced by disparities of 10, 7, and 3 minutes of arc in that order. Different groups of observers judged depth at different convergence distances (60, 90, 130, 200, and 300 cm) to control for relative cue size; that is, the stimulus subtending a larger visual angle at closer distances.

Fried used three viewing conditions, only two of which concern us. In the reduced cue condition the observer viewed the experimental stimuli in the dark. In the second condition the stimuli were seen in an illuminated field with familiar objects and a checkerboard floor. The viewing distances and disparities were the same in both conditions; measurements of perceived depth were always made with a reduced cue field. Ten observers participated under both conditions at each of the five viewing distances. The difference between the results of the two viewing conditions was significant only for 300 cm, but for both conditions veridicality of depth perception was suggested for viewing distances up to 200 cm. Figure 4-5 shows perceived depth values for 10 minutes of disparity. Fried believed that these results were more important than those of the two smaller disparity conditions, for there were no previous depth measurements to affect the judgment.

None of the studies reviewed here has actually determined whether a given extent in depth appears the same when seen at different distances. One experiment, however, examines depth perception for a constant physical depth presented at different distances. Heine (reported in Ogle, 1953; Von Kries, 1962) arranged three vertical rods to form an equilateral triangle on a horizontal plane. (The exact size of the triangles is not reported.) These rods, viewed in a lighted room, appeared as an equilateral triangle only when seen at a distance of 33 to 75 cm. At distances shorter than 33 cm the depth between the base and apex of the triangle was greater, whereas at distances of more than 75 cm the depth appeared to be less. The range of veridical depth perception is considerably smaller than that of Fried (1973) or Foley and Richards (1972), although the reason is unclear.

Perceived Depth at Far Viewing Distances

The results of the experiments described so far suggest that depth constancy does exist. In all these experiments stimuli have been presented at small viewing distances. A number of experiments in which stimuli were presented at greater distances are examined below. Their results indicate an absence of depth constancy.

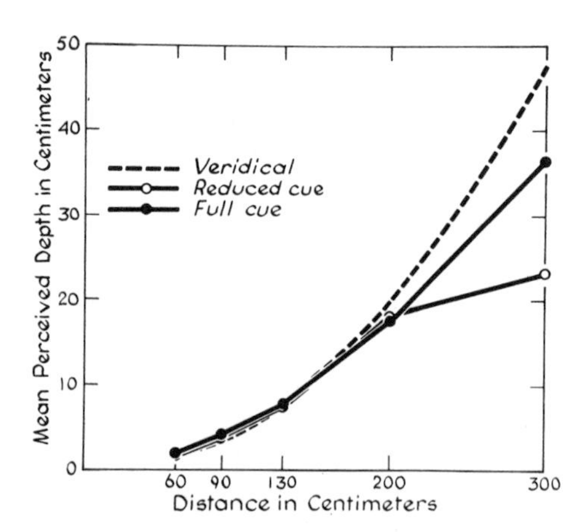

Figure 4-5 Fried's results for 10 minutes of arc disparity seen at five distances for two viewing conditions. Mean perceived depth closely approximates veridicality up to a viewing distance of 200 cm. (Adapted from Fried, 1973.)

Gogel (1960c) has reported an experiment that shows a failure of depth constancy for stimuli at viewing distances greater than 150 cm. He had observers view a 10.2 x 2.54 cm piece of cardboard presented at 160, 320, and 480 cm in illuminated visual alley. A stereoscopic disk at an infinite accommodative distance was reflected off a beam splitter interposed between the observer and the visual alley. For each position of the rectangle the observer had two tasks. The first was to place the disk at the perceived distance of the rectangle, the second, to place it at a position such that the perceived depth between it and the rectangle was equal to the perceived horizontal size of the rectangle. The binocular disparity necessary to match the size of the rectangle could be calculated from the difference between those two settings. Observers also made verbal estimates of the distance and size of the rectangle for each of the three distances. If depth constancy existed at those distances, a constant amount of depth should have been produced to match the constant size of the rectangle. However, although there were accurate verbal estimates of the size of the rectangle, the failure of the predictions from depth constancy were notable. The mean depth produced by observers to match the size of the rectangle increased linearly with the distance of the rectangle.

Foley's (1967a) results also indicate a failure of depth constancy. Three observers bisected the distance between themselves and a reference light by using a stereoscopic light for which they could vary disparity. The apparatus was similar to that used in the Foley study previously described. The reference light was placed at five different distances which ranged from beyond optical infinity (i.e., the eyes were divergent when the light was fixated) to a convergence angle appropriate for a viewing distance of about 400 cm. The results indicated that each of the three observers needed a fairly constant amount of disparity to perform this task, regardless of the convergence distance to the farthest light. This finding differs from Foley (1967b), which indicated that a larger disparity was needed with a decrease in convergence distance. The discrepancy in the results of the two studies is probably due to the difference in the range of convergence distances.

Several other studies by Gogel indicate that perceived depth is nonveridical at far viewing distances (see Gogel, 1960a, b, 1964, 1972); for example, Gogel (1960a) showed that perceived depth for a constant disparity may differ with different reference stimuli even though the perceived distance is the same. Two stimulus arrays were shown to two groups of observers. In both groups a ring was presented at a distance of 457 cm and flanked on either side by transilluminated transparencies of a playing card placed at a distance of 305 cm. The size of the playing card differed for the two groups of observers: one group saw a card of normal size, the other, a card twice the normal size. The observers in both groups estimated the distances of the ring and the cards on the basis of verbal reports and by throwing darts to each of the perceived distances. If depth perception were determined solely by disparity and convergence distance (or perceived distance), we would expect that the perceived distances between the cards and the ring would be the same for both conditions. Although there was no significant difference in the perceived distances of the cards in the two groups of subjects, the subjects viewing the double-sized cards saw the rings as being nearer than the subjects viewing the normal cards. This result indicates that factors other than disparity and convergence or perceived distance can affect stereoscopic depth perception.

Summary and Conclusion

Stereoscopic depth constancy or veridical stereoscopic depth perception occurs only at near viewing distances. The exact range, however, in which it occurs is difficult to ascertain. Fried (1973) found that veridical depth perception occurred up to 200 cm. Other results also suggest veridical depth perception for stimuli within this range, but Heine's experimental results indicate that depth constancy holds only between 33 and 75 cm. Despite the ambiguity of the exact range, the existence of depth constancy over only a limited range is not surprising; other constancies also have their limits. The fact that there is a limit should be of theoretical importance. Any theory dealing with depth constancy must take into account that depth constancy fails at certain viewing distances and that it occurs only in a certain range of viewing distances.

THEORIES OF STEREOPSIS AND DEPTH CONSTANCY

The fact that depth constancy occurs, albeit for a limited range of distances, clearly indicates that somehow absolute distance information can be processed with disparity information; that is, different calibrations of disparity occur as a function of absolute distance. Even the data that show deviation from perfect depth constancy also indicate a different calibration of disparity for different absolute distances. The hypotheses or theories to be examined later in this section are concerned with how the visual system can provide different calibrations of disparity for different viewing distances. Before discussing them, however, we pause for a brief consideration of some theories of stereopsis that do not include postulates of the relation between perceived depth and distance.

Kaufman (1974) has provided a comprehensive review of numerous theories of stereopsis in which he concludes that all have failed to take into account the inverse square law. They are concerned with problems such as how the visual system achieves single vision and how to define the necessary stimulus for depth perception but they have not addressed themselves to the question of depth constancy.

In our view most of the theories can be modified to explain or account for depth constancy by the addition of a simple postulate. As an example, we consider the fusion or projection theory, which is the most popular in terms of its apparent number of adherents, having been proposed independently by many (e.g., Kepler, cited in Kaufman, 1974; Boring, 1933; Charnwood, 1951; Dodwell & Engel, 1963; and Linksz, 1952). An attractive feature of this theory is that it can explain both single vision and depth perception. The mechanism by means of which fusion operates involves a cortical combination of the two visual inputs. Visual inputs from the two retinae project back through a hypothetical network in the cortex until a match of identical inputs from the two eyes is obtained. Figure 4-6 is a schematic representation generally used to illustrate projection. Two images of object A falling on corresponding points (3,3) on the retinae project back to the reference plane F at a. Two other images of the more

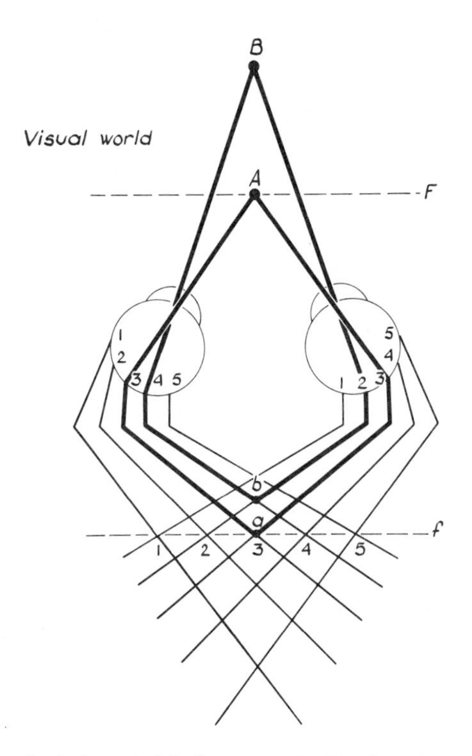

Figure 4-6 A hypothetical cortical fusion network. The dotted line represents the plane of fixation on the cortex. A stimulus *(B)* not on the fixation plane projects to a different point in the cortex than a point *(A)* on the fixation plane. (A similar figure has appeared in many sources. This one is adapted from Gregor, 1973.)

distant, nonfixated object B falling on disparate points (4,2) project back to a different position b. All we need for the fusion theory to account for depth constancy is an additional postulate that states that projections are different for different absolute distances of stimuli or for different convergence states. If a different convergence state determines a different projection, the outcome of this projection will be like that of our demonstration (Figure 4-2); that is, as convergence increases, the same disparity will project smaller depth values.

The same sort of simple additional postulate or modification can be suggested for other theories; for instance, in Sperling's (1970) theory of stereopsis convergence and accommodation play an important role. It would be a simple matter to incorporate the idea of different calibration values of disparity for different oculomotor states into his theory.

Why such a simple postulate has not been included in most theories is not clear. Perhaps, it is such an obvious point that theorists may have expected the reader to add the simple postulate. Whatever the reason, several hypotheses or theories are directly relevant to depth constancy, and we now turn our attention to them. We discuss four hypotheses that differ in their accounts of the calculation of different calibrations of disparity for different distances and what determines them. The four hypotheses are (a) perceived distance-squared calibration, (b) oculomotor adjustment calibration, (c) perceived distance-linear calibration, and (d) perceived size calibration.

Perceived Distance-Squared Calibration Hypothesis

In this chapter we have made several references to size constancy. The present hypothesis comes closest to the usual hypothesis for size constancy, namely, the size-distance invariance hypothesis, which states that for a given proximal size (visual angle) there is a unique ratio between perceived size and perceived absolute distance. This hypothesis does not specify what determines perceived size and perceived absolute distance, but presumably for a given retinal size perceived absolute distance determines perceived size in the absence of additional information. Likewise, the perceived distance-squared calibration hypothesis states that for a given disparity perceived absolute distance determines perceived depth. In our conceptual framework different calibrations of disparity are determined by perceived absolute distance. Presumably the cues that determine perceived absolute distance in depth perception are the same as those that determine perceived absolute distance in size perception. The perceived distance-squared calibration hypothesis states that, whatever the cues processed to determine absolute distance, perceived depth is determined by the perceived absolute distance of the fixated stimulus.

One of the earliest statements of this hypothesis was formulated by Von Kries (1962).in 1911:

> . . . if F denotes the real distance of fixation and E the real distance of another point, and if F' and E' denote their apparent distances, respectively, then the relative parallax between F' and E' must be the same as the relative parallax between F and E.

The term "relative parallax" refers to diparity. Von Kries states the relation between perceived and real distance as follows:

$$\frac{1}{F'} - \frac{1}{E'} = \frac{1}{F} - \frac{1}{E}$$

This mathematical statement is equivalent to our statement of the inverse square law except that the critical variable here is perceived absolute distance. However, because, on the surface, the two statements look different, the equivalence is shown.

$$\frac{E' - F'}{E'F'} = \frac{1}{F} - \frac{1}{E}$$

$$E' - F' = E'F'\left(\frac{1}{F} - \frac{1}{E}\right)$$

The left term now represents perceived depth. Multiplying the denominators and numerators of the two fractions by I (the interocular distance), we have

$$E' - F' = E'F'\left(\frac{I}{FI} - \frac{I}{EI}\right)$$

Note that $I/F = \alpha$ and $I/E = \beta$, where α and β are the angles in radians subtended by two lines of sight in Figure 4-1. Therefore we have

$$E' - F' = \frac{E'E'}{I}(\alpha - \beta)$$

If $E' \simeq F'$, we have

$$E' - F' = d' = \frac{F'^2}{I}\,\delta$$

which is equivalent to the inverse square law except for the substitution of perceived absolute distance for the physical distance. When Von Kries' hypothesis is restated as in (3), it should be obvious that it states that disparity is calibrated by perceived absolute distance squared.

More recent statements of the apparent distance-squared hypothesis are those of Wallach and Zuckerman (1963), Epstein (1973), and Foley (in press). Wallach and Zuckerman's hypothesis is logically identical to Von Kries' in the sense that they predict that the perception of depth should follow the inverse square law. They do not, however, make it explicit that perceived absolute distance is involved in the calibration of disparity information (compensation is their term). Wallach and his associates manipulated viewing distance, and their results can be interpreted as being due to the differences in accommodation and convergence in the different viewing conditions. Thus their oculomotor adjustment calibration hypothesis can be considered in the next section but because they stated that accommodation and convergence served as reliable cues for distance and spoke in terms of "registered" distance rather than "registered" convergence and accommodation we infer that it is concerned with perceived absolute distance.

Epstein's statement of the present hypothesis is couched in terms of a broader hypothesis about visual preception, which he labeled "the taking into account" hypothesis. This hypothesis treats, in addition to depth constancy, such phenomena as the perception of direction, orientation, size, speed, and brightness, for all of which the proximal stimulus alone provides insufficient information for veridical perception. Epstein states that perception of depth is determined by the visual system taking into account perceived distance—in this case the inverse square of the perceived distance.

Foley (in press), who has elaborated on this hypothesis more than anyone else, suggests that the reference point in visual space which calibrates disparity need not be the fixation point nor need it be a visual stimulus at all. The perceived depth of a point in visual space is related to the binocular disparity from the reference point and the perceived distance of the reference point. He has shown that if this hypothesis is correct we can infer perceived absolute distance from relative distance tasks such as the apparent frontoparallel plane (Ogle, 1962), the apparent equidistant circle (Foley, 1966), and apparent distance bisection (Foley, 1967b). The distance functions inferred from these different tasks are in good agreement. Perceived distance, thus inferred, exceeds physical distance is less than the physical distance. There is no a priori justification for identifying these inferred distances with perceived distances in the phenomenological sense. However, measures of perceived distance, using a manual pointing response, also show that for near targets perceived distance exceeds physical dis-
sponse, also show that for near targets perceived distance exceeds physical distance with only convergence as a cue (Foley & Held, 1972). Gogel and Tietz (1973) reached a similar conclusion with a more indirect measure based on perceived motion concomitant with lateral head motion.

Foley's analysis implies that veridical depth perception should occur around 250 cm. This seems discrepant with the studies we have cited (i.e., Fried, 1974; Fry, 1950; Heine, cited in Ogle, 1953; and Ogle, 1953). Perhaps this discrepancy emerges because Foley's analysis is concerned largely with data in which convergence is the only cue to distance. Foley and Held (1972) have found that although there is nonveridical pointing at near visual targets when convergence is the only cue to distance, pointing at targets is closer to veridicality when several cues are available (e.g., accommodation, relative size, perspective). Much of the evidence we have discussed involves experiments in which several cues to absolute distance have been available; for example, cues of accommodation and relative size were present in the experiments of Heine (cited in Ogle, 1953) and Fry (1950).

The specification of perceived absolute distance is a problem for the present hypothesis, as recognized by Von Kries:

> It is true, it can only be regarded as an approximation, as is self-evident, because the subjective values of the perceived depth cannot generally be determined with extreme accuracy. Attention must also be called to the fact that an accurate test is very difficult, because the perceived distance of the point of fixation can never be certainly and exactly told. Accordingly, the observational data by which the assumption may be tested are not only limited, but frequently difficult to interpret correctly.

Related to this difficulty is the unclear meaning of "perceived distance." Operationally, we might define perceived distance as whatever distance an observer reports. Epstein, however, suggests that the term perceived distance should *not* refer exclusively to "immediately available, conscious, perceptual experience." He argues that in "many instances the factor that interacts with the retinal counterpart to determine the focal perceptual output is not consciously noted or reported by the observer. It is only registered in the visual system." The obvious pitfall of including the registered perceived distance in the formulation is the difficulty of falsifying the hypothesis; that is, when the data deviate from the prediction, we can postulate a different registration or perceived distance from that reported. Having noted the shortcoming, Epstein has suggested guidelines for resorting to the distinction between perceived and registered distance. (For details see Epstein, 1973.)

Despite the weaknesses we have discussed, the present hypothesis is consistent with much of the data. This hypothesis predicts depth constancy for an experimental situation in which the perceived absolute distance is veridical. Unfortunately four of the experiments that showed veridical depth perception or depth constancy did not require observers to report perceived absolute distances (Fried, 1974; Fry, 1950; Heine, 1900, cited in Ogle, 1953; and Ogle, 1953). If an assumption were made that perceived absolute distance was veridical, these results would be consistent with the present hypothesis. Moreover, those experimental results at a far viewing distance that demonstrated a failure of depth constancy could be due to the perceived absolute distance not being veridical. Thus failures of depth constancy are not necessarily inconsistent with this hypothesis. Foley (1972) found perceived depth to be predictable from the present hypothesis up to at least 2° of disparity, even though perceived absolute distance was found to be nonveridical, by using a pointing response measure of visual localization. Thus the results of experiments that have found both veridical and nonveridical depth perception are consistent with this hypothesis. In those experimental situations in which depth perception is found to be veridical it is necessary that veridical perceived absolute distance be demonstrated.

Oculomotor Adjustment Calibration Hypothesis

This hypothesis is similar to the first hypothesis discussed–if oculomotor adjustment (convergence and accommodation) determines perceived absolute distance. The present hypothesis, however, restricts the determinant of different calibrations to oculomotor adjustment only. It differs from the preceding hypothesis in that it is not perceived absolute distance but the oculomotor adjustment per se that calibrates disparity.

The most explicit hypothesis that postulates the calibration of disparity as dependent on oculomotor adjustment is Richards' physiological model of stereopsis (1971a, 1975). In this model depth is signaled by the pooled activity of binocular units which have encoded retinal disparity. The evidence for the existence of such units is provided by Barlow, Blakemore, and Pettigrew's electrophysiological study (1967). Given Hubel and Weisel's (1969) discovery that the lateral geniculate body (LGB) projects to staggered contralateral slabs in area 17, Richards postulated that the cortical binocular units are driven by a pair of staggered contralateral and ipsilateral projection slabs which receive signals from the LGB. Richards postulates three types of binocular unit: one that signals crossed disparity, another that signals uncrossed disparity, and a third that signals no disparity. The perceived depth depends on a comparison of the pooled output from one class of disparity detector unit with the activity of another class. Richards suggested that the three types of binocular unit are *not* directly driven by the retinal signal but that their activity can be modulated at the level of the LGB. Retinal disparity information may thus be effectively transformed by the LGB before the level of binocular encoding. The modulation at the LGB that trans-

*Finding nonveridicality of distance perception for these short viewing distances in Foley's study does not necessarily contradict our assumption of veridicality in the studies cited above, for in his experiment there was the possibility of a cue conflict between accommodation and convergence. However, Foley and Held (1972), in an experiment that dealt with distance perception, have argued that such a cue conflict is not sufficient to explain the magnitude of nonveridical distance pointing. If an experiment did find veridical depth perception and nonveridical distance perception, the present hypothesis would be disconfirmed.

forms the effective retinal information is determined by the state of the oculomotor system.

The choice of the LGB as the site of disparity calibration has a physiological basis. First, the activity of geniculate cells in the cat and monkey is known to alter when the eyes make vergence movements (Feldman & Cohen, 1968; Richards, 1968). Second, different layers in the LGB have different cell sizes. Because an extraocular input correlated with convergence may affect the cells of various sizes differently, there may be differences in information flowing from the LGB to the cortex as a function of convergence.

In Richards' model the speculation concerning the degree of transformation of disparity that takes place at the LGB is not based on evidence from physiological data; that is, at present physiological knowledge does not suggest the extent of the transformation that takes place at the LGB; he relies instead on psychophysical data.

What is crucial to the oculomotor calibration hypothesis is that the visual system monitors the oculomotor state for information about absolute distance. The viability of this hypothesis is dependent on evidence that convergence and/or accommodation can be cues to absolute distance. Although there has been no systematic attempt to determine whether accommodation by itself is a cue to absolute distance, several attempts have been made to determine whether convergence (often with accommodation) is an effective cue to distance. The results are equivocal. After reviews of the literature, Osgood (1953) and Woodworth and Schlosberg (1954) tentatively concluded that convergence may serve as a cue to distance in the absence of other cues; Foley (in press) concluded that it is a cue to distance, although short distances are overestimated and large distances underestimated. Graham (1965) and Hochberg (1971) suggest that convergence is only a minor cue to distance; Irvine and Ludvigh (1936) and Ogle (1962) concluded that convergence is not a cue to distance or at best an unreliable one. The fact that these reviews indicate the ambiguous status of convergence as a cue might be considered the *coup de grace* for the oculomotor adjustment calibration hypothesis. However, the question whether convergence serves as a cue to distance can be divided into two parts: (a) does oculomotor adjustment provide distance information for the visual system and (b) is the distance information provided by oculomotor adjustment, if any, used to make distance judgments? The studies that assess convergence as a cue to the perception of absolute distance have been directed only to the second question, whereas it is the first question that is important to the present hypothesis. The question is whether information from oculomotor adjustment is used in processing depth by the visual system as distinct from whether oculomotor adjustments affect the perception of the absolute distance of the fixation point.

Many studies indicate that oculomotor adjustment provides distance information for the visual system. The occurrence of accommodation-convergence micropsia of linear size and angular size (e.g., Biersdorf, Ohwaki, & Kozil, 1963; Heinemann, Tulving, & Nachmias, 1959; Komoda & Ono, 1974; Leibowitz, Shina, & Hennessy, 1972; Ono, Muter, & Mitson, 1974; Wallach & Floor, 1971) supports this interpretation. Other perceptual phenomena have been found to vary with oculomotor adjustment; for example, both visual acuity (Freeman, 1933; Luckiesh & Moss, 1933) and stereoscopic acuity (Amigo, 1963) appear to decrease with increased fixation distance; Ricco's summation area decreases in size with increasing convergence (Richards, 1967) and CFF increases (Harvey, 1970). These studies imply that the visual system is processing the distance information provided by the oculomotor adjustment. Therefore the oculomotor adjustment calibration hypothesis should not be rejected because of data from experiments dealing only with distance judgments.

Two attractive features of Richards' model are that it provides a bridge between physiology and psychophysics and that it can be investigated by an difficulty is that a given perceived ratio can arise from two sets of different clear whether the compensatory calibration takes the form of a square or linear relationship to convergence and accommodation distance. One source of evidence comes from the data cited to support the perceived distance-squared and distance-linear calibration hypotheses. Because the data were generated by experiments that manipulated oculomotor adjustment, the data showing relations between perceived depth and oculomotor adjustment support the present hypothesis.

Because we have concluded that depth constancy occurs under near viewing conditions, it is worthwhile to postulate that the calibration of disparity is related to the inverse square of the distance signified by accommodation and convergence. Then depth constancy is predicted for an experimental situation in which accommodation and convergence are correctly monitored by the visual system. Because Wallach and Zuckerman (1963), Fried (1974), and Foley and Richards (1972) manipulated oculomotor adjustment and found veridical depth perception, this postulate is strongly supported. The fact that perceived absolute distances were not measured in these experiments is not critical for testing this postulate, although without them the choice between the oculomotor and perceived distance-squared calibration hypotheses would be difficult to make. The oculomotor adjustment calibration hypothesis does not require veridical distance perception for veridical depth perception, although it does require that the processing of information about the oculomotor state be correctly monitored.

If the oculomotor state is incorrectly monitored, this hypothesis predicts that depth perception will not be veridical and that depth constancy will not occur. This can explain the finding that depth constancy fails at far viewing distances and that a change due to adaptation can affect depth perception. When the viewing distances become large, the difference in convergence angles becomes small for a given change in absolute distance; for example, although there is almost a 4° change in convergence angle when convergence changes from a fixation distance of 100 to 50 cm, there is only a $.25^\circ$ change when fixation goes from a 300 to a 250 cm distance. The failure of depth constancy at far viewing distances may be due to difficulty for the visual system in discriminating small changes in convergence angle. Moreover, because the experimental evidence for a change in depth perception due to adaptation comes from manipulating oculomotor adjustment, we can simply postulate a new registration of the oculomotor state to account for the results of Wallach, Frey, and Bode (1972) and Wallach and Frey (1972).

The modified hypothesis can explain some of the failures and occurrences of depth constancy. This hypothesis, however, implies that for a constant disparity a change in perceived depth is associated solely with a change in oculomotor state. Because oculomotor adjustment is not the only factor that determines perceived depth, this hypothesis is not complete (e.g., see Gogel, 1964).

No Calibration or Perceived Distance-Linear Calibration Hypothesis

In contrast to the two hypotheses we have already considered, the present hypothesis does not predict the existence of depth constancy. It stems from Luneburg's theory which has as its starting point the finding that under some viewing conditions perceived space is non-Euclidian. (For details see Luneburg, 1947, 1950; Blank, 1953, 1957, 1959.) A hypothesis formulated by Hardy, Rand, Rittler, Blank, and Boeder (1953) in the context of the Luneburg theory states that "the perceived ratio of radial distance for any point of a stimulus to that of the point of perceived greatest radial distance depends only on the difference in convergence between the two points, independent of the stimulus." Two points should be noted about this hypothesis: (a) the hypothesis is concerned with perceived ratios of distances, not perceived distance or perceived depth in terms of scalar units, and (b) if the difference in convergence in the hypothesis is equated with binocular disparity, the disparity between a stimulus and the farther stimulus determines a ratio of perceived distance for the two stimuli. Foley (1967a) states this hypothesis in the following form:

$$\frac{d'}{D'_n} = K\delta$$

where δ is disparity between a near and a far point, d' is the perceived depth between those two points, D'_n is the perceived distance to the near point from an observer, and K is an observer constant. A prediction from this hypothesis is that for any pair of points in space, if the disparity between near and far stimuli remains constant, the perceived ratio of distances remains constant.*

Because the hypothesis makes predictions in terms of perceived ratios, it is difficult to restate it in terms of our conceptual framework for the calibration of disparity (the reason why the heading for this section is so cumbersome.) The difficulty is that a given perceived ratio can arise from two sets of different perceived distances; for example, the same ratio would apply to 10 and 5 cm as to 20 and 10 cm. If an assumption is made that the visual system does not process absolute distance or depth in terms of scalar units but only the ratio of distances, the hypothesis in terms of our framework would state that there is no calibration of disparity as a function of different viewing distances. If, however, an assumption is made that the perceived absolute distance varies as a function of the actual distance, the hypothesis would state that there would be differential calibrations for different viewing distances. This calibration would take the form of

$$d' = KD'_n \delta$$

in terms of the distance of the near stimulus or

$$d' = D'_f \frac{K\delta}{1 - K\delta}$$

in terms of the distance of the far stimulus. The value of D'_n and D'_f would be the perceived distance, which is the determinant of the calibration.

The present hypothesis based on either assumption would not predict the existence of depth constancy. The hypothesis would predict for a constant disparity a linear relationship between depth and viewing distance rather than a squared relationship. Because of this relationship, Foley has named the hypothesis the depth/distance invariance hypothesis. One of Foley's studies (1967a) was conducted to test its prediction, contrasting it with the prediction based on Von Kries' hypothesis. Foley found that the disparity necessary for an observer to produce a given ratio of depth to distance did not vary with the angle of convergence to the far point in a given configuration. He also found that the perceived distance to the far point varied as a function of convergence distance for two out of three observers, which provides some support for the linear calibration hypothesis.

*Research based on the Luneberg theory has often defined disparity in relation to the point in visual space with the smallest convergence angle, the far point, in contrast to disparity relative to the fixation point. One result of this definition is that for the Luneberg theory disparity does not vary with eye movements, whereas with the usual definition disparity varies with each new point of fixation.

Although Foley (1967a) attempted explicitly to test the implication of this hypothesis and provided confirming evidence, he found that the hypothesis did not hold for a situation in which the range of convergence angles was greater (1967b). He found (1967b) that the disparity necessary for a given depth-distance ratio varied with the magnitude of the convergence angle to the farthest point in visual space. Furthermore, our examination of experimental evidence clearly indicates that the present hypothesis should be rejected for near viewing distances. When evidence supports this hypothesis, the experimental conditions usually involve relatively far distances. At present, there are no provisions within the hypothesis to account for the experimental results that show depth constancy at near viewing distances. Whether the hypothesis can be modified within Luneburg's theoretical framework to account for the results is unclear.

Perceived Size Calibration Hypothesis

The last hypothesis to be considered differs from others in that it states that depth perception is not directly dependent on absolute distance information. According to Gogel (1960abc, 1963c, 1964), the perception of depth from disparity is related to the perceived size per unit of visual angle of stimuli adjacent to the disparate stimulus. Simply put, the hypothesis is

$$d' \simeq \frac{S'\delta}{c\theta}$$

where S' represents perceived size, and θ, the visual angle subtended by the stimulus.* Note that unlike the preceding three hypotheses this equation does not contain terms for absolute distance—perceived or actual. Gogel (1960a) noted that the only term in the equation that can be affected by perceived distance is S'. He emphasized, however, that perceived distance is not the only factor that can determine S', for example, object familiarity can affect S'. Thus the critical variable in the hypothesis is perceived size. To state the hypothesis in terms of our conceptual framework, the ratio of perceived size to visual angle of the fixated stimulus determines the calibration of disparity.

Experimental evidence consistent with the perceived size calibration hypothesis was obtained by Gogel (1960abc, 1964). Gogel's (1960a) results were discussed earlier in this chapter but are also restated briefly here. Gogel found that when a normal-sized and double-sized playing card were perceived at the same

*Our statement of Gogel's hypothesis is incomplete. For a more detailed exposition, including a more general equation than the approximation given here, see Gogel (Reference note 3, 1960abc, 1963). The predictions from the equation noted here and the more general equation may differ. although for small disparities the predictions are close.

absolute distance a ring with a constant disparity was perceived at different depths from the cards. The fact that both convergence distance and perceived distance were the same for the two cards strongly suggests that perceived size determined the calibration of disparity. From the standpoint of Gogel's hypothesis, an unfortunate aspect of the experiment was that reports of the perceived size of the playing cards were not obtained; what the perceived size per unit of visual angle was for this experimental situation is not known.

Gogel (1964) did measure perceived size and distance. A binocularly viewed three-dimensional configuration was seen against a monocularly viewed panel at distances of 548.6 or 182.9 cm. Perceived depth and perceived size increased as perceived distance to the stimuli increased, even though the accommodation and convergence cues from the binocular configuration remained constant. The results of these two experiments clearly indicate that calibration is not solely determined by convergence (Gogel, 1960a, 1964) or perceived absolute distance (Gogel, 1960a) and therefore contradicts the predictions from the three other hypotheses.

Although the perceived size calibration hypothesis is explicit in stating that perceived size and visual angle determine perceived depth for a given disparity, it can also predict relationships between perceived distance and depth for certain situations. If a special case is considered, namely, when S' is solely determined by the perceived absolute distance, the present hypothesis provides the same prediction as the perceived distance-linear calibration hypothesis. If S' is solely determined by perceived distance and if the size-distance invariance hypothesis holds, then

$$S' = kD'\theta$$

Substituting $\theta D'$ for S' in Gogel's formulation, we have

$$d' = \frac{kD'\delta}{c}$$

which is identical with the prediction of the perceived distance-linear calibration hypothesis with the assumption that k/c is equal to K. As noted in the preceding discussion, this prediction implies that depth constancy does not exist. If only this special case is considered, the perceived distance-linear calibration hypothesis and the present hypothesis must stand or fall on the same evidence. Thus Gogel's hypothesis would be supported by results obtained when the viewing distance is relatively far and not supported when the viewing distance is relatively near. Gogel (Reference Note 2) suggests that one possible reason for the support of depth constancy may be that at relatively near distances the relation between perceived and physical depth from binocular disparity changes from one in which physical depth is overestimated to one in which physical depth is underestimated (Gogel, Reference Note 3). Depth intervals in the vicinity of this transition distance may be perceived fairly veridically. For a similar view see Foley (in press).

Gogel has treated both size and depth perception as dependent on perceived distance. He described perceived depth from disparity as being dependent on perceived egocentric distance (1973) and has shown (1972) that the scalar perception of depth is determined by what he calls the egocentric reference distance, which is believed to be determined by the specific distance tendency and oculomotor cues. Perhaps Gogel's original hypothesis should not be treated as a competing hypothesis but rather as a hypothesis supplementary to the three already discussed. Perceived size may be an additional factor in the calibration of disparity—in addition to perceived distance or oculomotor state.

Summary and Conclusion

The perceived distance-squared calibration and the oculomotor distance-squared calibration hypotheses can predict the existence of depth constancy. Of the two, the perceived distance-squared calibration hypothesis predicts depth constancy for an experimental situation in which the perceived absolute distance is veridical. The oculomotor distance-squared calibration hypothesis also predicts depth constancy for an experimental situation in which accommodation and convergence are correctly monitored by the visual system. In contrast, the perceived distance-linear calibration and perceived size calibration hypotheses do not predict an inverse square relation with distance, hence do not predict the existence of depth constancy. The last two hypotheses do not predict depth constancy for all conditions, whether or not the perceived absolute distance is veridical. They do predict that for a given disparity there is a constant perceived ratio between depth and perceived distance. Thus among the four hypotheses two predict depth constancy, the other two do not.

The experiments we have examined also divide themselves into two sets; one contains data that confirm the existence of depth constancy, the other does not. The difference between the two sets is related to the use of near or far viewing distances. Given the evidence, two different conclusions are possible. The first is that one of the first two hypotheses is correct and that the last two should be rejected. The failure of depth constancy at far viewing distances can be explained by one of the two by postulating an incorrect "registration" of perceived distances or oculomotor states at far viewing distances. The converse of this conclusion, namely, that one of the last two is correct and that there is incorrect registration of perceived distance or size at near distances, is another logical possibility; but because at near viewing distances the perceived distance and size are more likely to be veridical, it is difficult to defend this position. Presumably at near viewing distances it is less likely that an incorrect registration will occur because of more available cues. The second possible conclusion is an eclectic one; namely, that one of the first two holds for near viewing distances and one of the last two holds for far viewing distances. This conclusion would necessitate a postulate concerning some mechanism that would allow the visual system to switch the calibration of disparity from square to linear as a function of viewing distance. This would not, however, be an efficient way for the visual system to operate. The most appropriate concluding statement is that to choose between the two possibilities or to choose among the four hypotheses necessitates much more experimental work, sophomoric as it may sound.

GENERAL SUMMARY AND CONCLUSION

Binocular or retinal disparity per se provides insufficient information for stereoscopic depth constancy. For depth constancy to occur the visual system must calibrate disparity information as a function of viewing distance. Our review of the literature showed that depth constancy can occur up to 200 cm, which indicates that such modification of calibration is made by the visual system. Four hypotheses concerning how different calibrations for different distances take place were examined. The four hypotheses were (a) perceived distance-squared calibration hypothesis (b) oculomotor adjustment calibration hypothesis, (c) perceived distance-linear calibration hypothesis, and (d) perceived size calibration hypothesis. At present there are few experiments which attempt to test among the hypotheses.

Not much attention has been paid to depth constancy in relation to the attention paid to size constancy. Most studies of depth perception are neither directly addressed to the question whether depth constancy occurs nor to the question of the conditions under which it occurs. Furthermore, many theories of steropsis do not give consideration ot the problem of depth constancy. It is hoped that the present relative lace of interest will change and that this chapter will serve as an inpetus for further work on depth constancy.

REFERENCE NOTES

1. Foley, J. M. *Converging on an index of perceived distance.* Talk given before the American Academy of Optometry, 1972.

2. Gogel, W. C. Personal communication, July 17, 1975.

3. Gogel, W. C. *The perception of shape from binocular disparity cues.* Experimental Psychology Department, U.S. Army Medical Research Laboratory. Fort Knox, Kentucky. AMRL Report No. 331, 1958.

REFERENCES

Amigo, G. Variation in stereoscopic acuity with observation distance. *Journal of the Optical Society of America*, 1963, **53**, 630-635.

Barlow, H. B., Blakemore, C., & Pettigrew, J. D. The neural mechanism of binocular depth descrimination. *The Journal of Physiology*, 1967, **193**, 327-342.

Beverley, K. I. & Regan, D. Evidence for the existence of neural mechanisms selectively sensitive to the direction of movement in space. *The Journal of Physiology*, 1973, **235**, 17-29.

Biersdorf, W. R., Ohwaki, S. & Kozil, F. J. The effect of instruction and oculomotor adjustments on apparent size. *American Journal of Psychology*, 1963, **76**, 1-17.

Blank, A. A. The Luneburg theory of binocular space perception. In S. Koch (Ed.), *Psychology: A study of a science* (Vol. 1). New York: McGraw-Hill, 1959.

Blank, A. A. The geometry of vision. *British Journal of Physiological Optics*, 1957, **14**, 154-169; 222-235.

Blank, A. A. The Luneburg theory of binocular visual space. *Journal of the Optical Society of America*, 1953, **43**, 717-727.

Boring, E. G. *The physical dimensions of consciousness.* New York: Century, 1933.

Charnwood, J. R. B. *Essay on binocular vision.* London: Hatton, 1951.

Dodwell, P. C. & Engel, G. R. A theory of binocular fusion. *Nature*, 1963, **198**, 39-40.

Epstein, W. The process of "taking-into-account" in visual perception. *Perception*, 1973, **2**, 267-285.

Epstein, W. & Davies, N. Modification of depth judgment following exposures to magnification of uniocular image: Are changes in perceived absolute distance and registered direction of gaze involved? *Perception & Psychophysics*, 1972, **12**, 315-317.

Epstein, W. & Morgan, C. L. Adaptation to uniocular image magnification: Modification of the disparity–depth relationship. *American Journal of Psychology*, 1970, **83**, 322-329.

Feldman M. & Cohen, B. Electrical activity in the lateral geniculate body of the alert monkey associated with eye movements. *Journal of Neurophysiology*, 1968, **31**, 455-466.

Foley, J. M. Distance perception. In R. Held, H. Leibowitz, & H. L. Teuber (Eds.), *Handbook of Sensory Physiology* (Vol. 8). Berlin: Springer-Verlag, in press.

Foley, J. M. Disparity increase with convergence for constant perceptual criteria. *Perception & Psychophysics*, 1967b, **2**, 605-608.

Foley, J. M. Binocular disparity and perceived relative distance: An examination of two hypotheses. *Vision Research*, 1967a, **7**, 655-670.

Foley, J. M. Locus of perceived equidistance as a function of viewing distance. *Journal of the Optical Society of America*, 1966, **56**, 822-827.

Foley, J. M. & Held, R. Visually directed pointing as a function of target distance, direction, and available cues. *Perception & Psychophysics*, 1972, **12**, 263-268.

Foley, J. M. & Richards, W. Effects of voluntary eye movement and convergence on the binocular appreciation of depth. *Perception & Psychophysics*, 1972, **11** 423-427.

Freeman, E. Anomalies of visual acuity in relation to stimulus-distance. *Journal of the Optical Society of America*, 1933, **22**, 285-292.

Fried, A. H. Convergence as a cue to distance. (Doctoral dissertation, New School for Social Research, 1973). *Dissertation Abstracts International*, 1974, **34**, 3247B. (University Microfilms No. 74-146.)

Fry, G. A. Visual perception of space. *American Journal of Optometry and Archives of American Academy of Optometry*, 1950, **27** 531-553.

Gogel, W. C. The organization of perceived space. I. Perceptual Interactions. *Psychologische Forschung*, 1973, **36**, 195-221.

Gogel, W. C. Scalar perceptions with binocular cues of distance. *American Journal of Psychology*, 1972, **85**, 477-497.

Gogel, W. C. The measurement of perceived size and distance. In W. D. Neff (Ed.), *Contributions to sensory physiology* (Vol. 3). New York: Academic, 1968.

Gogel, W. C. Perception of depth from binocular disparity. *The Journal of Psychology*, 1964, **67** 379-386.

Gogel, W. C. The visual perception of size and distance. *Vision Research*, 1963, **3**, 101-120.

Gogel, W. C. Perceived frontal size as a determiner of perceived binocular depth. *The Journal of Psychology*, 1960a, **50**, 119-131.

Gogel, W. C. The perception of shape from binocular disparity cues. *The Journal of Psychology*, 1960c, **50**, 179-192.

Gogel, W. C. The perception of a depth interval with binocular disparity cues. *The Journal of Psychology*, 1960b, **50**, 257-269.

Gogel, W. C. & Tietz, J. D. Absolute motion parallax and the specific distance tendency, *Perception & Psychophysics*, 1973, **13**, 284-292.

Graham, C. H. Visual space perception. In C. H. Graham (Ed.), *Vision and visual perception.* New York: Wiley, 1965.

Gregor, G. P. H. *Binocular single vision, fusion and suppression: An experimental study.* Unpublished master's thesis, York University, 1973.

Gulick, W. L. & Lawson, R. B. *Human stereopsis.* New York: Oxford University Press, 1976.

Hardy, L. H., Rand, G., Rittler, M. C., Blank, A. A., & Boeder, P. *The geometry of binocular space perception.* New York: Knapp Memorial Laboratories, Institute of Ophthalmology, Columbia University College of Physicians and Surgeons, 1953.

Harvey, L. O., Jr. Critical flicker frequency as a function of viewing distance, stimulus size and luminance. *Vision Research*, 1970, **10**, 55-63.

Heinemann, E. G., Tulving, E., & Nachmias, J. The effect of oculomotor adjustments on apparent size. *American Journal of Psychology*, 1959, **72**, 32-45.

Hering, E. *Spatial sense and movements of the eye* (C. A. Radde, Trans.). Baltimore: American Academy of Optometry, 1942. (Originally published, 1879.)

Hochberg, J. Perception II. Space and movement. In J. W. Kling & L. A. Riggs (Eds.), *Woodworth & Schlosberg's experimental psychology* (3rd ed.). New York: Holt, Rinehart & Winston, 1971.

Holway, A. H. & Boring, E. G. Determinants of apparent visual size with distance variant. *American Journal of Psychology*, 1941, **54**, 21-37.

Howard, I. P. & Templeton, W. B. The effect of steady fixation on the perception of relative depth. *Quarterly Journal of Experimental Psychology*, 1964, **16**, 193-203.

Hubel, D. & Wiesel, T. Anatomical demonstration of columns in the monkey visual cortex. *Nature*, 1969, **221**, 747-750.

Irvine, S. R. & Ludvigh, E. Is ocular proprioceptive sense concerned in vision? *Archives of Ophthalmology* (New York), 1936, **15**, 1037-1049.

Kaufman, L. *Sight and mind: An introduction to visual perception.* New York: Oxford University Press, 1974.

Komoda, M. K. & Ono, H. Oculomotor adjustments and size-distance perception. *Perception & Psychophysics*, 1974, **15**, 353-360.

Kries, J. von. Notes on perception of depth. In H. von Helmholtz, *Treatise on physiological optics* (Vol. 3) (J. P. C. Southall, Ed. and Trans.). New York: Dover, 1962. (Originally published, 1925.)

Lawson, R. B. & Gulick, W. L. Apparent size and distance in stereoscopic vision. In J. C. Baird (Ed.), Human Space Perception: Proceedings of the Dartmouth Conference. *Psychonomic Monograph Supplements,* 1970, 3(13, Whole No. 45).

Leibowitz, H. W., Shina, K., & Hennessy, R. T. Oculomotor adjustments and size constancy. *Perception & Psychophysics*, 1972, **12**, 497-500.

Linksz, A. *Physiology of the eye* (Vol. II). New York: Grune & Stratton, 1952.

Luckiesh, M. & Moss, F. K. The dependence of visual acuity upon stimulus distance. *Journal of the Optical Society of America*, 1933, **23**, 25-29.

Luneburg, R. K. The metric of binocular visual space. *Journal of the Optical Society of America*, 1950, **40**, 627-642.

Luneburg, R. K. *Mathematical analysis of binocular vision.* Princeton, New Jersey: Princeton University Press, 1947.

Ogle, K. N. The optical space sense. In H. Davson (Ed.), *The eye* (Vol. 4). New York: Academic, 1962.

Ogle, K. N. Precision and validity of stereoscopic depth perception from double images. *Journal of the Optical Society of America*, 1953, **43**, 906-913.

Ono, H. Some thought on different perceptual tasks related to size and distance. In J. C. Baird (Ed.), Human Space Perception: Proceedings of the Dartmouth Conference. *Psychonomic Monograph Supplements,* 1970, 3 (13, Whole No. 45).

Ono, H., Mitson, L., & Seabrook, K. Changes in convergence and retinal disparities as an explanation for the wallpaper phenomenon. *Journal of Experimental Psychology*, 1971, **91**, 1-10.

Ono, H., Muter, P., & Mitson, L. Size-distance paradox with accommodative micropsia. *Perception & Psychophysics*, 1974, **15**, 301-307.

Osgood, C. E. *Method and theory in experimental psychology.* New York: Oxford, 1953.

Richards, W. Visual space perception, In E. C. Carterette & M. P. Friedman (Eds.), *Handbook of perception* (Vol. 5.). New York: Academic, 1975.

Richards, W. Anomolous stereoscopic depth perception. *Journal of the Optical Society of America*, 1971a, **61**, 410-414.

Richards, W. Size-distance transformations. In O.-J. Grusser & R. Klinke (Eds.), *Pattern recognition in biological and technical systems.* Berlin: Springer-Verlag, 1971b.

Richards, W. Spatial remapping in the primate visual system. *Kybernetik*, 1968, 4, 146-156.

Richards, W. Apparent modifiability of receptive fields during accommodation and convergence and a model for size constancy. *Neuropsychologia*, 1967, **5**, 63-72.

Shipley, T., & Rawlings, S. C. The nonius horopter–II: An experimental report. *Vision Research*, 1970, **10**, 1263-1299.

Sperling, G. Binocular vision: A physical and a neural theory. *American Journal of Psychology*, 1970, **83**, 461-534.

Swenson, H. A. The relative influence of accommodation and convergence in the judgment of distance. *Journal of General Psychology*, 1932, **7**, 360-380.

Wallach, H. & Floor, L. The use of size matching to demonstrate the effectiveness of accommodation and convergence as cues to distance. *Perception & Psychophysics*, 1971, **10**, 423-428.

Wallach, H. & Frey, K. J. Adaptation in distance perception based on oculomotor cues. *Perception & Psychophysics*, 1972, **11**, 77-83.

Wallach, H., Frey, K. J., & Bode, K. A. The nature of adaptation in distance perception based on oculomotor cues. *Perception & Psychophysics*, 1972, **11**, 110-116.

Wallach, H., Moore, M. E., & Davidson, L. Modification of stereoscopic depth perception. *American Journal of Psychology*, 1963, **76**, 191-204.

Wallach, H. & Zuckerman, C. The constancy of stereoscopic depth. *American Journal of Psychology*. 1963. **76**, 404-412.

Woodworth, R. S. & Schlosberg, H. *Experimental psychology*, New York: Holt, 1954.

CHAPTER 5

THE METRIC OF VISUAL SPACE

WALTER C. GOGEL
University of California, Santa Barbara

This chapter examines the evidence for a two-factor theory of spatial perception (Gogel, 1972b). The theory distinguishes between absolute and relative cues or other absolute and relative sources of perceptual information. It is asserted that these cues or other sources of information are often in conflict and that the perception of metric space is a product of the rules by which the visual system integrates this often conflicting information. In the final portion of this chapter some of the implications of this theory for the achievement of spatial constancy are considered.

THE MEASUREMENT OF ABSOLUTE AND RELATIVE CUES

Absolute cues or other absolute factors are interactions between the observer and the stimulus object that contribute to the perceived characteristics of an

The preparation of this chapter was supported by PHS Research Grant MH 15651 from the National Institute of Mental Health.

object, whereas relative cues involve perceptual interactions between two or more stimulus objects. The distinction between absolute and relative cues can be clarified by considering the conditions under which each is characteristically measured. Although absolute cues can contribute to a perception when relative cues are present, their contribution is most easily demonstrated in the absence of relative cues. Thus absolute cues can be identified by measuring the perceived characteristic of an object in the absence of other objects; that is, with the object presented in an otherwise homogeneous visual field. As an example the perceptual effect of absolute cues of distance or of absolute cues of orientation is most readily investigated by measuring the perceived distance or perceived orientation of an object presented in an otherwise homogeneous visual field. The perceptual effect of absolute cues of motion is most readily investigated by measuring the perceived velocity and direction of motion of an object in the absence of any other objects.

It is easier to measure the effect of absolute cues in the absence of relative cues than the converse. Because the absolute cues associated with each object are independent of the number of objects present, it is unlikely that a situation can be produced in which the effects of relative cues can be measured in the total absence of absolute cues. It is possible, however, to examine the effectiveness of absolute cues compared with relative cues. For this purpose consider the situations illustrated in the three parts of Figure 5-1. The upper portion of A illustrates the situation in which a single point of light (Point 1) moving up and down, as shown by the arrows, is presented in an otherwise dark visual field. Assume that under these conditions the observer is able to provide a report of motion that varies in some consistent manner with the physical motion. This could be demonstrated, for example, by presenting different groups of observers with different amplitudes or directions of motion and finding that the average response varies systematically as a function of the different conditions. If this occurs, it can be concluded that absolute cues of motion are available to specify the apparent motion of the single point of light. Indeed it is likely, as indicated in the lower portion of A, that the observer will perceive Point 1 moving vertically in agreement with its physical motion. Relative motion cues can be investigated by introducing another point of light (Point 2) in the vicinity of Point 1 with a motion that is different from that of Point 1. This is illustrated in the upper portion of B. Point 2 moves horizontally and Point 1 continues to move vertically with the phase of motion of Point 2 in relation to Point 1 indicated by the arrows. Suppose that the apparent motion of Point 1 is identical before and after Point 2 is added to the display. In this case there is no need to assume that a relative motion cue exists between Points 1 and 2. The apparent motion of Point 1 can continue to be explained by the cue of absolute motion.

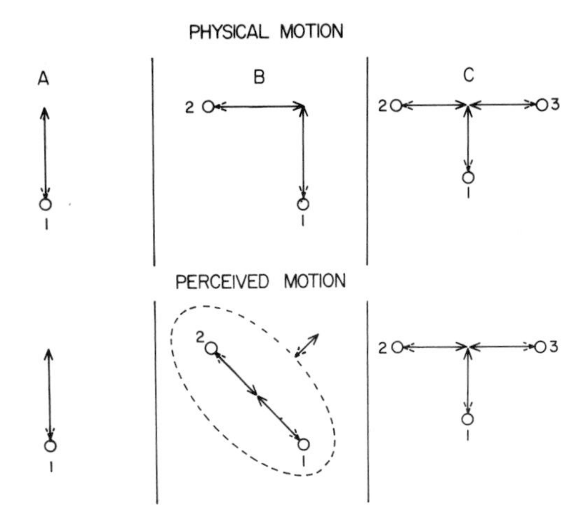

Figure 5-1 Front view drawings illustrating a situation (*A*) involving only an absolute cue of motion and situations (*B* and *C*) in which both absolute and relative cues are present.

Also, suppose that Point 2 is seen moving along the same (horizontal) path, whether Point 1 is present or absent. In this case, because the apparent motion of neither point is affected by the presence of the other point, there would be no reason to assume that any relative motion cues are involved in the perception of motion associated with the display in B. Relative motion cues usually are effective under these conditions. Rather than Point 1 appearing to move vertically and Point 2 appearing to move horizontally, with the two points simultaneously present, both points will appear to move toward and away from each other on a diagonal between upper left and lower right, as shown in the lower portion of B. This perceived motion of one object in relation to the other is called perceived relative motion and can be attributed to relative motion cues occurring between the two points. It is also possible that Points 1 and 2 will appear to move as a pair in a direction between lower left and upper right with this perceived motion also shown in the lower portion of Figure 5-1B and called perceived common motion (Johansson, 1964). Of importance to the present discussion, the perception of motion of Point 1 in the absence of Point 2, as shown in the lower portion of Figure 5-1A, indicates that a cue of absolute motion is available and the change in the path of apparent motion of Point 1 produced by the introduction of Point 2, shown in the lower portion of Figure 5-1B, indicates that a cue of relative motion has been added. The change in the perception of motion of Point 1 as a function of the addition of Point 2 is called induced motion; Point 2 is called the induction object and Point 1, the test object. The lower portion of Figure 5-1B illustrates a case in which the induction between the two objects is equal and opposite and either point could be called the test or induction object, depending on the purposes of the experiment. It is not always true, however, that the effect of relative cues between the test and induction objects is equal and opposite; for example, induced motion is often demonstrated by the situation in which a moving frame results in an apparent motion of an enclosed stationary point (Duncker, 1939). In this situation the effect on the apparent motion of the point produced by introducing or removing the frame is greater than the effect on the apparent motion of the frame produced by introducing or removing the point. The full procedure for examining the distinction between absolute and relative cues and for comparing the contribution of each is independently to measure perceived object characteristics as each of the two objects of the display is added or removed.

Whenever the perceptual outcome from one source of information acting alone is different from that obtained from another source acting alone, the two sources of information when presented simultaneously are in conflict. To achieve a single unitary perception in such cases the visual system must ignore one of the sources or combine the two by differentially weighing the contribution of each. As will become evident, the second is the alternative often preferred by the visual system. The perceptual outcome is neither that expected from one nor the other of the conflicting sources but is some compromise between the outcome expected from each. Also, informational conflicts in visual perception occur frequently; for example, according to the above discussion, whenever two or more objects move so that the distance between them changes, absolute and relative cues of motion will be in conflict.

In addition to conflicts between absolute and relative cues, conflicts also occur between different relative cues. Figure 5-1C provides an example of a conflict between two different relative cues of motion. The physical motions of Points 1 and 2 in Figure 5-1C are similar to those in Figure 5-1B, but in Figure 5-1C unlike Figure 5-1B, another point of light, Point 3, has been added. The physical motion of Point 3 is horizontal and opposite in phase to that of Point 2. The relative motion of Point 1 with respect to Point 3 is equal but opposite to the relative motion of Point 1 with respect to Point 2. In other words, the relative motion cues from Points 2 and 3 with respect to Point 1 are in conflict. Because the magnitudes of these conflicting cues are equal but opposite, they will cancel and, as shown in the lower portion of Figure 5-1C, Point 1 will usually appear to move vertically. If Points 2 and 3 are called induction objects, with Point 1 the test object, Figure 5-1C illustrates the situation in which two induction objects produce an opposite but equal induction effect in the test object. Induction conflicts from relative motion cues will occur whenever the direction or magnitude of the relative motion of the test object with respect to an induction object differs for the different induction objects. It should also be noted that although the apparent motion of Point 1 is the same in A and C the cause of this motion is different in the two situations. In A only the absolute motion cue was available. In C the absolute motion cue is in agreement with the opposed but balanced relative motion cues that result from the equal but opposite motion of Points 2 and 3. As discussed, opposing induction effects (opposed relative cues) can occur with respect to a variety of perceived object characteristics and under a variety of conditions. The visual system must integrate these frequently occurring conflicts of information from relative cues in order to achieve a unitary perception. Considerable evidence is available that in many situations this process consists not in the suppression of one of the sources of information but in the differential weighting of the conflicting sources.

The distinction between absolute and relative cues is often applied to the perception of distance. An example of the difference between an absolute and relative cue of perceived distance can be discussed with the help of Figure 5-2. Suppose that the position of the observer's eyes is represented by A and B and that Points *e*, *f*, and *g* are three possible positions of points of light located at different distances in the median plane of the observer. Consider the case in which only one point of light, for example, Point *e*, is presented in an otherwise dark visual field. Assume that under these conditions Point *e* is perceived at some specific (nonrandom) distance from the observer. Because this perceived distance involves an interaction between the observer and the object that can be measured in the absence of other objects, the cues or other factors determining this perception are absolute cues or absolute factors. Because the perception is that of distance from the egocenter of the observer, this distance is often called

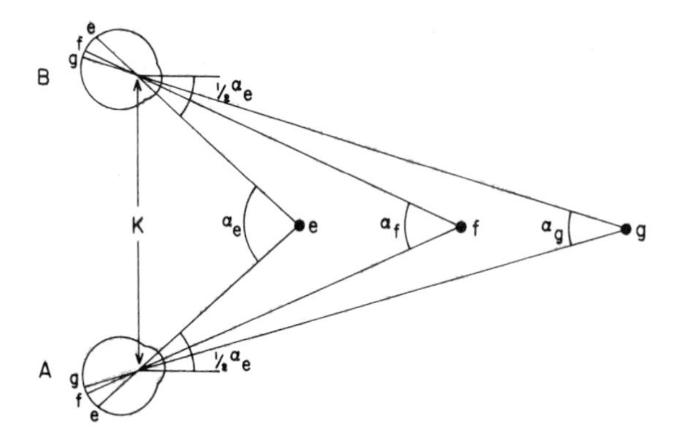

Figure 5-2 A top view schematic diagram that is useful in the discussion of the major cues of egocentric and exocentric distance.

perceived egocentric distance. Similarly, if Point *f* were presented alone, it would also be perceived at some egocentric distance, but its perceived egocentric distance would differ from that of Point *e*, depending on whether the absolute cues to the distance of Point *e* and Point *f* were the same or different. Presenting Points *e* and *f* simultaneously, however, could result in a perception of the distance of either or both that would be different from the perceived distance that occurred when each point was presented singly. In this case relative cues must be postulated to explain the different perceptions occurring from the successive and simultaneous presentations of Points *e* and *f*. A likely cue of absolute distance in this example of binocular observation is the convergence of the eyes, whereas the relative distance cue is that of binocular disparity. The perceived distance between objects produced by a relative cue is called perceived exocentric distance. The procedure of postulating a relative cue to be present only if the addition of a second object produces a perception that cannot be accounted for by absolute cues dictates that absolute and relative cues will have different psychophysical functions. It follows that the simultaneous presence of absolute and relative cues in perceived distance as in motion perception is necessarily a frequent source of cue conflicts.

OTHER ABSOLUTE AND RELATIVE FACTORS

Absolute and relative cues are important determiners of spatial perceptions; but not all absolute or relative factors that modify perceptions are cues that can be defined in terms of the conditions of stimulation. It has been found that as spatial cues are increasingly reduced or restricted perceptions tend to resemble certain specifiable conditions. These tendencies for perceptions to approach certain states rather than to become random as stimulus conditions are reduced are called observer tendencies. Several observer tendencies have been identified in the perception of distance. One called the equidistance tendency (EDT) states that with the reduction of relative distance cues objects or parts of objects will tend to appear at the same distance with the effectiveness of this tendency increasing as the directional separation between the objects or parts is decreased (Gogel, 1965a). The EDT is a factor in relative (exocentric) distance perceptions and can be in agreement or conflict with relative distance cues. Another tendency called the specific distance tendency (SDT) states that with the reduction of cues of egocentric distance objects will tend to appear at a near distance (about 2 m) from the observer (Gogel & Tietz, 1973). The SDT is a factor in egocentric distance perceptions and can be in agreement or in conflict with absolute distance cues.

METRIC AND SCALAR TASKS

Evidence is presented to support the conclusion that conflicts from different sources of perceptual information are ubiquitous. These conflicts can occur between absolute and relative cues, between different absolute or between different relative cues, and between cues and observer tendencies and must be integrated by the visual system to produce a stable unitary perception with metric qualities. Metric perceptions are capable of quantitative variation. It is clear that spatial perceptions are metric. One object is perceived to be twice the size of another and at three times its distance. Another object is perceived to be 10 cm high and 300 cm from the observer. A rod is perceived tilted 15° from the vertical or as forming a 30° angle with another rod. An object is perceived moving horizontally at a particular velocity or parallel to and with twice the velocity of another object. All are examples of metric perception. Although numbers are often used to describe these perceptions, the perceptions do not depend on the observer having a concept of number. This is obvious in that many animals are capable of metric perceptions; for example, the ability of a rat to modify the force of his jump as a function of the distance between one stand and another (Russell, 1932) implies that the rat has a metric perception of distance.

The kind of metric information that must be available in order for the observer to respond differentially to size, distance, orientation, and velocity depends on the judgment or task required (Gogel, 1968; Ono, 1970). One type of judgment, called a scalar judgment, requires that the observer respond to the perceived absolute magnitude of the object characteristic. A report that an object is 10 cm high and 300 cm from the observer or is tilted 15° from the vertical, to the extent that it is based on perceptual information requires that the perceptions contain scalar information. A task that utilizes scalar aspects of the perception is, for example, a verbal estimate of a size or distance in feet or inches. Examples of nonverbal responses that require scalar information are ballistically reaching for an object, jumping over a puddle, or visually deciding that an object is sufficiently small to be grasped by the hand.

Another type of judgment or task utilizes only relative (nonscalar) information in the perception. In this task the observer is asked to indicate the size, distance, orientation, or velocity of one object in relation to another. This is the type of task that has been studied most frequently; for example, in studies of size constancy the procedure used most commonly is to obtain judgments of the perceived size of an object at one distance in relation to other objects at other distances. Information that is capable of supporting judgments of the relative characteristics of objects but not judgments of absolute magnitudes is nonscalar. A basic difference between scalar and nonscalar information is that a task requiring the former uses "some unit not simultaneously in the modality in which the judgment is being made. For the estimating human it is the memory of the ruler or metric stick. For the jumping animal it is a learned or innate relation between muscular effort and perceived distance" (Gogel, 1968, p. 126).

It is clear that a theory of spatial perception that considers only the relative aspects of perceptions is incomplete. If perceptual information in regard to the three-dimensional world were completely nonscalar, coherent behavior based on perceptions would be impossible. If the human or animal is to move about in the three-dimensional world, picking up objects or avoiding them during locomotion, etc., it is necessary that information be available in regard to absolute sizes, distances, orientations, and velocities of objects and not only information that compares one object with another. The need to understand the processes in the perceptual system that provide scalar as well as nonscalar information is also obvious in the study of perceptual development. Behaviors associated with scalar information often involve the coordination of motor and visual information. Jumping over a ditch requires the integration of information regarding body weight, force of the jump, and perceived scalar distance. Ballistically reaching for an object requires the integration of information regarding the length of the arm, the velocity of the arm motion, and the visual perception of scalar distance. Clearly much of the information involving motor responses to visual stimuli will change with growth (see Held, 1965; Bower, 1974). This includes the oculomotor sources of scalar information; for example, the convergence of the eyes. With growth the interpupillary distance and thus the amount of eye convergence required to fixate an object at a given distance will increase. Convergence as a source of scalar distance information must be continuously recalibrated as the interpupillary distance of the observer changes with growth. Clearly a theory of spatial perception must account for the ability of the observer to perform scalar as well as nonscalar tasks. It seems reasonable that the important sources of information for scalar judgments originate in absolute cues or other absolute sources of perceptual information.

ABSOLUTE AND RELATIVE SOURCES OF INFORMATION IN PERCEIVED DISTANCE

This discussion is concerned with the reasons, both experimental and theoretical, for differentiating between absolute and relative factors in spatial perception. The phenomena are divided (somewhat arbitrarily) into two categories, one of which concerns the research on distance perception. In this case the perception of distance is the dependent variable. The other category, which also involves perceived distance but not always or even usually as the dependent variable, is concerned with induction effects. Examples of absolute and relative cues in each of these categories were given at the beginning of this chapter in relation to Figures 5-2 and 5-1, respectively. The need to differentiate between absolute and relative sources of information is considered first for the perception of distance.

Absolute and Relative Cues of Distance

Absolute cues of distance are cues that determine the apparent distance of an object from the observer (a perception of egocentric distance). The major absolute cues of distance are the oculomotor cues of convergence and accommodation, the cues of familiar size and absolute motion parallax. Absolute cues of distance are of special significance in that they are potential sources of scalar information. Relative cues of distance indicate that one object is more distant than another (a perception of exocentric distance) without specifying the apparent distance of either object from the observer. The major cues of exocentric distance are binocular disparity, relative (retinal) size, and relative motion parallax. If a relative cue is continuously distributed in depth, it is sometimes referred to as a gradient (Gibson, 1969); for example, the perspective cue formed by a distribution of objects of constant physical size extending to the horizon can be regarded as an extension of the relative size cue to distance. Gradients will not be considered to be qualitatively different from cues. Although the overall perceptual effect of a cue gradient may be larger than that obtained with the minimal stimulus conditions defining the cue, there is no reason to suppose that the basic processes in cues and gradients differ essentially.

THE GEOMETRY OF DISTANCE CUES

The basic experimental evidence for the need to distinguish between absolute and relative cues to distance is that the results from many investigations of the perception of distance cannot be understood without postulating both kinds of factors. The basic theoretical evidence for the need to make this distinction is that the two kinds of factors differ at the level of the proximal stimulus. Both forms of evidence are considered in relation to Figure 5-2.

The stimulus definitions of absolute and relative cues differ as illustrated in Figure 5-2. Consider the absolute cues of convergence, accommodation, motion parallax, and familiar size. Convergence and absolute motion parallax are illustrated by the situation in which a single point of light is presented at one of the distances e, f, or g. If A and B are the right and left eye of the observer, the convergence of the eyes required to fixate the point is given by α. For values of α that are not too large, α in radians is equal to K/D, where K is the distance between the eyes and D is the distance of the point from the observer. The same equation can be used to express the relation between physical distance D

and accommodation in diopters in which K is equal to unity and D is expressed in meters. Also α in the expression $\alpha = K/D$ can be used to express the cue of absolute motion parallax. In this case instead of representing the observer's eyes A and B are the two successive positions of the observer's head and K is the magnitude of lateral head motion. Figure 5-2 can also represent the familiar size cue to egocentric distance if K is used to indicate the width of the familiar object viewed from either positions e, f, or g. In this case α is proportional to the size of K on the eye. Thus the equation $\alpha = K/D$ can represent all of the major absolute cues of distance (Gogel, 1974a). As indicated by this equation, all of these absolute cues are inversely related to the physical distance of the object from the observer.

Figure 5-2 can also be used to illustrate the relative cues of distance by considering the case in which at least two of the points of light are presented simultaneously. The radial distance between any two points will be called d, the difference in the convergence to the nearer and farther points will be called γ; that is, $a_e - a_f = \gamma_{ef}$ and D, as before, is the distance of a point from the observer. For situations in which D, is large compared with d (as is usual) it can be shown that, as a good approximation, $\gamma = Kd/D^2$, where D is the average value of D to each of the points (Graham, 1965). If A and B are the two eyes of the observer, γ is proportional to the difference in horizontal extent between the points on the two eyes. This is the cue of binocular disparity. If A and B are two positions from which the objects are viewed with head motion, γ is proportional to the relative eye turning or relative motion of the objects on the eye if the direction of the eye is fixed. In this case γ defines the cue of relative motion parallax. Finally, if two objects of the same physical size (K) are presented at different distances from the observer, γ represents the difference between their angular sizes on the two eyes. This is the relative size cue of distance that occurs if two objects of the same shape and physical size are presented at different distances, either simultaneously or successively, or if an object of constant physical size is moved to different distances while being viewed by the observer. If the objects are familiar, γ also represents the relative size cue of distance associated with a familiar size. It follows that γ represents the major relative cues of distance defined by differences in visual angles expressed in radians. For a constant physical depth d these cues decrease inversely as the square of the distance from the observer.

Clearly, absolute and relative cues are distinguished by their relation to physical distance, with the absolute varying inversely with physical distance and the relative (for a constant d) varying inversely as the square of distance. Also, they are distinguished in terms of their definitions as stimuli (α and γ). Although α and γ are measured in the distal world, they are proportional to and therefore represent proximal stimuli. In terms of proximal stimuli absolute cues defined by α and relative cues defined by γ differ in some important respects. Absolute cues of distance are defined less in terms of retinal position or retinal extent than relative cues. The absolute cue of convergence requires nonvisual information regarding the position of the eyes in the head. The accommodative cue of absolute distance probably requires information regarding the state of ciliary muscles. Absolute motion parallax requires information regarding the changes in the position of the eye or eyes fixating the object as the head is moved, and the familiar size cue to egocentric distance, in addition to information regarding retinal size, requires that familiar characteristics be available to determine the perceived size of the object. It seems likely that it is the presence of these nonretinal components of information in absolute cues that make it possible for absolute cues to introduce scalar information into the perception. The proximal stimulus for relative cues of depth (d) on the other hand is differences in retinal extent between the two eyes (binocular disparity), differential motion across the retina (relative motion parallax), and changes or differences in retinal size (relative size cue).

EVIDENCE FOR ABSOLUTE AND RELATIVE CUES OF DISTANCE

Theoretical Evidence for Absolute and Relative Cues of Distance. The distinction between absolute and relative cues in proximal stimulation and in their different relations to physical distance provides a necessary but not a sufficient reason for differentiating between these two kinds of cue in perception. Geometrically, the two kinds of cue are redundant. Perhaps they are also redundant perceptually so that, for example, if egocentric cues of distance are available, the perception of exocentric distance would be explained as a difference between two egocentric perceptions of distance. Conversely, if relative distance cues were available throughout the visual field, perhaps in the form of cue gradients, it might seem to be unnecessary to consider the contribution of egocentric cues, in which case, the perceptions of egocentric distance would be the result of a process of summing perceived exocentric distances. The latter possibility was proposed in a paper concerned with absolute and relative cues to the perception of size, shape, and distance (Gogel, 1963). There are, however, a number of reasons for rejecting both possibilities. One limitation of any attempt to explain egocentric distance perceptions completely in terms of the perceptual summation of distance from exocentric cues is the evidence that identifies egocentric cues to distance as present in situations in which all exocentric cues of distance are excluded. This evidence is considered later in this chapter. Also opposed is the difficulty of explaining scalar perceptions in terms of exocentric cues. Finally, as will be discussed, egocentric cues or perceived egocentric distance are necessary to transform (calibrate) the exocentric cue to a perceived scalar distance.

The reasons for concluding that perceived distance cannot be explained entirely in terms of egocentric cues are compelling. The most important concerns the general difference in sensitivity between these two kinds of cue. It is not surprising that the exocentric cues which involve comparisons between retinal extents or retinal motions are more precise than egocentric cues which involve oculomotor adjustments, absolute retinal or eye motions, or remembered size. The experimental evidence available seems to support this expected difference in precision; for example, the perception of depth from binocular disparity is usually more precise than the perception of depth from successive presentations of different values of convergence. Indeed, under optimal conditions the cue of binocular disparity can achieve a threshold value of several seconds of arc in the perception of the equidistance of objects. Also, as an examination of the evidence regarding egocentric cues shows, most egocentric cues are not capable of mediating perceived egocentric distances beyond several meters from the observer. Thus *egocentric* cues cannot account totally for the egocentric perceptions that obviously extend to greater distances than those within which direct egocentric cues would be effective. This is illustrated in Figure 5-2. As will be discussed, a relative cue requires calibration by egocentric cues in order to specify a perception of exocentric distance. Assume that the perception of egocentric distance (D'_e) to the first point e calibrates the exocentric cues ($\alpha_e - \alpha_f$) associated with the depth ef. The perceived depth (d'_{ef}) resulting from this calibration process can then be added to D'_e to determine D'_f, that is, $D'_f = D'_e + d'_{ef}$. In a similar manner the perceived egocentric distance D'_f would then calibrate the perceived depth (d'_{fg}) associated with the depth extent fg. Probably a more exact expression of this process would involve a process of summation (integration) in which the D' to a point would determine the $\Delta d'$ associated with an increment of depth at that distance, and so on. In any event, it is likely that by this process of summation the increased precision of exocentric cues of distance, compared with direct egocentric cues, is utilized to extend the egocentric scalar aspect of perceptions throughout the distance of the visual field. Indeed it is highly probable that *perceptions* are never exocentric without also being egocentric and never relative without also being scalar. Although it is essential to distinguish between egocentric and exocentric and scalar and nonscalar cues, if more than one object is present at different distances, these different cues merge in their contribution to the perception of a distance and the resulting perception simultaneously contains qualities that are egocentric, exocentric, relative, and scalar.

The results from distance responses can be misinterpreted if the distinction between absolute and relative cues of distance is not understood. One source of not infrequent misinterpretation is present because relative cues of distance can occur between successive as well as between the simultaneous presentations of objects (Gogel, 1968; Gogel & Sturm, 1971). To illustrate this problem consider the following experimental situation. Two luminous squares of the same physical size are viewed monocularly through a restrictive aperture and are presented successively, one at 3 and the other at 6 m from the observer in an otherwise dark environment. Suppose that the purpose of the experiment is to evaluate the distance cues available in this situation. Suppose also that absolute cues of distance are not available, but because the far square is smaller on the retina than the near square the size cue to exocentric distance is present between the successive presentations. The observers are randomly assigned to two groups. To balance the orders of presentation one group is presented first with the far and then with the near square, whereas the other group is presented first with the near and then with the far square. To measure perceived distance the experimenter obtains verbal reports from the observer of the distance of a square on each presentation. In the absence of absolute cues of distance the average reports from the first presentations of the two groups will be the same or nearly the same; for example, on the first presentations the squares, regardless of their retinal sizes, might be seen at 2 m. On the second presentations, however, the relative size cue between first and second presentations will make the square presented second appear at half or twice the distance of the first. Thus on second presentations the group presented with the near square first will report the far square to be at twice the apparent distance of the square presented first; that is, at 4 m. The group presented with the far square first, because of the relative size cue between the successive presentations, will report the near square to be at one-half the apparent distance of the square presented first; that is, at 1 m. Thus the group receiving the far-near order of presentation will make reports of 2 and 1 m, respectively, whereas the group receiving the near-far order of presentations will give reports of 2 and 4 m, respectively. If the reports from the two presentations for each group are averaged, the group presented with the far square first will average 1.5 m. The group presented with the near square first will average 3 m. From these average reports the experimenter may conclude, erroneously, that the absolute size of the retinal image, that is, the size of a single retinal image by itself, is an effective cue to distance. Only if the first and second reports are analyzed separately by the experimenter would the differences in the reports of distance be identified correctly as due to the relative size cue resulting from changes in the retinal size of the object between the successive presentations. Even if this particular error in interpretation is not made, the psychophysics of the relative size cue could be misinterpreted. Because the reports from the second presentations have a ratio of 4 to 1 (4 to 1 m) the experimenter may conclude, again erroneously, that the ratio of perceived distances from the relative size cue is greater than would be expected from the ratio of retinal sizes. The lack of absolute distance cues on the first presentations resulted in an exaggeration of the differences between the reports of distance on the second presentations. This illustrates the need to differentiate and the difficulty in differentiating between absolute and relative cues of distance. Conversely, differences or lack of differences between the perceptions associated with first and subsequent responses to different stimulus conditions can provide evidence of the identification of distance cues. In a recent study Mershon and King (1975) used the difference between responses to first and second presentations of auditory stimuli to demonstrate that auditory intensity is a cue to exocentric but not to egocentric distance. It seems that the distinction between absolute and relative distance cues is as important in audition as it is in vision.

Experimental Evidence for Absolute and Relative Cues of Distance. From the point of view of this chapter the experimental evidence concerning the validity of the different cues of distance, as described in relation to Fig.5-2, is of interest in two respects. First, of course, this evidence is relevant to the need to differentiate between absolute and relative cues of distance. Second, egocentric cues of distance are potential sources of scalar information; for example, if the observer has information regarding the distance between the nodal points of his eyes and the convergence required to place the object in the center of each fovea (fixation), sufficient information is available to specify the apparent scalar distance of the object. Similarly, information concerning the state of the ciliary muscles necessary to produce a sharp image on the retina is potentially capable of supplying scalar information about the distance of the object. Thus oculomotor cues are possible sources of scalar information required for the scalar aspects of perceptions; or the cue of absolute motion parallax can produce scalar information if the observer can utilize simultaneously the information regarding the amount of head movement in the frontoparallel plane and the amount of eye turning necessary to fixate the object despite the head motion. Familiar size is potentially a particularly effective source of scalar information that can be readily applied to all portions (including distant portions) of the visual field. If an object has a particular perceived size because it is familiar, this familiar size, together with the retinal size, can possibly specify not only the scalar distance of the object from the observer but also a scale for calibrating the entire visual field.

Research in regard to whether the accommodation and convergence of the eyes to an object presented in a homogeneous surround will determine a perception of egocentric distance has a long and varied history (Woodworth, 1938; Linksz, 1952). Considerable evidence has been accumulating in support of the conclusion that oculomotor cues can be consequential cues of perceived egocentric distance (Foley, 1976; Foley & Held, 1972; Gogel, 1972c; Gogel & Tietz, 1973; Richards & Miller, 1969; Wallach & Frey, 1972; Wallach & Smith, 1972). There are some limitations, however, on the effectiveness of these cues. Oculomotor cues are not likely to be effective beyond about 3 m from the observer; they are often clearly demonstrated only for closer distances, are not highly precise as cues even at the distances at which they are effective, and can vary markedly in effectiveness between observers. An illustration of the validity of oculomotor cues and their lack of precision at distances not too close to the observer is found in a study by Gogel (1972c) in which a single point of light was presented at two different convergence distances to two different groups of observers. The verbal reports of distance were significantly different in the expected direction for the different groups when the distances were 5 and 25 ft but not when the distances were 10 and 20 ft.

Absolute motion parallax as a cue to distance is studied by measuring the perceived distance to a single object with no cues available to this distance except the one provided by the motion of the head in a frontoparallel plane. The cue of absolute motion parallax under these conditions for a lateral motion of the head is illustrated by Figure 5-3. Positions 1 and 2 represent the extreme positions of the head as it is moved laterally. The solid rectangle represents the physical location of the stimulus object, the lines of regard from the position of the observer to the solid rectangle are construction lines that illustrate the visual directions of the object with ϕ the change in visual direction as the head is moved from Position 1 to Position 2. The only purpose of the dashed lines is to improve the perspective of the drawing. They would not be present in an actual experiment. The angle ϕ can represent the change in the direction of gaze if the rectangle is fixated as the head is moved or the angular motion of the

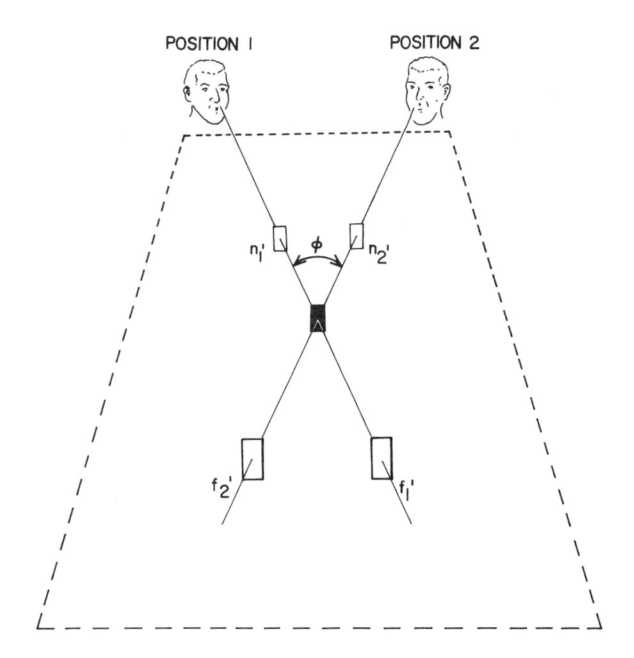

Figure 5-3 A perspective top-view drawing illustrating the apparent motion of an object associated with lateral motion of the head as a function of the difference between the physical and perceived distance of the object.

rectangle on the eye if the gaze is always straight ahead. The open rectangles at n' and f' represent two possible apparent distances of the rectangle. If the rectangle appears at n' it will appear to move in the same direction as the head motion (between n'_1 and n'_2). If the rectangle appears at f' it will appear to move in a direction opposite that of the head motion (between f_1 and f_2). An apparent motion of an object with or against the direction of motion of the head (an apparent concomitant motion of the object) will occur if the object is incorrectly localized in apparent distance. Thus an apparent concomitant motion of the object indicates at least a partial failure of the absolute motion parallax cue to specify the distance of the object. A total failure of this cue would occur if the direction and magnitude of the concomitant motion were entirely predicted by the perceived distance of the object, measured in the absence of head motion. At least some amount of failure of the distance cue of absolute motion parallax was found in a study by Gogel and Tietz (1973) in which points of light were presented one at a time in an otherwise dark visual field at different distances (30, 91, 181, 457, and 883 cm) as the head was moved a lateral distance of 13.5 cm. The single points of light were observed monocularly or binocularly. Because there is a tendency, in the absence of distance cues, for objects to appear at a relatively near distance (the SDT), it was expected that the apparent distance of a point of light physically located at distances other than the SDT would be modified in its direction. Thus the perceived distance of a point of light at distances greater or less than that of the SDT should be perceptually underestimated or overestimated, respectively. It follows from Figure 5-3 and the SDT that distant points should appear to move with the head, whereas near points should appear to move against the head. This was the usual result obtained from the motion reports. The sources of egocentric information of distance available during the head motion were absolute motion parallax and oculomotor cues. An apparent motion of the point indicates that the point was not perceived at the distance (the physical distance) expected from absolute motion parallax and oculomotor cues. It can be concluded from this study that the perception of distance expected from absolute motion parallax together with oculomotor cues can be modified by the rather weak factor of the SDT. Verbal reports of distance were obtained in the study by Gogel and Tietz (1973) before moving the head. Under these conditions the only available egocentric information in regard to distance was that of the SDT and oculomotor cues. The verbal reports of distance (corrected for observer biases in applying a foot ruler) were consistent with the head motion data in indicating that apparent distance was overestimated for the near and underestimated for the far distances of the points. Of interest in evaluating oculomotor cues to egocentric distance, the study also provides evidence that accommodation and convergence (or accommodative-convergence) can support perceptions of egocentric distances at near distances (less than several meters from the observer).

Only a few studies have attempted to evaluate the distance cue of absolute motion parallax that results from head motion in the absence of other egocentric cues of distance. In an experiment by Eriksson (1972a, Experiment I) objects subtending a constant visual angle were presented monocularly one at a time at 2, 3, or 4 m from the observer in an otherwise dark visual field. Verbal reports of the distances of the objects were obtained while moving the head laterally through 5.5 cm. Although a significant difference in the verbal reports were obtained from the 2-m distance, compared with either of the other distances, there were indications that some observers had based their judgments on cognitive information that included the degree of apparent lateral motion of the objects. Eriksson concluded that the evidence for the validity of the absolute motion parallax cue was marginal in this experiment. It should be noted, however, that the objects were quite far from the observer and the constant visual angle of the successively presented objects provided misleading information that the objects were equidistant. It is concluded from a study by Johansson (1972) that absolute motion parallax can mediate the perception of egocentric distances near the observer. In this study a monocularly observed test figure of constant visual angle presented at 30, 60, 120, or 240 cm (Experiment I) or at 35, 60, or 120 cm (Experiment II) was viewed while moving the head through a distance of about 1 cm. Following the head motion, the figure was removed and its apparent distance was measured by moving a rod in distance in a full cue visual field (binocularly observed) to duplicate the apparent distance of the test figure. In both experiments the measured apparent distance closely approximated the physical distance of the test figure. Although the contribution of accommodation to these results was discounted by some preliminary observations, it is possible that the absolute accommodation to the test figure found in the studies by Foley (1975) and Gogel and Tietz (1973) as a consequence of continued observation of the test figure or perhaps the accommodative cue between the successive presentations of the test figure and measurement field contributed to the precision of the results.

Many studies have been concerned with familiar size as a cue to perceived egocentric distance. A comprehensive review of this research to about 10 years ago has been provided by Epstein (1967). The present discussion does not attempt to extend that review. It is suggested, however, that the issues involved in assessing the validity of the familiar size cue have changed somewhat. Although the validity of the function relating familiar size and apparent egocentric distance had been questioned (Gogel, 1964b), it is now clear that reported distance will vary in the expected direction as a function of the familiar size of the object (Gogel, 1969; Gogel & Mertens, 1967). It is also clear that quite large individual differences occur in this function (Epstein, 1963b). In order to be a cue to perceived distance it is necessary for familiar size to determine perceived size. Although known (assumed or familiar) size can modify the report of distance, there is some question whether familiar size is able to determine perceived size (Epstein, 1967, p. 51). It is apparent from a study by Epstein (1961) that reported size is not always specified by familiar size. In this study different sizes of playing cards were presented under reduced conditions of observation. Despite a constant familiar size, the reported size of the card sometimes varied with its retinal size. Similar results were obtained in a study by Gogel and Mertens (1967) and in another by Gogel (1969) in which observers judged the perceived size and distance of a playing card under otherwise reduced conditions as a function of its retinal size. The reports of apparent size varied with retinal size, that is, the playing card sometimes appeared as nonnormal in size even though the reports of distance varied in the directions expected if familiar size were a valid cue to perceived egocentric distance. It may seem from these results that familiar size is often an adequate cue to egocentric distance despite its frequent failure to determine perceived size. It has also been suggested that this seeming inconsistency in the results occurs because of the intrusion of cognitive factors in the reports of distance (Gogel, 1974a). As in the case of a point of light (Gogel & Tietz, 1973), possibly a familiar object presented monocularly in a dark surround will appear at a distance of about 2 or 3 m from the observer (the SDT), regardless of its physical distance. If this occurs, it follows in agreement with the size-distance invariance hypothesis that the familiar object will appear as larger or smaller than normal, depending on whether the perceived distance of the object is greater or smaller than expected from its familiar size. In this explanation it is asserted that the perception of the familiar object as larger or smaller than normal will result in the observer concluding (and reporting) that the familiar object is at a greater or smaller distance, respectively, than the distance at which it is perceived. This tendency has been postulated in other studies (Joynson & Kirk, 1960; Kaufman & Rock, 1962). Consistent with it are several studies which have indicated that observers expect distant objects to appear smaller than near objects of the same physical size (Carlson, 1960, 1962; Epstein, 1963a). Thus it is possible that under otherwise reduced conditions of observation reported distance will differ as a function of familiar size not because the familiar objects appear at different distances but because they appear off-sized as a result of the SDT and so are cognitively judged as located at different distances than the distance at which they appear. Therefore an adequate test of familiar size as a cue to perceived egocentric distance requires a test procedure that is sensitive to perceived distance without involving cognitive effects.

A procedure that seems to meet these requirements is illustrated in Figure 5-4, which is similar to Figure 5-3 in that only one object (the solid rectangle) is present in the field of view and the observer moves his head laterally between Positions 1 and 2 while gazing at the rectangle. Unlike Figure 5-3, however, in Figure 5-4 the solid rectangle is rigidly attached to a rod extending from the

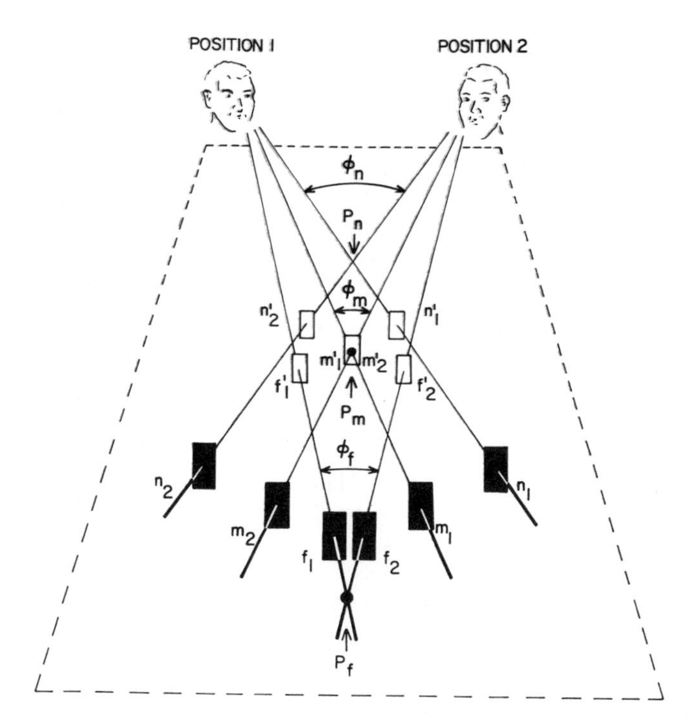

Figure 5-4 A perspective top-view drawing illustrating the measurement of perceived distance by the adjustment of the pivot distance until the apparent concomitant motion of the object disappears.

observer to the object. This rod can be pivoted anywhere along its length with three alternative pivot distances, P_n, P_m, and P_f. In an actual apparatus the rod would be below the level of the rectangle and the observer's eyes and would not be visible to him. The apparent distance of the rectangle, indicated by the open rectangles in Figure 5-4, is shown arbitrarily as located closer to the observer than its physical distance. This apparent distance is assumed to be constant for all positions of the head and all pivot distances. Consider the case in which the pivot distance is at P_f. The rectangle will appear to move in the direction of the head (between f'_1 and f'_2) as the head is moved laterally between Positions 1 and 2. Consider the case in which the pivot distance is at P_n. The rectangle will appear to move against the head (between n'_1 and n'_2) as the head is moved. Only when the pivot distance is at the apparent distance of the rectangle will the rectangle appear to be stationary (at m'_1 or m'_2) as the head is moved. If the observer registers the lateral motion of his head and the angle ϕ at the pivot distance correctly, the distance of the pivot that will result in the object appearing stationary is also the apparent distance of the object from the observer. An apparatus based on these principles was constructed to allow continuous variation in the pivot angle ϕ (Gogel & Newton, 1976). The pivot distance associated with the pivot angle at which no object motion was perceived with the lateral motion of the head is considered to be a "pure" measure of the apparent distance of the object; that is, a measure that will not be modified by cognitive factors.

This apparatus was applied to an examination of the familiar size cue to perceived egocentric distance (Gogel, 1976). The familiar objects (color transparencies) most frequently used in the study were of a guitar simulating a distance of 1236 cm and a key simulating a distance of 63 cm (a ratio of about 20 to 1). These were physically at 133 cm, were viewed monocularly through a lens that placed them accommodatively at infinity, and were presented one at a time in an otherwise dark visual field. The verbal reports were calibrated for observer differences in mentally applying a foot ruler to perceived distances. The brightness ratios of the familiar objects were varied between several experiments to control for a possible effect of apparent brightness on apparent distance. The results from the first and the subsequent presentations were analyzed separately in order to evaluate familiar size as a cue to egocentric as distinct from exocentric distance. Using the geometrical means of the data from the two simulated distances, the ratios of calibrated verbal reports were about 6 to 1, whereas the head motion procedure produced ratios of about 1.3 to 1. The greater obtained ratio of responses from the verbal reports than from the moving pivot procedure indicates that cognitive factors can strongly influence verbal reports of distance when the stimulus is a familiar object. It is also clear from the results obtained with the moving head procedure that the familiar size cue to distance was *not* completely ineffective. The measures of perceived distance to the key and guitar obtained from this procedure were significantly different and this difference occurred on the first presentations. It is apparent from these results that familiar size is a relatively ineffective but not negligible cue of egocentric distance.

Eriksson and Zetterberg (1975) examined the ability of observers to estimate verbally the distances of familiar objects or parts of familiar objects as they appeared in outdoor environments. The stimuli were viewed through an aperture in an enclosed automobile in which a head and chin rest was installed. The object distances varied from 3.3 to 150 m. The highly veridical reports of size and distance indicated to the experimenters that familiar size was an accurate cue of size and distance. An interpretation consistent with the results obtained from the study which employed the head motion procedure already discussed is that to a considerable degree the observers were using familiar sizes to determine the reports of size, regardless of the apparent sizes of the objects, and were also using off-sized perceptions to determine at least in part the reports of distance.

Consider the major cues of exocentric distance. Evidence is abundant that binocular disparity can be highly effective in determining that one object is behind or in front of another, particularly in situations in which the objects or contours are not too separated in depth or direction. Also there is considerable evidence for the validity of the relative size cue of distance (Gogel, 1964b; Gogel & Sturm, 1971). Although relative motion parallax is often assumed to be one of the effective exocentric cues, the evidence seems to be equivocal. In evaluating the relevant research regarding this cue, it is possibly important to consider again the distinction between cognitive and perceptual information. As shown in Figure 5-3, if the apparent distance of a physically stationary object is smaller or greater than its physical distance, it will appear to move with or against the head, respectively, as the head is moved in a frontoparallel plane. It follows from an extension of the principles illustrated in Figure 5-3 that if the perceived depth between two objects is less than the physical depth the physically more distant object will seem to move more in the direction of the head motion than the physically less distant object (Gogel & Tietz, 1974). Thus the perception that two objects move differently in this manner with head movement may be sufficient to indicate cognitively that one of the objects is more distant than the other. Because the exocentric distances are perceptually underestimated in many situations, the observer would have many opportunities to encounter this cognitive information before participating in the experiment. The knowledge might also be acquired during the experiment, particularly in one in which training trials are used (Dees, 1966; Ferris, 1972). Eriksson has investigated the cue of relative motion parallax in a series of studies. In one study (Eriksson, 1972a) evidence was provided in support of the cue of relative motion parallax, but reported distances, in addition to being variable, showed systematic underestimations of the physical depths and an appreciable number of the responses in the study were two-dimensional. In another study Eriksson (1972b) found that although moving the head laterally resulted in a clear increase in the reports of depth, compared with the results obtained when the head was stationary, the direction of the report was dominated by the cue of height in the visual field. In a third study Erkisson (1972c) found that relative motion parallax was dominated by the relative size cue to exocentric distance as well as by the vertical position cue. This is consistent with the study by Gogel and Tietz (1974) in which it was also found that the relative size (perspective) cue dominated the cue of relative motion parallax produced by lateral head motion. It seems that this cue is readily modified by other exocentric cues, including the usually rather ineffective cue of vertical displacement in the visual field. In a fourth study Eriksson (1974) examined the observer's ability to perceive the depth between objects that were distributed in distance as a result of walking toward them. Information contrary to the actual distances was provided by the constant retinal sizes of the objects and by their vertical positions. Despite these misleading cues, it was found that the perception of the relative order in distance was in agreement with the physical distribution of the objects. It may be unnecessary to attribute the results ob-

INTERACTIONS BETWEEN ABSOLUTE AND RELATIVE CUES OF DISTANCE

Two types of interaction between absolute and relative cues of distance are considered. One concerns the manner in which the two kinds of cue contribute redundant although not necessarily equivalent information to a perception. In this type of interaction the two kinds of cue are often in conflict. Indeed, the paradigm suggested for distinquishing between absolute and relative cues makes it clear that such conflicts are unavoidable. An absolute cue is identified by the perceptions associated with values of the stimulus when only one object is present in the field of view and the different values are presented to different groups of observers to avoid interactions between presentations. It is necessary to postulate a relative cue in addition to an absolute only if additional objects, presented simultaneously or successively, modify the perception from that obtained with the object in isolation. It follows that the psychophysics of absolute and relative cues must be different.

A second type of interaction between absolute and relative cues is in the production of perceptions for which both kinds of cue are necessary. An example of this interaction is implied in the conclusion that exocentric perceptions resulting from relative distance cues must sum with egocentric perceptions resulting from absolute distance cues if the perception of egocentric distance is to extend to the more distant parts of the visual field. Another example is found in the calibration of exocentric cues by egocentric cues (or by perceived egocentric distance). Consistent with Figure 5-2, exocentric cues by themselves are ambiguous in specifying a depth interval; for example, the physically equal depth intervals *ef* and *fg* in Figure 5-2 would produce different cue magnitudes (different proximal stimuli) on the eye; that is, $\alpha_e - \alpha_f \neq \alpha_g - \alpha_h$. Conversely, the same magnitude of exocentric cue will represent different physical depths at different physical distances from the observer. All the different distal stimuli that produce the same proximal stimulus on the eye are called equivalent configurations. Because it is consistent with the proximal stimulus for the perception to resemble any of the equivalent configurations, it follows that equivalent configurations are ambiguous in specifying a perception. An exocentric cue by itself defines a group of equivalent configurations and therefore is an ambiguous stimulus. Egocentric cues or egocentric perceptions are necessary to resolve this ambiguity by limiting the perception to one of the possible equivalent configurations. Also, because egocentric information is often scalar, this calibration process, in removing the stimulus ambiguity, specifies that the depth perception will be of an absolute magnitude.

Few studies have examined the cue conflicts implied in definitions of absolute and relative cues of distance. One study (Gogel, 1961) measured the relation between convergence and perceived distance as a function of the number of binocular objects (one, two, or three) distributed in stereoscopic distance. The binocular disparities between the objects were constant, the convergence to the configuration, variable. It was found that the function relating convergence and perceived distance was not identical for the different objects in the configuration. The significance of these results for the present discussion is that for the same egocentric cues of distance the perceived distance of an object can differ, depending on its position in a binocular configuration. Direct egocentric cues of convergence and the summed cues of egocentric distance involving binocular disparity are conflicting. This psychophysical difference between objects in configurations and objects presented alone is supported by a similar study (Gogel, 1972c; Experiment I). To the extent that these two studies represent a general relation between egocentric cues and exocentric perceptions, absolute and relative cues provide information that psychophysically is not identical and therefore to some degree is in conflict.

Studies concerned with the resolution of the stimulus ambiguity involved in exocentric cues also seem to be limited to the cue of binocular disparity. The research regarding the relation between egocentric cues and the perception of depth from binocular disparity has been summarized by Foley (1976). It seems that a given binocular disparity will produce a different perceived exocentric distance as a function of egocentric cues of perceived distance or the perceived distance of the configuration from the observer. According to a study by Gogel (1964a), it is the perceived distance to the configuration, regardless of the factors by which this perceived distance is achieved, that calibrates the binocular disparity cue.

Absolute and Relative Observer Tendencies in the Perception of Distance

Not all factors that contribute to the perception of distance have their origin in the conditions of stimulation, as shown by measuring spatial perceptions under conditions of stimulus ambiguity or stimulus reduction. It appears that the perception is often less ambiguous than would be expected from the conditions of stimulation or from the sensitivity of the visual system; for example, the visual field subjectively seems to be equally clear throughout its extent even though it is readily demonstrated that visual acuity deteriorates rapidly toward the periphery. Evidence that stimulus ambiguity does not necessarily produce perceptual uncertainty is found in the observer tendencies defined at the beginning of this chapter. Egocentric and exocentric perceptions of distance occur despite total cue reduction, a conclusion supported by evidence of the specific distance tendency (SDT) and equidistance tendency (EDT), respectively. It is also asserted that these tendencies, even under conditions in which some distance cues remain, can modify the perceived distances normally expected from egocentric and exocentric cues. It follows that observer tendencies and the usual sources of stimulus information (cues) often are in conflict.

THE EDT AND ITS INTERACTIONS WITH CUES OF EXOCENTRIC DISTANCE

EDT data have been summarized elsewhere (Gogel, 1965a, 1973). Of particular interest for the purposes of this chapter is the evidence that confirms the ability of the EDT to modify the perceived depth expected from exocentric cues of distance. Because the EDT results in objects or parts of objects appearing equidistant, it is in opposition (conflict) with cues that indicate that the objects are at different distances. Whether the EDT will significantly modify the perceived depth from exocentric cues depends on the effectiveness of the EDT compared with the exocentric cues. Because this effectiveness has been shown to vary with the directional separation of the objects (Gogel, 1965a; Lodge & Wist, 1968), the EDT's contribution to the perception is expected to increase as the objects under consideration are more adjacent in the visual field. The effectiveness of exocentric cues also varies inversely with the visual separation of the objects. Evidence for this is found in the decrease in the precision of cues that occurs with separation in the visual field (Hersch & Weymouth, 1948); for example, the precision with which two objects can be adjusted to apparent equidistance by using the binocular disparity cue is inversely related to the directional separation of the objects (Graham, Riggs, Miller, & Solomon, 1949; Ogle, 1950). It seems that the effectiveness of both the EDT and exocentric cues of distance varies inversely with the visual separation of the objects. Also, for a constant separation of objects the major exocentric cues may differ in their general precision or effectiveness with binocular disparity probably the most precise. It seems reasonable to conclude that the cue conflicts that inevitably occur between the EDT and cues of exocentric distance will be resolved differently, depending on the effectiveness of the cues and their change in effectiveness as a function of the separation of the objects in the visual field.

The ability of the EDT to modify the perceived depth from binocular disparity was tested by Gogel, Brune, and Inaba (1954). In this study the binocularly viewed objects were widely separated in direction to reduce the effectiveness of the stereoscopic cue. The effectiveness of the EDT was maintained at a high level by arranging the situation so that the EDT occurred between one of the binocularly viewed objects and a monocularly viewed, directionally adjacent object at a different perceived distance. Under these conditions the EDT markedly affected the perception obtained from the binocular disparity cue. There is evidence also that the EDT can modify the perception of depth from binocular disparity under less contrived circumstances. Harker (1962) found that a luminous line slanted in depth will appear modified in slant toward the frontoparallel plane. A substantial part of this effect is attributed to the EDT between the parts of the line. This interpretation is further supported by the results of a study by Howard and Templeton (1964) in which the binocular inspection of a stereoscopic depth modified the perception of stereoscopic depth presented subsequently. Also, as shown by Owens and Wist (1974), the contribution of the

EDT in relation to a binocular disparity can increase within a period of observation. In this study the perceived depth between binocularly observed stimuli presented at different distances in an otherwise dark surround was measured as a function of exposure time. In one condition two points of light were presented with different depth separations. In another the same points were connected by a thin luminous line. The addition of the line was expected to increase the contribution of the EDT to the perception. With the two points of light, the effect of the EDT, measured by the decrease in the apparent depth between the points, increased with increased time of exposure. As expected, adding the line resulted in a decrease in the perceived depth between the points that was more rapid and more extreme. The results are interpreted in terms of a change in effectiveness of both the binocular disparity cue and the EDT as a function of time of observation; the rate of change, however, is different for these two factors. It is plausible that the increase in the contribution of the EDT with the addition of the line can be attributed to the increase in the number of parts of the stimulus between which the EDT effects occurred. This interpretation is consistent with a study by Wist and Summons (1976) in which EDT effects were greater for objects of larger size (larger visual mass). The increase in the effectiveness of the EDT with visual mass can also be used to explain the binocular induction effect found by Werner (1938). According to binocular induction, a frame physically slanted in depth appears to be frontoparallel, whereas an enclosed bar (a less massive object) physically located in the frontoparallel plane appears to be slanted.

If the EDT can modify the perceived depth from binocular disparity, as the evidence indicates, it is even more likely to modify the perceived depth from exocentric cues less effective than binocular disparity. A study by Gogel and Mershon (1968) provides evidence that the EDT can modify the effect of the size cue to distance. Two objects at the same physical distance were viewed monocularly through two windows that were apparently but not physically equidistant. The effect of the EDT, despite the relative size cue between the objects, was to make each object appear at the distance of its directionally adjacent window with a resulting distortion in the perceived sizes of the objects. The similar distortion of size which occurs in the Ames distorted room can be explained in a similar manner. In a study by Gogel and Harker (1955) it was found that the effectiveness of relative size cues increased as the objects were increasingly separated in a frontoparallel plane. In a subsequent study (Gogel, 1956) it was concluded that this effect could be attributed to the decreasing contribution of the EDT compared with the relative size cue as the directional separation of the objects increased. Consistent to this interpretation, the impressive perceptions of depth found in wide-screen movies can probably be explained by the release of the relative size (perspective) cues from the restraining effects of the EDT. There is evidence also that the EDT can readily modify the perceived depth expected from a difference in the accommodative cue. This is indicated by experiments (Mershon & Gogel, 1970) in which monocular, in contrast to binocular, observation results in objects, physically at different distances, appearing, near or in the same frontoparallel plane.

It is clear that under conditions in which the EDT is in conflict with binocular disparity (Owens & Wist, 1974) or in which two EDT effects are in opposition (Lodge & Wist, 1968) the contribution of the EDT can vary with the observation period. It is probable, however, that the effect of the EDT is almost instantaneous under conditions in which the EDT clearly dominates the perception (Owens & Wist, 1974) or in which exocentric cues are reduced or eliminated. The rapid effect of the EDT in the absence of opposing cues is consistent with the perceptions obtained from monocular observation of objects attached to the Ames rotating trapezoidal window (Gogel, 1956). Another factor that can modify the relative effectiveness of the EDT is the visual fixation. Wist and Summons (1976) have found that visually fixating an object increases the effectiveness of the EDT generated between that object and a test object in relation to the effect of other nonfixated objects. As suggested by these authors, this effect of fixation on the EDT may indicate that the effectiveness of the EDT varies with attention.

Summing this research on the EDT, it can be concluded that spatial, temporal, and fixation (or attention) variables regulate the contribution of the EDT in relation to the contribution of exocentric cues of distance. It might seem that the synthesis of a perception of distance, taking all of these factors into consideration, would represent a difficult task for the visual system. In a complex visual field a variety of such factors would be available and would increase in their consequences as the visual field was reduced. It is possible, however, that all of these factors can be expressed in terms of the single variable of precision. The precision of a cue or of some other source of influence can be defined as the inverse of the variability of the perception resulting from that cue or source. The smaller the variability, the more precise the cue. It has been suggested that the visual system gives more weight to information as it is more precise (Taylor, 1962). The more precise the cue, the greater its effectiveness (contribution) in determining a perception in relation to the contribution of other conflicting factors. It is possible that only to the extent that temporal, spatial, and fixational factors modify the precision (or conversely the variability) of the perception associated with a cue or other factor will they also modify its effectiveness. If this interpretation is valid, the effect of a variety of variables on the contribution of a perceptual cue would be predicted by the total effect of all these factors on the variability or cue threshold. This is not to imply that cue precision is the *cause* of cue effectiveness but rather that both factors are likely to be reflections of the same underlying process.

Possible Explanations of the EDT. Kaufman (1974) has noted that the EDT can be demonstrated by the situation in which an afterimage in one eye is seen in the directional (and depth) vicinity of a binocularly viewed object. In this case the afterimage, like the fixated, binocularly viewed object, will be in sharp focus without diplopic images. Perhaps as a consequence, the observer will perceive the two objects as equidistant. Evidence against this explanation of the EDT can be found in experiments in which objects tended to appear equidistant despite differences in accommodative distances (Gogel, 1956; Gogel & Mershon, 1968) and, in unusual cases (Gogel, Brune, & Inaba, 1954), differences in convergence. Consider however, a somewhat related explanation. Suppose that the apparent distance of the object to which the observer is accommodating is registered by the state of contraction of the ciliary muscle. The question then is how this information is applied to objects in the visual field. If it were applied to all the objects, this would result in all appearing at the same distance according to the cue of accommodation. Similarly, the information concerning the state of the extraocular eye muscles involved in converging on an object might be applied equally to all the objects in the visual field so that according to this information, all would appear equidistant, regardless of their physical distances. If the oculomotor cues worked in this manner, in the absence of other cues, they would produce a perception of equidistance. Although it seems unlikely that this could explain the differential effect of the EDT on perception as a function of directional separation, it warrants consideration. This explanation also indicates the need to understand the process by which distance information from oculomotor cues is applied to objects distributed throughout the visual field.

THE SDT AND ITS INTERACTIONS WITH CUES OF EGOCENTRIC DISTANCE

The evidence for the specific distance tendency (SDT) has also been summarized (Gogel, 1973). The SDT is an observer tendency in the perception of egocentric distance and is in conflict with all egocentric information that would indicate that the object is not at a distance of several meters from the observer. Only evidence regarding the SDT pertinent to the possibility that it can modify the perception of distance from cues of egocentric distance is discussed here. The experiment by Gogel and Tietz (1973) is appropriate for this purpose. In that study observers moved their heads laterally while gazing (monocularly or binocularly) at a physically stationary point of light located at one of several physical distances in an otherwise dark visual field. According to the SDT, a point of light not at the distance of the SDT should appear displaced toward this distance, and as a result of this discrepancy between the physical and apparent distance of the point it should appear to move concomitantly with head motion. The direction of this motion is indicated in Figure 5-3. Using the frequency of occurrence of apparent concomitant motions of the object in particular directions, a distance was calculated such that if a point of light had been placed at that distance it would usually have appeared stationary as the head was moved. Although this distance (about 3 m from the observer) was somewhat greater than the measure of the SDT determined by the corrected verbal report of distance (about 2 m), the study provides evidence of the validity of the SDT. This study also suggests that the SDT can modify the apparent distance expected from oculomotor cues of egocentric distance. The oculomotor cues were accommodation and accommodative-convergence in the case of monocular observation and accommodation and fusional-convergence in the case of binocular observation. The apparent displacement of the points of light in the direction of the SDT occured despite these oculomotor cues.

Possible Explanations of the SDT. Owens (1974) has provided an interpretation of the interrelation of the SDT and oculomotor cues different from that we have given. Visual performance on a number of tasks seems to be optimal at some intermediate distance from the observer (Freeman, 1932; Schober, 1954). Recent studies with the laser optometer (Leibowitz & Hennessy, 1975) have verified that the accommodation of the eyes in the absence of visual stimulation (the resting state of accommodation) is to a relatively near distance. It is possible that this resting distance or distance of optimal function is related to the SDT. From a review of the literature Owens (1974) has concluded that the resting state of accommodation and convergence is to a relatively near distance and that errors in accommodation (and perhaps convergence) will occur in the direction of the resting state. These errors are particularly likely if the stimulus to accommodation, such as that from a point of light of rather low intensity presented in a dark surround, tends to be inadequate. Owens examined the relation between the SDT and the resting state of accommodation or convergence by correlating the measured distance of the SDT (using head motion) with the measured distance of each of these two resting states. Although no significant correlation was found between the SDT and the resting state of accommodation, a significant correlation occurred between the SDT and the resting state of convergence. If the SDT is produced by the resting state of convergence, it cannot be said to be in conflict with or to modify the perception of egocentric distance from oculomotor cues. Unfortunately this study, although of great interest, does not specify the relation between the SDT and the resting state of convergence. Theoretically the SDT is the physical distance at which the apparent distance of an object is accurate in the absence of all factors except the SDT. The perceptual bias introduced by the resting state of convergence theoretically is specified by the physical distance at which an object is accurately perceived in the absence of all distance factors except the resting state of convergence. Assume for the purposes of this discussion that these two distances are different. It will follow that if both the SDT and the resting state of convergence are effective factors the distance at which an object is accurately perceived, as measured by the head-motion technique, will be influenced by both. Thus to the extent that the resting state of convergence has perceptual consequences it must correlate with the SDT as measured by the head-motion technique, regardless of whether these two states are the same or different.

The SDT as a Source of Scalar Calibration. The SDT is not only a possible source of conflict with egocentric cues of distance but it also may provide a "yardstick" for the scalar calibration of cues of distance. It is possible that the SDT can produce a useful reference for the perception of egocentric distance from the absolute convergence of the eyes. This has been suggested by Von Hofsten (1974) in a study that supports the conclusion that the absolute value of convergence (or the double images associated with the convergence) is scaled perceptually by the resting state of convergence or perhaps equivalently by the SDT. Also, because apparent egocentric distance calibrates exocentric cues, it follows that the SDT under reduced cues of egocentric distance will calibrate exocentric distance cues such as binocular disparity. In order for this to occur, however, the binocular configuration must be positioned perceptually with respect to the distance defined by the SDT. Some evidence regarding this process is found in several experiments (Gogel, 1972c; Mershon & Lembo, 1976) which have indicated that with a reduction in egocentric cues of distance a binocular configuration extended in apparent depth will assume a position in relation to the SDT. This position is such that the far point of the configuration usually appears to be near the distance defined by the SDT; the other parts of the configuration extend from this apparent distance toward the observer. More research will be necessary, however, before the facts regarding this phenomenon become clear.

ABSOLUTE AND RELATIVE FACTORS IN PERCEPTUAL INDUCTION

Absolute and relative cues can be distinguished not only with respect to distance perceptions but also with respect to other perceived characteristics. In a variety of, and perhaps in all perceptions informational conflicts occur frequently between different relative cues and between absolute and relative cues. It is the purpose of this portion of the chapter to examine this evidence and to note some of its implications.

Many of the perceived characteristics of objects have been examined under conditions that produce induction effects. Induction effects can be said to occur whenever it is found that the presence of one object (an induction object) changes the perception of another object (the test object) from what would have occurred if the induction object had been absent. It will be noted that this definition of an induction effect is the same as that of a relative cue and that the terms induction effect and relative effect are used interchangeably. As in the measurement of a relative cue, the perception of the test object under induction conditions can be determined by absolute and relative factors. Adding the induction object provides relative cues to modify the perception from that produced by the absolute cue alone. In different situations the absolute and relative cues can be in agreement or conflict; for example, in Figure 5-1B the absolute and relative cues of motion would have been in agreement if the physical motions of Points 1 and 2 had been on the diagonal between upper left and lower right rather than vertical for Point 1 and horizontal for Point 2. If the absolute and relative cues are in agreement, the effect of the relative cues is not distinguishable from the absolute cues and an induction effect is not evident. Thus induction effects by definition imply a conflict between absolute and relative cues. To the degree that the absolute cues produce veridical perceptions inductions are nonveridical (illusions). An example of a well-known illusion that involves a conflict between absolute and relative cues is the rod-and-frame illusion in which a physically vertical rod appears tilted in a direction opposite to that of a surrounding tilted frame. Cues that in the absence of the frame would usually result in the rod appearing vertical (Rock, 1954) are probably the position of the image of the rod on the eye, the position of the eye in relation to the head and body, and the position of the head and body in space. These are absolute cues of the position of the rod, for they involve the interaction between the observer and the object. The addition of the tilted frame introduces visual interactions between the frame and rod that are in opposition to (in conflict with) the absolute cues of orientation. Illusions can also occur between conflicting relative cues; but whether the induction effects produce an illusion or reduce an illusion originally occuring with absolute cues, the basic processes are unchanged and act to integrate the contributions of the often conflicting sources of information. Predicting the changes in induction effects as a function of stimulus conditions requires a knowledge of how absolute and relative cues vary in effectiveness with the changes in the stimulus conditions. From the decrease in the precision of relative cues that occurs as the spatial separation of objects is increased, it might be expected that the effectiveness of relative cues would decrease with increasing separation of the objects. The evidence to support this expectation for both frontoparallel (directional) and depth separation of the objects is discussed. The stimulus or observer dimensions that determine changes in the effectiveness of absolute cues have received much less research attention.

Evidence for the Distinction Between Absolute and Relative Factors in Perceptual Induction

The presence of a reference line or other object near a moving object can lower the threshold for the perception of the motion. This has been found by Leibowitz (1955) to occur for a 16-second but not for a 0.25-second exposure.

Leibowitz explains this difference in terms of the distinction between a displacement and a motion threshold. Mates (1969), using an exposure duration of 6 seconds, found that the velocity threshold decreases as the number of reference lines increases. Shaffer and Wallach (1966) measuring the extent of motion required to reach threshold found that the presence of reference lines decreased the threshold for very short as well as longer exposure times. These researchers distinguish between subject-relative motion (absolute motion as used in this chapter) and object-relative motion (relative motion as used in this chapter). Although the latter has a lower threshold, correlations found between the thresholds from the two kinds of motion raised the question whether they were basically different. Kinchla and Allan (1969) have proposed a quantitative model of the discrimination of position and motion in terms of the response to absolute and relative information. The model is applicable theoretically to modalities other than vision and to continuous motion and static changes in position. Consistent with the model, it was found that observers used relative rather than absolute cues of displacement if the visual separation of objects was not large (less than 15°). The results support the two-factor theory of the visual perception of motion or displacement in which absolute cues are used if the separation is sufficiently large, whereas relative cues determine the perception at smaller separations. Kinchla and Allan (1969) note that it would be premature to conclude that the use of absolute and relative motion is mutually exclusive. The point of view expressed in this chapter is that relative and absolute cues are not used in an all-or-none fashion, but instead that both kinds of cue contribute simultaneously in a graded manner to the final perception. It is possible, of course, that under some conditions relative cues will be so much more effective than absolute cues that the contribution of absolute cues to the perception may be undetected. It would be expected, however, that the influence of both kinds of cue would be evident if the conditions were appropriately varied.

There is direct evidence that conflicting relative cues contribute to the perceived characteristics of objects in a weighted or graded manner and, by analogy at least, a similar process of resolving conflicts between absolute and relative cues would be expected. A study of induced motion by Gogel (1974b) provides some support for the notion that conflicting absolute and relative cues contribute to the perception of motion in a graded manner. The physical conditions of motions and the average results from this study are illustrated in Figure 5-5, in which the lettered distances *ab*, *cd*, and *ef* represent the location, direction, magnitude, and phase of the repetitive motions of points of light presented in an otherwise totally dark visual field. As one point moved from the position marked by one successive letter to the other and returned, the remaining point or points moved between the same successive letters. Two or three points of light were presented with different amounts of separation

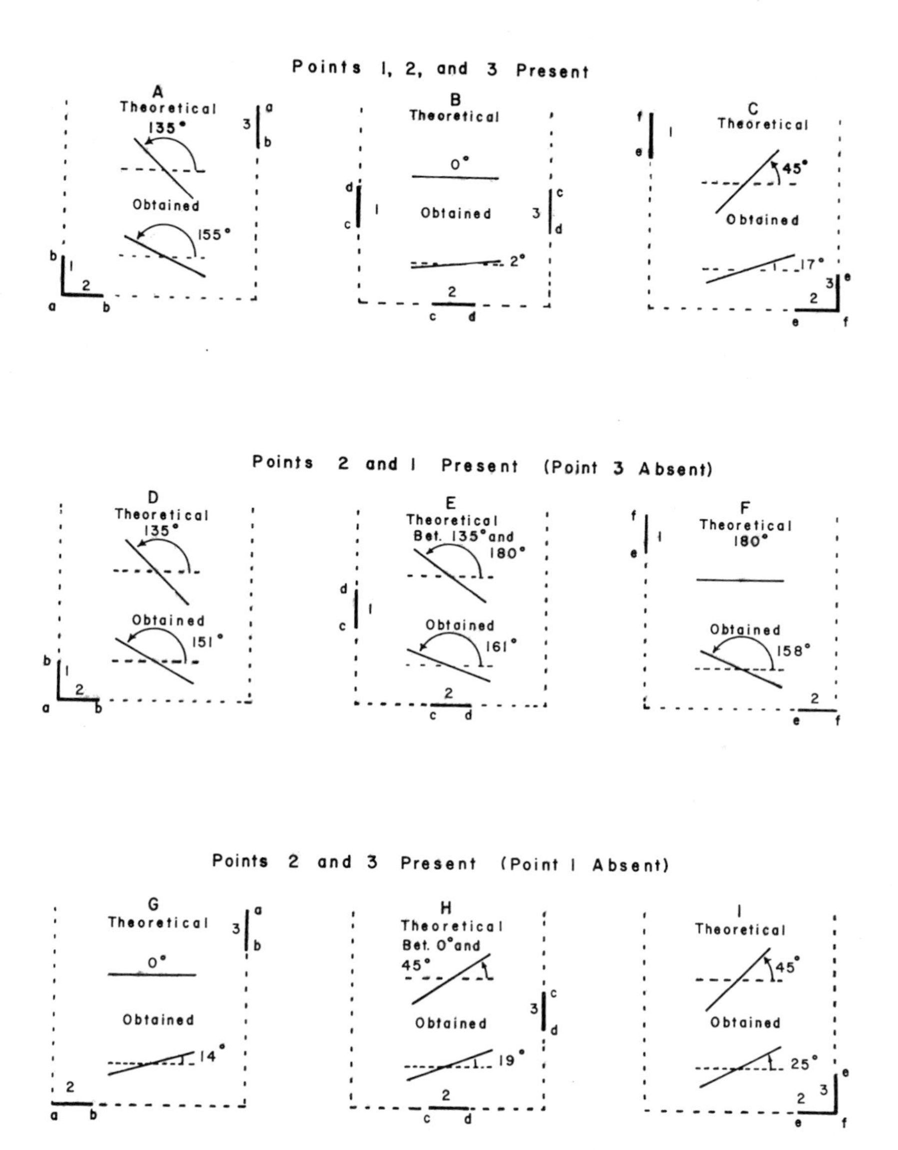

Figure 5-5 Theoretical limits and results obtained from situations in which three moving points (top row) or two moving points (middle and bottom rows) were presented simultaneously (from Gogel, 1974b).

between the points. The task was always to indicate the apparent direction of the path of motion of Point 2. In the situations in which only two points of light were presented (Points 2 and 1 or Points 2 and 3) the problem was to determine whether absolute and relative motion cues in conflict would add abgebraically in determining the perceived motion of Point 2. Consider, for example, the situations represented in diagrams *G*, *H*, and *I*. The absolute motion of Point 2 in each of these drawings was horizontal and, if the absolute motion completely determined the perceived motion, the perceived motion in these diagrams would also be horizontal. If the relative motion between Points 2 and 3 were completely dominant, Point 2 would be perceived as moving at 45°. The effectiveness of the relative motion cue is expected to decrease with increasing separation of the points, whereas the effectiveness of the absolute motion cue, because it is independent of the presence of other objects, should be independent of the separation of the points. The "theoretical" slants for the two-point situations of Figure 5-5 indicate the apparent motion expected from Point 2, if the cue of relative motion is completely dominant when the points are most adjacent and if the cue of absolute motion is completely dominant when the points are most displaced. From the results obtained neither the absolute nor the relative cue dominated completely in any of the situations shown in either the middle or lower row of diagrams of Figure 5-5. Also, in the lower (but not in the middle) row of diagrams the effectiveness of the absolute motion cue of Point 2 showed a statistically significant increase as the separation between the points was increased. The results from the lower diagrams are consistent with the conclusion that absolute and relative cues of motion contribute algebraically to the perception and that the effectiveness of the relative motion cue decreases with increasing separation of the points.

Effectiveness of Relative Cues as a Function of Object Separation

THE EFFECT OF SEPARATIONS IN A FRONTOPARALLEL PLANE

It has been proposed as a general principle (the adjacency principle) that the effectiveness of relative cues decreases as the perceived distance between the objects producing the relative cues increases. This adjacency effect has been found with either frontoparallel or depth separation (Gogel, 1970). An example of the effect of the separation of objects in a frontoparallel plane on a motion induction is shown in the upper row diagrams in Figure 5-5. In the situations represented by this row three moving points of light are presented simultaneously by using the conditions of separation between the paths of motion of the three points indicated in the diagrams. In A Point 2 is adjacent to Point 1 and displaced from Point 3. In C Point 2 is adjacent to Point 3 and displaced from Point 1; and in B Point 2 is equally separated from each of the other two points. The "theoretical" orientations indicate the apparent path of motion of Point 2 if this apparent motion were dominated by the adjacent point in A and C and equally affected by both induction points in B. More generally, if cue effectiveness and frontoparallel separation are inversely related, the apparent motion of Point 2 should be between upper left and lower right in A, horizontal in B, and between upper right and lower left in C. As indicated in Figure 5-5, these were the results obtained. It should be noted also that the "theoretical" apparent motions expected if the relative motion cue between Point 2 and the point most adjacent to Point 2 had been completely dominant were not achieved. Although the relative motions between adjacent points contributed most to the perception, the remaining motion cues also had some effect (Gogel, 1974b).

Frontoparallel separation can also modify relative cues of distance. Two experiments supporting this conclusion are illustrated by Figures 5-6 and 5-7. In Figure 5-6 A and D are normal-sized playing cards; B is a double-sized playing card and C is a gray square. The upper right portion of B and C were removed and the objects carefully positioned to provide an interposition cue that would result in D appearing in front of B. All four objects were viewed monocularly at the same physical distance in an otherwise dark visual field. If the interposition cue had not been present, the relative size cue would have caused D to appear more distant than B. Also, from the relative size cue A normally would appear at the same distance as D and behind B but, because of the misleading interposition cue between B and D, A could not appear to be both at the distance of D and more distant than B. If the relative size cue between A and B were dominant over that between A and D, as would be expected from the adjacency principle, A would appear behind not in front of B. This was the usual

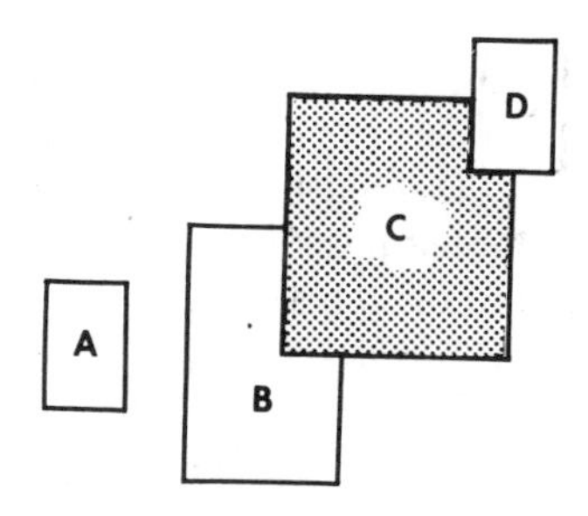

Figure 5-6 A schematic diagram of a display for testing the relation between the effectiveness of relative cues of distance as a function of the separation of the objects. Objects A and D are normal-sized playing cards, object B is a double-sized playing card, and object C is a gray card which, because of the interposition cue, results in object D appearing closer to the observer than object B (from Gogel, 1965b).

result showing that the relative size cue between the adjacent cards (A and B) was more effective than that between displaced cards (A and D) in determining the apparent position of A in the configuration (Gogel, 1965b). In addition, although B was more effective than D in determining the apparent distance of A, D also contributed to the perception, although to a lesser degree. This result is consistent with the graded contribution of the conflicting relative cues of distance to the perception. Also consistent with a contribution from both induction objects B and D, in a later study (Gogel, 1967) it was found that positioning A equally far from each of the other cards (near the lower right corner of C) resulted in more nearly equal contributions of the relative size cue from B and D to the perception.

Figure 5-7 is a top view illustration of a situation in which an Ames' trapezoidal window in a dark surround is physically oriented with its large end R at a greater distance from the observer than its small end L. Two luminous points A and B are located one on each side of the window at the same distance from the observer. All the objects are viewed binocularly. Despite the binocular observation, the trapezoidal shape of the window causes the left portion of the window L′ to appear to be more distant than the right portion R′. This error (illusion) in the perceived orientation of the window in depth produces a similar error in the perceived depth between A and B. Despite their physical equidistance, A appears at A′ and B at B′; that is, A appears to be considerably more distant than B. It seems that the perceived distance of each of the points is most determined by the binocular disparity between that point and the end of the window directionally adjacent to it; for example, A, which is physically behind the left end of the window and physically in front of the right end, correctly appears behind the left (adjacent) end of the window but incorrectly appears behind the right (separated) end.

Again, the effectiveness of an exocentric cue (in this case binocular disparity) is shown to be inversely related to the directional (frontoparallel) separation between the point and the different portions of the window. The conclusion that the binocular disparity between an object and the part of the window directionally adjacent to that object contributes most to the perception of the distance position of the object has been applied to the perceived path of motion of an object attached to the rotating trapezoid window (Gogel, 1956). The success of this application provides strong evidence that the adjacency effects are perceptual not cognitive. Also, as the physically equidistant pair of points A and B is physically (and stereoscopically) moved to increasing distances behind or in front of the window, the apparent distortion in distances between the points decreases markedly (Gogel, 1972a). This result indicates that depth separation as well as frontoparallel separation is an important factor in the effectiveness of the binocular disparity cue.

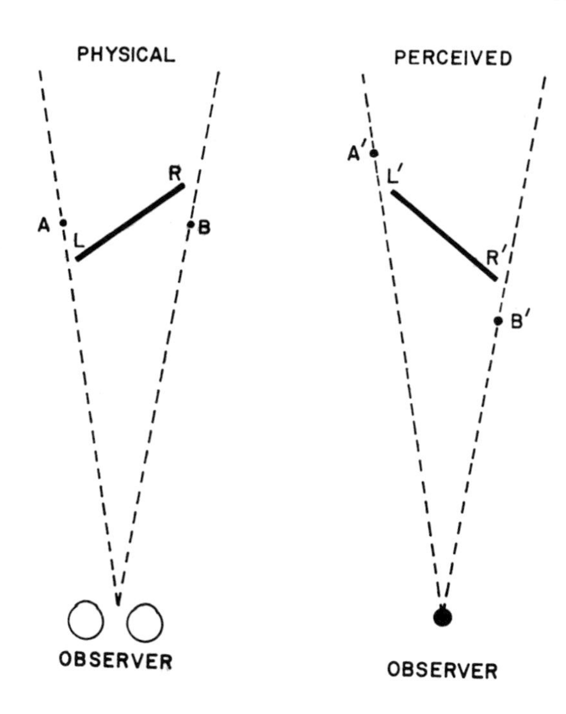

Figure 5-7 A top view schematic drawing that illustrates the difference between the physical positions of a trapezoidal window (LR) and two points (AB) and the perceived positions (L′ R′) and (A′ B′) of these objects. The error in the perceived orientation of the window has induced an error in the perceived positions of the points (from Gogel, 1973).

THE EFFECT OF SEPARATIONS IN DISTANCE

In discussing the induction effect from binocular disparity in Figure 5-7, it was noted that the magnitude of the induction decreased as the pair of points was moved together in front of or behind the trapezoidal window. Three additional studies that have examined the effect of separation in stereoscopic (perceived) distance on the direction or magnitude of induction effects are illustrated in the perspective drawings of Figure 5-8. The dashed lines are for illustrative purposes and do not represent stimuli used in the experiments. The induction effects considered in Figure 5-8 are a size induction based on the Ponzo illusion in A (Gogel, 1975), an orientation induction based on the rod-and-frame illusion in B (Gogel & Newton, 1976), and a motion induction in *C* (Gogel & Tietz, 1976). In each of the three drawings in Figure 5-8 two induction objects are shown, with one at a near and the other at a far distance. The induction objects

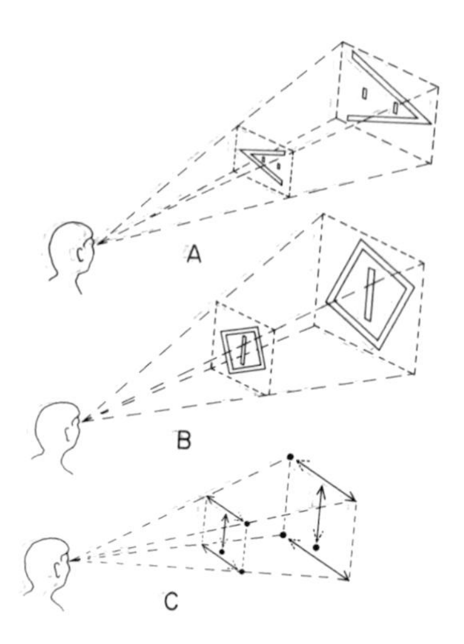

Figure 5-8 Perspective drawings of three experimental situations in which two opposite induction objects are presented at different stereoscopic distances and a test object is placed stereoscopically at one or the other or between the two in depth.

are two wedges of opposite orientation in A, two frames of opposite slant in B, and two pairs of points moving horizontally with the phase of each pair opposite at the two distances in C. The same visual angle is subtended by the near and far induction object in each of the three situations. Because the orientation of the induction objects is opposite at the near and far distances, the relative cues in the induction effects at these distances are in opposite directions. The test object in any presentation was at one distance only, but the experimenter was able to vary the stereoscopic distance of the test object between presentations. The visual angle of the test object at any distance was always constant. In A the test objects were two vertical bars of the same size at the same distance, one located near the apex of the wedge and the other displaced from the apex. According to the Ponzo illusion, the bar nearer the apex should appear to be the larger. In B, the test object was a physically vertical rod. According to the rod-and-frame illusion, the rod should appear to be tilted in the direction opposite to the physical tilt of the surrounding frame. In C the test object was the vertically moving point. According to the relative motion cue between the test and induction object, the test point should appear to move between upper left and lower right at the far plane and between upper right and lower left at the near plane. Also, in addition to presenting the two induction objects simultaneously, one induction object could be presented at a time, with the test object at the plane of the induction object or displaced from the induction object in distance in the situations represented by A and B. With only the near induction object present the displacement of the test object was behind the induction object. With only the far induction object present the displacement of the test object was in front of the induction object. According to the expected effect of distance separation on induction, displacing the test object behind or in front of the single induction object should decrease the magnitude of the induction. With both induction objects present the most effective relative cues should occur between the test object and the induction object nearest the test object in apparent distance. Thus in the situation illustrated by A the right bar should appear to be larger than the left bar at the far distance and smaller than the left bar at the near distance. In B the rod should appear to be tilted in a counter-clockwise direction at the far distance and in a clockwise direction at the near distance. In C the vertically moving point should appear to move between upper left and lower right at the far distance and between upper right and lower left at the near distance. In the studies associated with A and B the test object was sometimes positioned at a middle distance in addition to the near and far distances. In the study associated with C only the near and far distances were used and the vertical point was always at the distance of one of the pairs of horizontally moving points.

The results from these studies for the situations in which the two induction objects of opposite orientation were presented simultaneously at different distances unambiguously support the hypothesis that stereoscopic (perceived) separation in distance between a test and induction object reduces the effectiveness of the relative cue between these objects. This is indicated in that clear induction effects occurred despite the opposite orientations of the induction objects, with the direction of the induction determined by the induction object at the same apparent distance as the test object. The results from the situations in which the test object was stereoscopically displaced in front of the single induction object at the far distance clearly show the expected decrease in the magnitude of the illusion. Unlike the results obtained by using the trapezoidal window (Gogel, 1972a), displacing the test object behind the near induction object presented singly showed little or no decrease in the magnitude of the illusion in the situations of A and B. Three explanations have been considered to account for the asymmetrical results obtained from displacing the test object in front of the far induction object compared with an equal stereoscopic displacement behind the near induction object. One of these concerns the perceived size of the test object in relation to the induction object (Greene, Lawson, & Godek, 1972). Because the visual angle of the test object always remained constant, its perceived size in relation to the perceived size of the induction object varied inversely with its stereoscopic distance. This interpretation would predict an increase in the magnitude of the induction effect for distances beyond the induction object. Although this result occurred in the study of the Ponzo illusion by Greene et al. it did not occur in the studies illustrated by A and B. Also, the results obtained from the simultaneous presentations of the two induction objects are difficult to explain by this hypothesis.

A second possible explanation is that the adjacency effect is mediated by attention. In terms of this hypothesis, induction objects near the test object are unusually effective because they cannot be ignored. This would explain the asymmetrical distance results obtained from the single induction objects by making the reasonable assumption that it is easier to ignore the induction object for stereoscopic displacements of the test object in front of rather than behind the induction object. In the latter case the observer must look through the induction object to examine the test object visually. Direct evidence for this hypothesis is presented later, but it should be noted that, according to this hypothesis, when the two induction objects of opposite orientation are presented simultaneously the induction effect should be greater when the test object is at the distance of the near rather than the far induction object. Only in one of the situations in Figure 5-8 do the median results for the simultaneous presentations of the induction objects provide support for this explanation. In A the bar nearest the apex of the wedge was perceived to be 8% larger than the other bar at the near distance and 5% larger at the far distance. In B the induced tilt (without regard to sign) was 1.8° at the near distance and 4.2° at the far distance. In C the apparent tilt from the vertical of the path of motion induced in the test object independent of sign was 13° and 21° at the near and far distance. A third explanation which I have proposed assumes that (a) both absolute and relative cues can contribute in a graded manner to the perception of the test object, (b) the effectiveness of relative cues decreases with increasing separation of the test and induction object in distance, and (c) in some cases absolute cues decrease in their effectiveness with increasing distance from the observer. The results from the simultaneous presentations of the two induction objects and from the condition in which the test object was displaced *in front of* the single induction object clearly supports assumptions a and b. Although there is some limited support in the research for assumption c (Gogel, 1975), this assumption clearly requires further investigation.

ADDITIONAL EVIDENCE FOR THE DIFFERENTIAL CONTRIBUTION OF RELATIVE CUES AS A FUNCTION OF SPATIAL SEPARATION

The conclusion that the effectiveness of relative cues is inversely related to the apparent separation of the test and induction objects is called the adjacency principle. This principle is similar in some respects to the Gestalt law of proximity. It differs from the law of proximity in that (a) it applies to distance as well as frontoparallel separation, (b) it is specified in terms of perceived rather than either proximal or distal separation, (c) it is relevant to all the perceived characteristics associated with the relative cues and not only to perceptual grouping, and (d) it is concerned with the graded contribution of relative cues that occur between displaced as well as adjacent objects. Additional evidence for the adjacency principle has been summarized elsewhere (Gogel, 1973). It is clear that perceived spatial separation, for a variety of perceived characteristics, is an important dimension along which the effectiveness of many relative cues vary. Of importance to the present discussion, this evidence is consistent with the concept that the metric perception of size, slant, or motion is the result of a process by which weights are assigned by the visual system to various (often conflicting) sources of information. The adjacency principle is a statement of the manner in which weights are assigned to relative cues. The role of absolute cues in induction phenomena has been less studied, but, here also, as in the perception of distance, the evidence points to a graded process by means of which various sources of information, whether relative or absolute, are synthesized into metric perceptions.

THE ROLE OF ATTENTION IN ADJACENCY EFFECTS

The perceptual effects of spatial adjacency are consistent with the concept that the metric perception of size, slant, or motion is the result of a process by which weights are assigned by the visual system to various (often conflicting) sources of information. The weight assigned to a source determines the magnitude of the contribution of this source to the perception. One of the implications of identifying this weighting process is that the effect of a variety of observer or stimulus variables on perception can possibly be identified by their ability to modify the weights assigned to informational sources. One observer variable which has produced considerable controversy in the literature is that of set or attention. The basic question is whether the effect of set or attention is on perception or on the response with the perception unmodified (see Haber, 1966). On the basis of the evidence considered in this chapter, it is reasonable to expect that if set or attention has an effect on perception it occurs by increasing or decreasing the weight given by the visual system to the attended or ignored source of information, respectively. As mentioned before, one possible explanation of the adjacency principle is that objects adjacent to the test object tend to receive more of the observer's attention than objects displaced from the test object. The possible role of set or attention in the adjacency effect has been examined in three studies, one of which (Gogel, 1965b) was discussed in relation to Figure 5-6. As also mentioned, in the second study (Gogel, 1967), A of Figure 5-6 was positioned to be equally displaced from B and D rather than next to B. In these studies the contribution of set to the perception of the distance position of A was evaluated by determining whether the apparent distance of A was less for the instructions to judge A in relation to D than for the instructions to judge A in relation to B. It was found in both studies that set made some contribution to the apparent distance of A, but this contribution was much less than that of adjacency per se. The third study which examined the relation between set or attention and adjacency was considered in C of Figure 5-8. During part of this study the observers were instructed to look at the vertically moving test point and to indicate the slant of its apparent path of motion while simultaneously attending to one pair of induction points and ignoring the other. The results from this portion of the study showed substantial attention effects in which the apparent path of motion was modified in the direction expected from the induction object receiving the attention. The effects of both adjacency and attention were quite large and attention was able to account for about half the total effect of adjacency. In a second experiment in this study (see footnote 1, Gogel & Tietz, 1976) attention was varied in the situation represented by the upper portion of Figure 5-5. Although the adjacency effects previously obtained (Gogel, 1974b) were replicated closely, attention effects were not significant. It can be concluded that attention can modify the weights given to relative cues and under some circumstances (Figure 5-8C) can produce large modifications in the perception. Clearly the large changes obtained as a function of attention in the situation of Figure 5-8 C were perceptual. Thus the fact that objects adjacent to the test object would normally receive more attention than objects displaced from the test object contributes to the adjacency effect. It does not, however, seem that voluntary attention is able to account for all the adjacency results, even under the most favorable conditions tested, and under other circumstances it accounts for only a small proportion of these results.

THE TWO-FACTOR THEORY AND SPATIAL CONSTANCY

The Achievement of Distance Constancy

The experimental and theoretical distinction between absolute and relative cues of distance implies that the psychophysics of these cues differs, and because of this difference they are often in conflict. Further, the two-factor theory states that the resolution of these cue conflicts usually occurs not by the suppression of all but one of the conflicting cues but rather by a process in which weights are given differentially to the several cues before summing their contributions to the perception. It has been hypothesized that the weight given each cue in the resolution of these conflicts are directly related to the precision (threshold) of the perception resulting from that cue. Conflicts of distance cues, unlike those involved in induction effects, usually occur between absolute and relative cues rather than between relative cues. It would follow that much of the complexity required of the weighting process would be avoided in the perception of distance if only one type of cue (absolute or relative) were used by the visual sytem. Although a visual system designed to respond to absolute or relative distance cues but not to both would indeed be a simpler system, it would also be inadequate.

Consider the case of a visual system that responds only to absolute cues of distance. A response to absolute cues would permit the perceptions to be exocentric as well as egocentric, for the latter can mediate the former by a subtractive process. It would also permit the perception to be scalar because absolute cues of distance often provide scalar information. Distance perceptions from only egocentric cues, however, would be restricted to a few meters from the observer. Clearly, distance constancy under these circumstrances would be limited. On the other hand, a visual system that used only relative cues of distance would also be inadequate in that the perceptions of distance would be uncalibrated and unscaled. It will be recalled that the calibration of the relative cue of binocular disparity by perceived egocentric distance not only provided a scalar extension of the perception of distance but resulted in equal values of the relative cue being perceptually larger at greater distances from the observer. The calibration of a relative distance cue by an absolute distance cue avoids the restriction on perceived distance that would occur with increasing distances from the observer if perceived extent were proportional to cue magnitude. The calibration adjusts the perception in the direction of distance constancy. It seems that the occurrence of cue conflicts and the need to unify the contribution from diverse sources of distance information is the price the visual system pays for the achievement of a perceptual world extended and scaled in distance.

Underconstancy of distance is the condition in which perceived distance does not increase as rapidly as physical distance. From the preceding discussion both absolute and relative cues of distance are necessary in order to avoid a severe underconstancy of distance. Even in situations in which many absolute and relative cues of distance are available (full cue conditions) it cannot be assumed that perceived and physical distance will be proportional (perfect distance constancy), particularly with regard to distant portions of the visual field. Some amount of underconstancy in perceived distance is expected to occur with respect to far portions of an extended visual field despite full cue conditions of observation. One reason for expecting this is that the number and magnitude of distance cues are greater for a near than for a far depth interval; for example, all the absolute and relative distance cues will be above the threshold of discrimination for a depth interval extending from 3 to 6 ft. but few if any will be available for the same depth interval extending from 100 to 103 ft. Both the magnitude and number of distance cues are greater for the observer for near than for far portions of the visual field. Far portions of a visual field extended in distance approach the condition of complete cue reduction, which with increasing distance from the observer might be expected to result in a perceptual uncertainty, not in underconstancy, were it not for the EDT or the SDT. With increasing reduction of cues objects physically separated in depth will tend to be perceived at the same distance according to the EDT and at a distance near to the observer according to the SDT. From either the EDT or SDT depth segments distant in the visual field would tend to be perceptually underestimated in relation to depth segments nearer the observer. This effect of the observer tendencies to convert the expected perceived uncertainty from cue reduction found in distant portions of the visual field to distance underconstancy at least has the advantage of producing definite perceptions with systematic errors for which the observer can attempt some cognitive compensation. It has been suggested that the *over*-constancy sometimes found in distance judgments is the result of the observer's application of a cognitive overcorrection to the perceptual information obtained from more distant portions of the visual field (Gogel, 1974a; Rogers & Gogel, 1975). It is at least consistent with this interpretation that overconstancy of distance seems to be found only with more experienced observers, for example, with adults or older children (Wohlwill, 1963).

Perhaps one of the major advantages of the observer tendencies is that they provide a perceptual basis for the rapid utilization of cognitive information. Suppose, for example, that a distant familiar object is viewed under otherwise reduced conditions of observation. Suppose also that under these conditions the perceived size and distance of the object are mainly determined by the SDT. The perceived size, perceived distance ratio will be consistent with the size distance invariance hypothesis, and because the object appears to be smaller than normal (a ratio of perceived to familiar size less than unity) it will often be inferred (correctly) to be at a greater distance than the distance at which it is perceived. A similar advantage can occur from the EDT with familiar objects. By determining a perceived distance and thus a perceived size the EDT allows the observer, from the ratio of perceived to familiar size, to infer the correct distance of the object. It is likely that this cognitive system, based on perceptual information supplied at least in part by the observer tendencies, represents an extensive and possibly a rapid procedure by which the observer can modify spatial responses in the interest of achieving veridical responses. That this procedure can sometimes result in error does not detract from its usefulness under many circumstances.

The Achievement of Object Constancy

Object constancy is a special instance of an invariance hypothesis; for example, size constancy will occur according to the size-distance invariance hypothesis only if the perceived distance of the object is proportional to its physical distance, and shape constancy will occur according to the shape-slant invariance hypothesis only if the perceived slant of the object is proportional to its physical slant. Although there is considerable controversy in regard to the form and precision of these relations, it is clear that the perception of the third dimension is an important variable in the perception of size and shape. To the degree that the two-factor theory contributes to the understanding of the perception of distance, this theory contributes to the understanding of object constancy. Perhaps the theory presented here also adds to the understanding of object constancy by contributing to the analysis of an important precursor of object constancy; that is, the perception of an object as a unitary or organized event. Basic to the way in which an object maintains its perceptual identity despite changing conditions of stimulation is the way in which the object is perceptually segregated from the total distribution of stimulation.

Relative cues or induction effects by definition involve the perceptual interrelations of objects or parts of objects. These relative effects can be contrasted with the effects from absolute cues that specify the perceived characteristics of objects or parts independent of other objects or parts. Thus relative cues and their expression in induction effects are a measure of the perceptual interrelatedness of objects, whereas absolute cues indicate the characteristics of objects in isolation. Relative cues or induction effects, therefore, unlike absolute cues, are necessary for perceptual organization. Whenever relative cues or induction effects occur, the parts of the visual field involved in this interaction can be said to be perceptually interrelated or perceptually organized. With all other factors constant, the degree of perceptual organization can be assumed to increase as the amount of induction increases. Thus, according to the adjacency principle, greater perceptual organization will occur between spatially adjacent rather than spatially displaced objects.

A kind of perceptual organization related to figure perception is known as perceptual grouping. The Gestalt law of proximity has led to a number of demonstrations of perceptual grouping or figural organization; for example, a matrix of points will be perceptually organized into rows or columns, depending on whether the points are more widely separated vertically or horizontally. It should be noted, however, that superposed on the figural organization into rows or columns is a perceptual organization that permits the organized figures (rows or columns) to be perceived in relation to one another. The relative cues that specify the perceived positions of the points determine a perceptual organization that produces both the perception of figures and the perceptual relations between the figures. The perception of figures within the total perceptual organization occurs because of asymmetries in the spatial distribution of the points, or more precisely, because of asymmetries in the distribution of the effectiveness of the cues of position between the points.

The procedures for measuring induction effects as a function of spatial separation discussed in this chapter can be applied to the measurement of perceptual interrelatedness (perceptual organzation) across the visual field and to the local inhomogeneities in cue effectiveness essential to the perception of figures. Overall perceptual organization can be measured experimentally by measuring the reduction in the magnitude of induction effects as a single induction and test object are separated increasingly in apparent depth or direction.

The slope of this function may be expected to differ between observers and to constitute an important variable along which to measure perceptual development in the infant. The tendency toward strong or weak figural perception, on the other hand, may be experimentally determined by the use of two opposing induction objects in separated portions of the visual field with the induction objects located at different perceived distances from the test object. The greater the effect of the nearer induction object on the test object in relation to that of the more displaced induction object, the greater the tendency to form strong figural organizations. Again, this measure may constitute an important variable along which individual differences and perceptual development can occur. The comparative weight given absolute and relative cues in determining the magnitude of induction phenomena is another dimension along which individual and developmental differences can occur. It seems likely that this dimension would be related to the amount of overall perceptual organization measured by the decrease in the magnitude of the induction as a function of the perceived separation of a single induction and test object.

In conclusion, the two-factor theory of spatial perception may assist in identifying the process underlying figural organization or object identity. Conversely, the phonomemon of figural organization poses an interesting question for the two-factor theory. Absolute cues were defined as determining the perception associated with a single object presented in an otherwise homogeneous visual field. In the example in Figure 5-1 an absolute cue determines the apparent motion of Point 1 in A, whereas both absolute and relative cues of motion are available in B. This manner of defining an absolute cue, however, depends on the specification of the object as a single object. If Points 1 and 2 in B are considered together as a single object (with internally moving parts), the common motion illustrated by the single diagonal arrow in the lower portion of B can be defined as absolute motion. Similarly, if the configuration of two points represented by B move in a direction of common motion different from the common motion of another configuration presented at the same time, a still higher level of relative and common motion can be specified. It is obvious from this description that relative motion cues defined at one level produce a perceptual grouping (a sort of perceptual chunking) that permits the definition of higher levels of absolute and relative cues and results in a hierarchy of perceptual organization. Possibly it is this hierarchical organization that helps to prevent the perceptual world from being a confusing kaleidoscope of perceptually independent parts.

SUMMARY

This chapter has supported a point of view called the two-factor theory of spatial perception (Gogel, 1972b); a theory that distinguishes between absolute and relative determiners of perception. Absolute determiners contribute to the perception of the characteristics of an object independently of other objects. Relative determiners contribute to the perception of the characteristics of an object as a function of other objects in the visual field. The theoretical and experimental need to distinguish between absolute and relative factors has been examined in terms of cues and observer tendencies in the perception of distance and in induction effects. In the perception of distance the major reasons for distinguishing between absolute and relative factors are as follows.

1. The proximal definitions of the major distance cues of a particular kind (absolute or relative) are similar, whereas consistent differences in proximal definition occur between these two types of cue.
2. Considerable evidence exists that the absolute distance cues of convergence and accommodation can support the perception of egocentric distance at relatively near distances from the observer. The validity or magnitude of other absolute cues of distance is more in doubt.
3. Evidence for the validity and importance of relative distance cues is clear in the case of binocular disparity and relative size but less clear in the case of the relative motion parallax associated with head motion in a frontoparallel plane.
4. The perception of egocentric distance resulting from absolute sources of information calibrates or scales the perception of exocentric distance associated with relative cues.
5. The effect of relative cues summates with that of absolute cues in extending the perception of egocentric distance to distant portions of the visual field.
6. The perception of depth within a configuration containing relative cues of distance cannot be explained by egocentric or absolute cues as a subtractive process.
7. The distinction between absolute and relative distance factors is found not only in cues defined by proximal stimuli but also in observer tendencies as indicated by the SDT and EDT. Also, the need to distinguish absolute from relative factors is not limited to the perception of distance. The effect of relative cues on a variety of perceptions is often investigated in situations that can be classified as induction phenomena. The need to postulate absolute in addition to relative cues seems to be indicated in a number of induction phenomena but particularly in the perception of motion and position.

The experimental procedure for identifying an absolute and relative cue or factor dictates that absolute and relative cues are in conflict. Evidence that conflicting cues have a graded effect on the final perception is presented throughout this chapter. These conflicts can occur between absolute and relative cues, between different relative cues, or between absolute or relative cues and observer tendencies. A dimension that seems to be particularly important in specifying the manner in which conflicts between relative cues are resolved is that of perceived adjacency in a frontoparallel plane or in depth. The evidence is clear that for a variety of perceived characteristics, including the perception of size, distance, and motion, the effectiveness of cues between objects (relative cues) in determining perceived characteristics is inversely related to the perceived separation of the objects. One hypothesis that explains the effect of perceived adjacency is that sources of information are weighted inversely to their variability (directly to their precision). In addition to perceived adjacency, dimensions that can be significant in this weighting process are the time of observation and the attention of the observer.

Unfortunately the stimulus or the observer dimensions important for the weighting or precision of absolute factors have not been identified, although the distance of the object from the observer or the position of the stimulus on the retina could be a significant factor. In addition, perhaps both the time of observation and attention can modify the weight given absolute as well as relative cues. In any event, the processes of weighting different and often conflicting sources of information in the perception indicate the degree to which observer processes in addition to the stimulus are essential to an understanding of perception. The effect of attention on adjacency processes although quite small in some instances is appreciable in others. It seems that although perception is seldom if ever unresponsive or unrelated to stimuli, under some conditions it can be modified by processes internal to the observer. The weighting by the visual system of diverse sources of information is perhaps the stage or moment in the development of the percept at which perception is the most susceptible to those influences. A process is considered by means of which the SDT and EDT can provide a basis for the rapid assimilation of cognitive information into the response in an attempt by the observer to correct errors in perceptual information. Although the observer is not always successful in this attempt, as evidenced by overconstancy of size and distance, the use of inferential processes (probably without awareness) enables the observer to reduce the systematic perceptual errors that occur, particularly with regard to far distances.

The two-factor theory of spatial perception has relevance for understanding object and distance constancy. Relative cues or induction effects contrasted with absolute cues provide for the interrelation of parts of the visual field. Organization across the visual field is determined by the effectiveness of absolute compared with relative cues as a function of spatial separation. The tendency to perceive spatial inhomogeneities as figures or objects can be specified by the variation in effectiveness of opposing induction objects on a test object as a function of the relative separation of the induction objects from the test object. These two processes of field and figural organization provide the necessary conditions for the perception of object identity and the relation of the object to the remainder of the visual field.

REFERENCES

Bower, T. G. R. *Development in Infancy.* San Francisco: Freeman, 1974.

Carlson, V. R. Size-constancy judgments and perceptual compromise. *Journal of Experimental Psychology*, 1962, 63, 68-73.

Carlson, V. R. Overestimation in size-constancy judgments. *American Journal of Psychology*, 1960, 73, 199-213.

Dees, J. W. Accuracy of absolute visual distance and size estimation in space as a function of stereopsis and motion parallax. *Journal of Experimental Psychology*, 1966, 72, 466-476.

Duncker, K. The influence of past experience upon perceptual properties. *American Journal of Psychology*, 1939, 52, 255-265.

Epstein, W. *Varieties of Perceptual Learning.* New York: McGraw-Hill, 1967.

Epstein, W. Attitudes of judgment and the size-distance invariance hypothesis. *Journal of Experimental Psychology*, 1963a, 66, 78-83.

Epstein, W. The influence of assumed size on apparent distance. *American Journal of Psychology*, 1963b, 76, 257-265.

Epstein, W. The known size apparent distance hypothesis. *American Journal of Psychology*, 1961, 74, 333-346.

Epstein, W. & Baratz, S. S. Relative visual angle in isolation as a stimulus for relative perceived distance. *Journal of Experimental Psychology*, 1964, 67, 507-513.

Eriksson, E. S. Movement parallax during locomotion. *Perception & Psychophysics*, 1974, 16, 197-200.

Eriksson, E. S. *Movement parallax and distance perception* (Report No. 117). Uppsala, Sweden: University of Uppsala, Department of Psychology, 1972a.

Eriksson, E. S. *The effects of movement parallax on vertically separated stimuli at different distances* (Report No. 119). Uppsala, Sweden: University of Uppsala, Department of Psychology, 1972b.

Eriksson, E. S. *Movement parallax, anisotrophy, and relative size as determinants of space perception* (Report No. 131). Uppsala, Sweden: University of Uppsala, Department of Psychology, 1972c.

Eriksson, E. S. & Zetterberg, P. *Experience and veridical space perception* (Report No. 169). Uppsala, Sweden: University of Uppsala, Department of Psychology, 1975.

Ferris, S. H. Motion parallax and absolute distance. *Journal of Experimental Psychology*, 1972, 95, 258-263.

Foley, J. M. Primary distance perception. In R. Held, H. Leibowitz, & H. L. Teuber (Eds.), *Handbook of Sensory Physiology*, (Vol. 3), Berlin: Springer-Verlag, 1976, in press.

Foley, J. M. & Held, R. Visually directed pointing as a function of target distance, direction and available cues. *Perception & Psychophysics*, 1972, 11, 423-427.

Freeman, E. Accommodation, pupillary width and stimulus distance. *Journal of the Optical Society of America*, 1932, 22, 729-734.

Gibson, J. J. The stimulus variables for visual depth perception. In P. Tibbetts (Ed.), *Perception: Selected Readings in Science and Phenomenology*, Chicago: Quadrangle, 1969.

Gogel, W. C. An indirect method of measuring perceived distance from familiar size. In preparation, 1976.

Gogel, W. C. Depth adjacency and the Ponzo illusion. *Perception & Psychophysics*, 1975, 17, 125-132.

Gogel, W. C. Cognitive factors in spatial responses. *Psychologia*, 1974a, 17, 213-225.

Gogel, W. C. Relative motion and the adjacency principle. *Quarterly Journal of Experimental Psychology*, 1974b, 26, 425-437.

Gogel, W. C. The organization of perceived space I: Perceptual interactions. *Psychologische Forschung*, 1973, 36, 195-221.

Gogel, W. C. Depth adjacency and cue effectiveness. *Journal of Experimental Psychology*, 1972a, 176-181.

Gogel, W. C. Factors in perceptual organization. *Abstract Guide of the XXth International Congress of Psychology*, 1972b, Tokyo, 154-155.

Gogel, W. C. Scalar perception with binocular cues of distance. *American Journal of Psychology*, 1972c 85, 477-498.

Gogel, W. C. The adjacency principle and three-dimensional visual illusions. *Psychonomic Monograph Supplement*, 1970, 3 (Whole No. 45).

Gogel, W. C. The absolute and relative size cue to distance. *American Journal of Psychology*, 1969, 82, 228-234.

Gogel, W. C. The measurement of perceived size and distance. In W. D. Neff (Ed.), *Contributions to Sensory Physiology* (Vol. 3). New York: Academic, 1968.

Gogel, W. C. Cue enhancement as a function of task set. *Perception & Psychophysics*, 1967, 2, 455-458.

Gogel, W. C. Equidistance tendency and its consequences. *Psychological Bulletin*, 1965a, 64, 153-163.

Gogel, W. C. Size cues and the adjacency principle. *Journal of Experimental Psychology*, 1965b, 70, 289-293.

Gogel, W. C. Perception of depth from binocular disparity. *Journal of Experimental Psychology*, 1964a, 67, 379-386.

Gogel, W. C. Size cue to visually perceived distance. *Psychological Bulletin*, 1964b, 62, 217-235.

Gogel, W. C. Visual perception of size and distance. *Vision Research*, 1963, 3, 101-120.

Gogel, W. C. Convergence as a cue to the perceived distance of objects in a binocular configuration. *The Journal of Psychology*, 1961, 52, 303-315.

Gogel, W. C. Relative visual direction as a factor in relative distance perception (Whole No. 418). *Psychological Monographs*, 1956, 70, 1-19.

Gogel, W. C., Brune, R. L. & Inaba, K. *A modification of the stereopsis adjustment by the equidistance tendency*. (Report No. 157) U.S. Army Medical Research Laboratory, 1954, 1-11.

Gogel, W. C. & Harker, G. S. The effectiveness of size cues to relative distance as a function of lateral visual separation. *Journal of Experimental Psychology*, 1955, 50, 309-315.

Gogel, W. C. & Mershon, D. H. Perception of size in a distorted room. *Perception & Psychophysics*, 1968, 4, 26-28.

Gogel, W. C. & Mertens, H. W. Perceived depth between familiar objects. *Journal of Experimental Psychology*, 1968, 77, 206-211.

Gogel, W. C. & Mertens, H. W. Perceived size and distance of familiar objects. *Perceptual & Motor Skills*, 1967, 25, 213-225.

Gogel, W. C. & Newton, R. E. An apparatus for indirect measurement of perceived distance. 1976, 43, 295-302.

Gogel, W. C. & Newton, R. E. Depth adjacency and the rod-and-frame illusion. *Perception & Psychophysics*, 1975, 18, 163-171.

Gogel, W. C. and Sturm, R. D. Directional separation and the size cue to distance. *Psychologische Forshung*, 1971, 35, 57-80.

Gogel, W. C. & Tietz, J. D. Adjacency and attention as determiners of perceived motion. *Vision Research*, 1976, 16, 839-845.

Gogel, W. C. & Tietz, J. D. The effect of perceived distance on perceived movement. *Perception & Psychophysics*, 1974, 16, 70-78.

Gogel, W. C. & Tietz, J. D. Absolute motion parallax and the specific distance tendency. *Perception & Psychophysics*, 1973, 13, 284-292.

Graham, C. H. Visual space perception. In C. H. Graham, N. R. Bartlett, J. L. Brown, Y. Hsia, C. G. Miller, & L. A. Riggs, *Vision and Visual Perception*. New York: Wiley, 1965.

Graham, C. H., Riggs, L. A., Miller, C. G. & Solomon, R. L. Precision of stereoscopic settings as influenced by distance of target from a fiducial line. *The Journal of Psychology*, 1949, 27, 203-207.

Greene, R. T., Lawson, R. B., & Godek, C. L. The Ponzo illusion in stereoscopic space. *Journal of Experimental Psychology*, 1972, 95, 358-364.

Haber, R. N. The nature of the effect of set on perception. *Psychological Review*, 1966, 73, 335-350.

Harker, G. S. Apparent frontoparallel plane, stereoscopic correspondence and induced cyclorotation of the eyes. *Perceptual & Motor Skills*, 1962, 14, 75-87.

Held, R. Plasticity in sensory motor systems. *Scientific American*, 1965, 213, 84-94.

Hersch, J. J. & Weymouth, F. W. Distance discrimination. II: Effect on threshold of lateral separation of the test objects. *Archives of Opthalmology*, 1948, 39, 224-231.

Hochberg, C. B. & Hochberg, J. Familiar size and the perception of depth. *The Journal of Psychology*, 1952, 34, 107-114.

Hochberg, J. & McAlister, E. Relative size vs. familiar size in the perception of represented depth. *American Journal of Psychology*, 1955, 68, 294-296.

Howard, I. P. & Templeton, W. B. The effect of head fixation on the perception of relative depth. *Quarterly Journal of Experimental Psychology*, 1964, 16, 193-203.

Johansson, G. *Monocular movement parallax and near space perception* (Report No. 132). Uppsala, Sweden: University of Uppsala, Department of Psychology, 1972.

Johansson, G. Perception of motion and changing form. *Scandinavian Journal of Psychology*, 1964, 5, 181-208.

Joynson, R. B. & Kirk, N. S. An experimental synthesis of the Associationist and Gestalt accounts of the perception of size (Part III). *Quarterly Journal of Experimental Psychology*, 1960, 12, 221-230.

Kaufman, L. *Sight and Mind*. New York: Oxford University Press, 1974.

Kaufman, L. & Rock, I. The Moon Illusion, I. *Science*, 1962, 136, 953-961.

Kinchla, R. A. & Allan, L. G. A theory of visual movement perception. *Psychological Review*, 1969, 76, 537-558.

Leibowitz, H. W. Effect of reference lines on the discrimination of movement. *Journal of the Optical Society of America*, 1955, 45, 829-830.

Leibowitz, H. W. & Hennessy, R. T. The laser optometer and some implications for behavioral research. *American Psychologist*, 1975, 30, 349-352.

Linksz, A. *Physiology of the Eye* (Vol. 2), *Vision*. New York: Grune & Stratton, 1952.

Lodge, H. & Wist, E. R. The growth of the equidistance tendency over time. *Perception & Psychophysics*, 1968, 3, 97-103.

Mates, B. Effect of reference marks and luminance on discrimination of movement. *The Journal of Psychology*, 1969, 73, 209-221.

Mershon, D. H. & Gogel, W. C. The effect of stereoscopic cues on perceived whiteness. *American Journal of Psychology*, 1970, 83, 55-67.

Mershon, D. H. & King, L. E. Intensity and reverberation as factors in auditory perception of egocentric distance. *Perception & Psychophysics*, 1975, 18, 409-415.

Mershon, D. H. & Lembo, V. L. Scalar perceptions of distance in simple binocular configurations. *American Journal of Psychology*, 1976, in press.

Ogle, K. N. *Researches In Binocular Vision*. Philadelphia: Saunders, 1950.

Ono, H. Some thoughts on different perceptual tasks related to size and distance. In J. C. Baird, (Ed.), Human Space Perception: Proceedings of the Dartmouth Conference (Whole No. 45). *Psychonomic Monograph Supplement*, 1970, 31.

Ono, H. Apparent distance as a function of familiar size. *Journal of Experimental Psychology*, 1969, 79, 109-115.

Owens, D. A. An investigation of the relationship between the specific distance tendency and the oculomotor system. Pennsylvania State University: Masters thesis, 1974.

Owens, D. A. & Wist, E. R. The temporal course of the relationship between retinal disparity and the equidistance tendency in the determination of perceived depth. *Perception & Psychophysics*, 1974, 16, 245-252.

Redding, G. M., Mefferd, R. B. & Wieland, M. A. Effect of observer movement on monocular depth perception. *Perceptual & Motor Skills*, 1967, 24, 725-726.

Richards, W. & Miller, F., Jr. Convergence as a cue to depth. *Perception & Psychophysics*, 1969, 5, 317-320.

Rock, I. The perception of the egocentric orientation of a line. *Journal of Experimental Psychology*, 1954, 48, 367-374.

Rogers, S. P. & Gogel, W. C. Relation between judged and physical distance in multicue conditions as a function of instructions and tasks. *Perceptual & Motor Skills*, 1975, 41, 171-178.

Russell, J. T. Depth discrimination in the rat. *Journal of Genetic Psychology*, 1932, 40, 136-161.

Schober, H. Ueber die akkomodations ruhelage. *Optik*, 1954, 11, 282-290.

Shaffer, O. & Wallach, H. Extent-of-motion thresholds under subject-relative and object-relative conditions. *Perception & Psychophysics*, 1966, 1, 447-451.

Shinkman, P. G. Visual depth discrimination in animals. *Psychological Bulletin*, 1962, 59, 489-501.

Taylor, M. M. Figural after-effects: a psychophysical theory of the displacement effect. *Canadian Journal of Psychology*, 1962, 16, 247-277.

Von Hofsten, C. The role of convergence in visual space perception (Report No. 168). Uppsala, Sweden: University of Uppsala, Department of Psychology, 1974.

Wallach, H. & Frey, K. J. Adaptation in distance perception based on oculomotor cues. *Perception & Psychophysics*, 1972, **11**, 77-83.

Wallach, H. & Smith, A. Visual and proprioceptive adaptation to altered oculomotor adjustments. *Perception & Psychophysics*, 1972, **11**, 418 413-416.

Wern^r, H. Binocular depth contrast and the conditions of the binocular field. *American Journal of Psychology*, 1938, **51**, 489-497.

Wist, E. R. & Summons, E. Spatial and fixation conditions affecting the temporal course of changes in perceived relative distance. *Psychological Research*, 1976, in press.

Wohlwill, J. F. The development of "overconstancy" in space perception. In L. P. Lipsett, & C. C. Spiker, (Eds.), *Advances in Child Development and Behavior* (Vol. 1). New York: Academic, 1963.

Woodworth, R. S. *Experimental Psychology*. New York: Holt, 1938.

CHAPTER

6

ANALYSIS OF CAUSAL RELATIONS IN THE PERCEPTUAL CONSTANCIES

TADASU OYAMA
Chiba University, Japan

Perceptual constancies are labeled according to abstracted perceptual properties; for example, size, shape, and lightness. A more instructive labeling would be one that retained the context of the perceptual discrimination; for example, size at a distance, shape at an orientation, and lightness under illumination. This revision would call attention to the fact that several aspects of the object are experienced simultaneously. Of course, this fact is widely recognized by investi-

I am indebted to Takayuki Mori and Takeshi Watase of Chukyo University, Peter K. Kaiser of York University, Aiko Kozaki of Tokyo Woman's Christian College, and Kaoru Noguchi of Chiba University for their kind permission to use their experimental data for the causal inference presented in this chapter and for their helpful comments; to Ken Goryo and Muneo Mitsuboshi for their assistance in computer programming; and to Katsuo Taya and Kazuo Ikeda for their assistance in data processing.

gators and has been the basis for the description of various percept-percept relationships in the analysis of constancies. The size-distance invariance hypothesis is a familiar example. For most investigators these statements are not intended merely to describe relationships among percepts (responses) but as assertions concerning underlying causal chains as well; for example, perceived distance, in interaction with visual angle, causes a specific perceived size to occur. In this chapter such claims are examined. A theory and method of casual inference is applied to the data of experiments on size, shape, and lightness constancy. The results of this analysis suggest that the standard form of the causal account typified by the size-distance example ought not to be accepted as generally valid.

PERCEPTUAL CONSTANCIES AND INVARIANCE HYPOTHESES

Size Constancy

Under normal conditions for discrimination of distance, if two objects placed at different distances subtend the same visual angle, the far object will appear larger than the near object. However, if these two objects are luminous disks presented on a very dark background and are observed monocularly, they will appear to be the same size and at the same distance, although they are physically different sizes and at different distances. Sometimes two objects of the same size at the same distance (of course subtending the same visual angle) appear to be different sizes and at different distances: the far-looking object appears larger and the near-looking object appears smaller. The apparent sizes of objects increase or decrease nearly proportionally with their perceived distances. As these examples indicate, size constancy presupposes distance perception. A formal statement of the relationship between apparent size and apparent distance is called *size-distance invariance:* "A retinal projection or visual angle of given size determines a unique ratio of apparent size to apparent distance" (Kilpatrick & Ittelson, 1953; Epstein, Park, & Casey, 1961). This relation was originally suggested by Koffka (1935) and has been examined by many experiments (Epstein, 1967, 1973).

Shape Constancy.

The shape of an object whose projection changes because of changes in orientation in relation to the observer will remain relatively constant in appearance. Two shapes that project identical retinal shapes are discriminated perceptually; for example, a trapezoid presented upright and a square placed with its top slanted away may project identical retinal images but they are perceived differently, a trapezoid and a square, when the observer sees them in a full-cue situation. A given retinal image can be produced by an infinite number of combinations between physical shapes and physical slants. Theoretically, the observer has an infinite number of possibilities in selecting a shape-slant combination for a given retinal shape. Actually he perceives only one shape, at one slant, at a time. Perception of a correct shape is usually accompanied by perception of a correct slant, and perception of an incorrect shape is accompanied by perception of an incorrect slant. Perceived shape and perceived slant are coupled together so that if one changes the other changes also, as pointed out by Koffka (1935). Drawing an analogy to the *size-distance invariance* hypothesis, Beck and Gibson (1955) extended Koffka's idea by proposing a shape-slant invariance hypothesis: "A retinal projection of a given form determines a unique relation of apparent slant to apparent shape." Several experimental investigations support this hypothesis but a few do not (Epstein & Park, 1963; Kaiser, 1967).

Lightness Constancy

Lightness of an object tends to remain constant even though the intensity of light reflected by the surface of the object changes a great deal because of changes in illumination. This constancy was described by Helmholtz (1925):

> A grey sheet of paper exposed to sunlight may look brighter than a white sheet in the shade; and yet the former looks grey and the latter white, simply because we know very well that if the white paper were in the sunlight, it would be much brighter than the grey paper which happens to be there at the time (p. 131),
>
> . . . we are accustomed and trained to form a judgment of colours of bodies by eliminating the different brightness of illumination by which we see them . . . (p. 287).

Since Helmholtz's day this phenomenon has often been considered in relation to the perception of illumination. Koffka (1935) also suggested an invariant relation between perceived surface lightness (whiteness in his terminology) and perceived illumination (brightness); "a combination of whiteness and brightness, possibly their product, is an invariant for a given local stimulation under a definite set of total conditions" (p. 244). It may be called the *lightness-illumination invariance* hypothesis, but only a smaller number of experiments than those concerned with the size-distance and shape-slant invariance hypotheses (Beck, 1972) have been conducted to test it.

Invariance hypotheses such as those suggested in relation to the three main perceptual constancies, size, shape, and lightness, can also be formulated for relations between visual direction and apparent eye position, between apparent object orientation and apparent body position, between stereoscopic depth and apparent distance, and between apparent speed and apparent distance of moving object (Epstein, 1973). All these invariant relations contribute to the stability of our perceptual world.

POSSIBLE CAUSAL INTERPRETATIONS OF INVARIANT RELATIONSHIPS

What causal interpretations should be assigned to the relationships described in the invariance hypotheses when they are found in perceptual constancies and some other perceptual phenomena? Mathematical expressions of the invariance hypotheses are useful in considering possible causal interpretations of these relations. The size-distance invariance hypothesis can be expressed as follows:

$$\frac{S'}{D'} = K\theta \qquad (1)$$

or in a more generalized form

$$\frac{S'}{D'} = f(\theta) \qquad (2)$$

where S', D', and θ indicate perceived size, perceived distance, and visual angle, respectively, and K is a constant. The shape-slant invariance hypothesis applied to a trapezoidal object slanted around the horizontal axis is expressed by

$$\frac{h' \cos \phi'}{w'} = a\frac{h}{w} \qquad (3)$$

where h', w', ϕ', h, w, and a indicate, respectively, perceived height, perceived width, perceived slant, retinal height, retinal width, and a constant. The lightness-illumination invariance hypothesis is expressed by the following equation:

$$R'E' = bI \qquad (4)$$

where R', E', I', and b represent perceived lightness (or matched reflectance), perceived illumination (or matched illuminance), retinal illuminance, and a constant.

In all these equations the right side indicates some retinal variable (visual angle, retinal height-width ratio, or retinal illuminance) and a constant. If a certain value is given on the right side, the ratio or product of perceptual variables on the left side is uniquely determined. It chould be noted that *only* the ratio (S'/D' or $h' \cos \phi'/w'$) or product ($R'E'$) is uniquely determined, leaving individual variables (S', D', h'/w', θ', R', and E') undetermined. An infinite number of combinations of these values produces a given ratio or product; for example, in the size-distance invariance hypothesis a given visual angle determines a unique ratio of perceived size to perceived distance, but it cannot determine perceived size and perceived distance themselves. The observer needs other information to determine these individual variables uniquely. Such information comes from an outer source (physical stimulus) or an inner source (expectation, memory, innate or learned tendency). "Depth" cues such as convergence, accommodation, and linear perspective are examples of outer information and known or assumed size and specific distance tendency are examples of inner information. The retinal variables can be called the first independent variable and the other inner or outer information can be called the second independent variable. The terms "first" and "second" are used only for convenience and do not mean "primary" and "secondary" in determining perceptual variables. If many sources of information exist, they may constitute the third, fourth, and fifth independent variables, and so on. In the present discussion, however, we consider only the first and second independent variables for the sake of simplification.

When the first and second independent variables (e.g., visual angle and stimulus cue) are given in a size constancy situation, two dependent (perceptual) variables, S' and D', will be determined. Which of the two variables is determined first, S' or D'? At least, three different cases can be considered.

Case 1. Perceived distance is determined first by the second independent variable X, and perceived size is then proportionally determined by two variables, visual angle (the first independent variable) and the already determined perceived distance. Mathematical expressions of these relations are

$$D' = f_1(X) \tag{5}$$

and

$$S' = K\theta D' \tag{6}$$

Case 2. Perceived size is determined first by the second independent variable and perceived distance, by visual angle and the already determined perceived size:

$$S' = f_2(X) \tag{7}$$

and

$$D' = \frac{S'}{K\theta} \tag{8}$$

Case 3. Perceived size and distance are independently determined by one or two independent variables, and the form of the functions relating the two dependent variables to the second independent variable is identical. Consequently, a proportional relation helds between the two dependent variables S' and D' when the first independent variable θ is kept constant:

$$S' = K\theta\, f_3(X) \tag{9}$$

and

$$D' = f_3(X) \tag{10}$$

then

$$\frac{S'}{D'} = \frac{K\theta\, f_3(X)}{f_3(X)} = K\theta \tag{11}$$

Figure 6-1 shows causal relations schematically in these three cases. In Cases 1 and 2 perceived distance or perceived size is an intervening variable for the other perceptual variable, perceived size or perceived distance. In Case 3 such causal relations between the two perceptual variables do not exist. It is important that, in spite of these differences, the last equations of the three cases (6), (8), and (11) are mathematically identical to one another and to (1).

Is it possible to discriminate among the three cases on the basis of experimental results? The answer to this question is no if (5) to (11) are deterministic equations without random variations. The answer will be yes if these equations are probabilistic and random variations of the intervening variable are transmitted to the dependent variable. Fortunately most perceptual processes can be regarded as probabilistic. Consequently we can discriminate the three causal

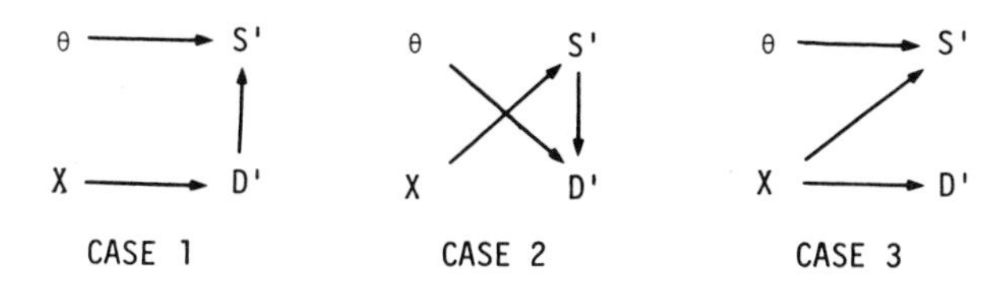

Figure 6-1 Schematic representations of three possible causal explanations of the size-distance invariance hypothesis. θ, X, S', and D' represent visual angle, stimulus cue, perceived size, and perceived distance.

models in the problem of perceptual constancies. The logic and procedure of the causal inference are discussed in the next section.

More complicated causal relations can be considered in the interpretation of the invariant relationship. In some of them both dependent variables are determined by two independent variables, and in others a dependent variable is determined by two independent and one intervening variables, as shown in Figure 6-4. The complicated cases are discussed in the next section.

Concerning the shape-slant invariance hypothesis and the lightness-illumination invariance hypothesis, we can also discuss three kinds of causal relations analogous to those just illustrated for the size-distance invariance hypothesis. In the shape-slant invariance hypothesis applied to a slanted trapezoid perceived height-width ratio (h'/w') and perceived slant are two dependent variables, whereas the retinal height-width ratio (h/w) is the first independent variable. A depth cue or the known or assumed shape will become the second independent variable.

In the lightness-illumination invariance hypothesis perceived lightness and perceived illumination are the related dependent variables and retinal illuminance, the first independent variable. What kind of outer information is used as the second independent variable? What information specifies the lightness of surface, the level of illumination, or both? The retinal illuminance of the other parts of the visual field or its ratio to that of the focal part is used as the second independent variable. The retinal illuminance ratio between the object image and its surround is one of the important determinants of perceived lightness of the object (Wallach, 1948). The ratio between the object image and even its indirect surround also effectively determines the perceived lightness of the object when the retinal illuminance of the direct surround is kept constant (Hsia, 1943; Oyama, 1968). The highest retinal illuminance in the visual field has been considered as one of the most important determinants of perceived illumination (Beck, 1959, 1961, 1972; Oyama, 1968).

If Case 1 or 2 is proved to be suitable for a situation of size constancy, for example, by the causal analysis of experimental results, the "taking-into-account" explanation suggested by Woodworth (1938) and Epstein (1973) can be applied. When Case 1 is fitted, perceived distance is determined first by an independent variable (e.g., a depth cue) and this perceived distance is taken into account by the observer to determine perceived size. When Case 2 is fitted, perceived size is determined first by an independent variable (e.g., known or assumed size) and perceived size is taken into account for determination of perceived distance. The "taking-into-account" explanation cannot, however, be applied to Case 3, in which perceived size and perceived distance are directly and independently determined by the independent variables. The observer does not need to take a perceived property into account to determine another perceptual property.

An important aspect of the "taking-into-account" explanation is the indifference to information source of the effect of the taken-into-account property on the to-be-discriminated property (Epstein, 1973). In Case 1, for example, whenever visual angle is constant, whether perceived distance is determined by convergence or by experience, the same perceived distance always produces the same perceived size. The situation is quite different for Case 3. Each dependent variable is connected with independent variables by a different function, although some of the functions may have an identical form. The perceived properties may vary for different sources of information. Of course, we can hypothesize some intervening variables between the independent and dependent variables and assume that the dependent variables are the same as far as the intervening variables are kept constant, regardless of the change on independent variables. This kind of intervening variable in Case 3 is different from the taken-into-account perceptual property in Cases 1 and 2. The former is only a hypothetical construct and not directly observable, whereas the latter have double characters, a dependent variable and an intervening variable, at the same time, and is directly observable. The causal analysis that we are going to discuss is applicable only to observable variables.

CAUSAL INFERENCE BY PARTIAL CORRELATIONS

"Even in the first course in statistics, the slogan 'Correlation is no proof of causation!' is imprinted firmly in the mind of the aspiring statistician or social scientist," said Simon (1954) in the beginning of his paper in which he proposed a statistical method to discriminate between "true" and "spurious" correlations. Simon's method was developed by Blalock (1962) on a more generalized form of causal inference by partial correlations. Applying the Simon-Blalock method, we can discriminate between direct and indirect causal relations among three or more variables.

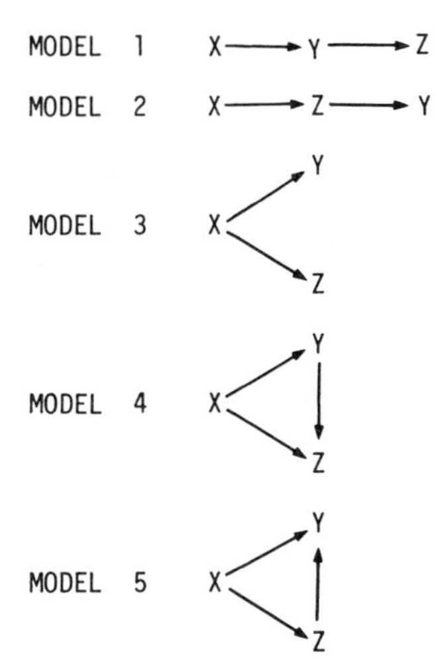

Figure 6-2 Schematic representations of the five causal models among three variables in which X is always the only independent variable.

According to Simon's three-variable model, five causal relations are expected among three variables, X, Y, and Z, if we assume that X is always an independent variable and that every relation between two variables is linear. The five causal relations are shown schematically in Figure 6-2.

In Model 1 relationships among the three variables are represented by the following equations,

$$Y = aX + u \tag{12}$$

$$Z = bY + v \tag{13}$$

where a and b are constants and u and v are random variables. As shown in these equations, X determines Y and Y determines Z. From (12) and (13) the following relations can be derived for a large sample:

$$r_{XZ} = r_{XY} \cdot r_{YZ} \tag{14}$$

From (14) the partial correlation between X and Z with Y held constant can be derived as zero,

$$r_{XZ \cdot Y} = 0 \tag{15}$$

because the numerator of the partial correlation is $r_{XZ} - r_{XY} \cdot r_{YZ}$. The details of mathematical derivation are given in Simon's paper.

Similarly, in Model 2, in which X determines Z and Z determines Y,

$$Z = a'X + u' \tag{16}$$

and

$$Y = b'Z + v' \tag{17}$$

where a' and b' are constant and u' and v' are random variables.

Hence

$$r_{XY} = r_{XZ} \cdot r_{YZ} \tag{18}$$

and

$$r_{XY \cdot Z} = 0 \tag{19}$$

In Model 3 Y and Z are independently determined by X and there is no direct causal relation between Y and Z.

$$Y = a''X + u'' \tag{20}$$

and

$$Z = b''X + v'' \tag{21}$$

where a'' and b'' are constants and u'' and v'' are random variables.

Then

$$r_{YZ} = r_{XY} \cdot r_{XZ} \tag{22}$$

and

$$r_{YZ \cdot X} = 0 \tag{23}$$

In Model 4 Y is determined by X and Z, by both X and Y:

$$Y = cX + s \tag{24}$$

and

$$Z = dX + eY + t \tag{25}$$

where c, d, and e, are constants and s and t are random variables. In Model 5 Z is determined by X, whereas Y is determined by X and Z:

$$Z = c'X + s' \tag{26}$$
$$Y = d'X + e'Z + t' \tag{27}$$

where c', d', and e' are constants and s' and t' are random variables. In Models 4 and 5 we cannot derive the simple relations found in Models 1 to 3. No partial correlation becomes zero.

A partial correlation is a net correlation between two variables when the influence of one or more additional variables has been eliminated. This correlation becomes zero, as indicated in (15), when, as in Model 1, the effect of X on Z operates only through the change on Y whose influence has been eliminated in this correlation. Similarly, the partial correlation between X and Y with Z held constant becomes zero in Model 2 because the effect of X on Y operates only through the change on Z whose influence has been eliminated. In Model 3 both Y and Z depend directly on X and the correlations between Y and Z result from the joint causal effects of X on both variables. This correlation is "spurious" as shown by the zero partial correlation between them in which X is held constant.

Simon (1954) presented two examples to illustrate the causal relations corresponding to Models 1 and 3 in this chapter.

For Model 1

> X is the percentage of female employees who are married, Z is the average number of absences per week per employee, Y is the average number of hours of housework performed per week per employee. A high (positive) correlation, r_{XZ} was observed between marriage and absenteeism. However, when the amount of housework, Y, was held constant, the correlation $r_{XZ \cdot Y}$ was virtually zero. . . .marriage results in a higher average amount of housework performed, and this, in turn, in more absenteeism (p. 469).

(Notations of variables were adapted to those in this chapter.)

For Model 3

> Y is the percentage of members of the group that is married, Z is the average number of pounds of candy consumed per month per member, X is the average age of members of the group. A high (negative) correlation, r_{YZ}, was observed between marital status and amount of candy consumed.

> . . . However, when age was held constant, the correlation $r_{YZ \cdot X}$, between marital status and candy consumption was nearly zero. . . .the correlation between marital status and candy consumption is spurious, being a joint effect caused by the variation in age (p.469).

Figure 6-3 shows the causal relations found in these examples. It should be noted that the partial correlation method can discriminate only between direct and indirect relations but cannot decide the direction of causality. Exactly the same partial correlations can be obtained from three different causal relations: (a) X determines Y and then Y determines Z, (b) Z determines Y and then Y determines X, and (c) Y independently determines X and Z. The partial correlation $r_{XZ \cdot Y}$ will be zero in all three cases. "Common sense" or other scientific knowledge indicates which is the most suitable for a given situation. In usual experimental situations causal arrows start from experimentally controlled stimulus variables, never from subjects' responses. If X represents the experimental variable, (a) should be the most suitable causal model for the experimental data in which $r_{XZ \cdot Y}$ was nearly zero.

Although Simon (1954) presented causal models for only three-variable cases, Blalock (1962) extended the use of partial correlation to the inference of causation in four-variable cases. He presented the following equations:

$$X_1 = e_1 \tag{28}$$
$$X_2 = b_{21}X_1 + e_2 \tag{29}$$
$$X_3 = b_{31\cdot 2}X_1 + b_{32\cdot 1}X_2 + e_3 \tag{30}$$
$$X_4 = b_{41\cdot 23}X_1 + b_{42\circ 13}X_2 + b_{43\ 12}X_3 + e_4 \tag{31}$$

MARRIAGE ⟶ HOUSEWORK ⟶ ABSENTEEISM
(X) (Y) (Z)

MODEL 1

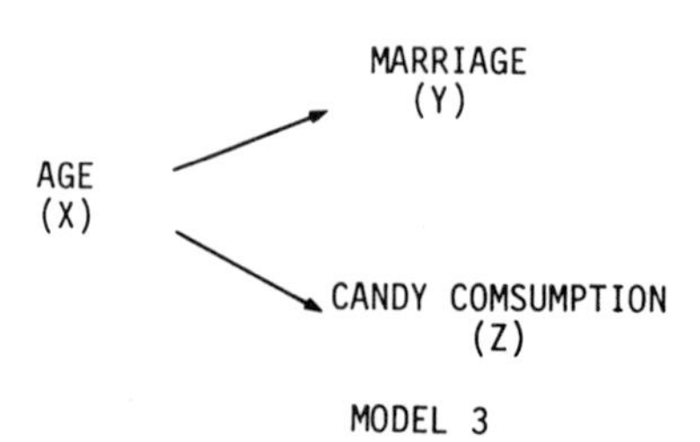

MODEL 3

Figure 6-3 Schematic representations of the causal relations in two examples described in Simon (1954).

These equations indicate that X_1 depends causally only on outside variables, the effects of which are represented by e_1, but X_2 depends on X_1 as well as outside factors whose effects are represented by e_2, X_3 depends on both X_1 and X_2, and finally X_4 depends on all of the remaining X's. This system, called "recursive," is contrasted with a possible set of equations whose causation may be reciprocal in which X_1 depends on X_2 and vice versa.

In (28) to (31) some of bs can be equal to zero. A zero b indicates that there is no direct causal link between the two variables concerned. Zero bs are shown by the fact that the corresponding partial correlations become zero. If $r_{13 \cdot 2} = 0$, then $b_{31 \cdot 2} = 0$, and there is no direct link between X_1 and X_3. Similarly, if $r_{24 \cdot 13} = 0$, then $b_{42 \cdot 13} = 0$, and this means that X_2 is no longer a direct

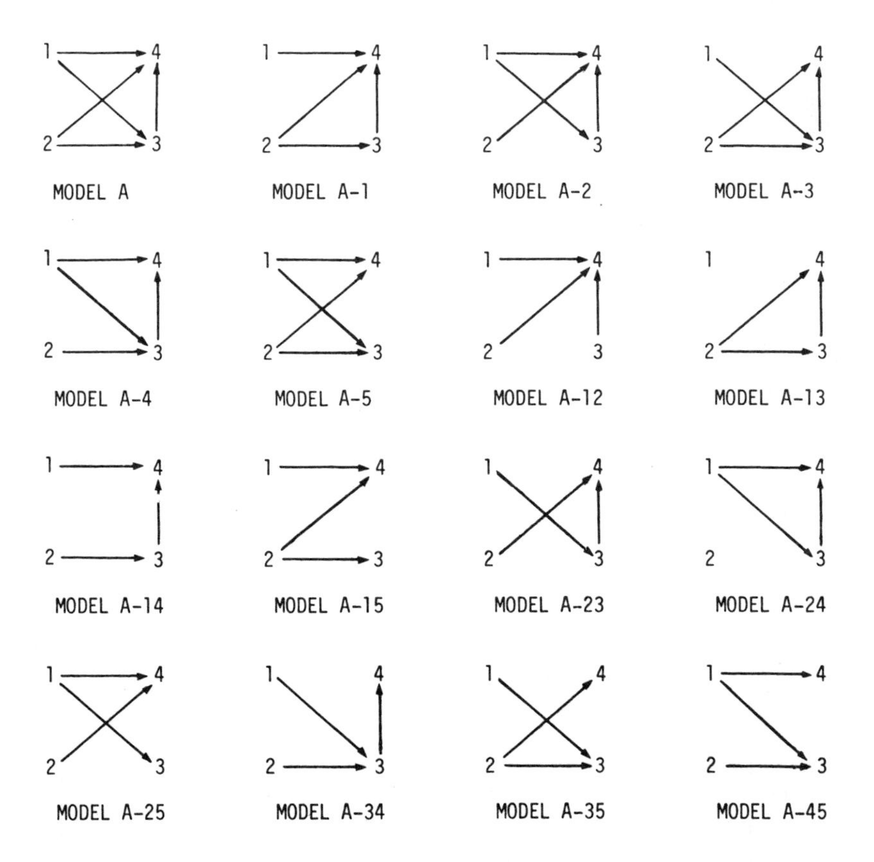

Figure 6-4 Sixteen causal models in A-series of the four-variable cases described by Blalock (1962). Reprinted from *The American Journal of Sociology*, 1962, **68**, 182-194, Chart 1 on pp. 186-187 by permission of The University of Chicago Press. Copyright 1962, The University of Chicago Press.

Table
1

DISTRIBUTIONS OF ZERO CORRELATIONS IN THE A-SERIES OF BLALOCK'S FOUR-VARIABLE CASUAL MODELS

Models	A	A-1	A-2	A-3	A-4	A-5	A-12	A-13	A-14	A-15	A-23	A-24	A-25	A-34	A-35	A-45
r_{12}	0*	0*	0*	0*	0*	0*	0*	0*	0*	0*	0*	0*	0*	0*	0*	0*
r_{13}		0*					0*	0*	0*	0*						
r_{14}								0*							0*	
r_{23}			0*				0*				0*	0*	0*			
r_{24}												0*				0*
r_{34}																
$r_{12.3}$		0	0				0	0	0	0	0	0	0			
$r_{12.4}$								0							0	0
$r_{13.2}$		0					0	0	0	0						
$r_{13.4}$								0								
$r_{14.2}$								0							0	
$r_{14.3}$								0			0*			0*		
$r_{23.1}$			0				0				0	0	0			
$r_{23.4}$												0				
$r_{24.1}$												0				0
$r_{24.3}$									0*			0		0*		
$r_{34.1}$													0*			0*
$r_{34.2}$										0*					0*	
$r_{12.34}$								0	0		0	0				
$r_{13.24}$								0		0						
$r_{14.23}$				0*				0			0			0	0	
$r_{23.14}$												0	0			
$r_{24.13}$					0*				0			0		0		0
$r_{34.12}$						0*				0			0		0	0

*Zeros with asterisks indicate the zero correlations described in Blalock's (1962) chart; the other zeros are automatically derived.

cause of X_4. Thus in four-variable cases not only first-order partial correlations, in each of which one variable is held constant, but also second-order partial correlations, in each of which two variables are held constant, should be obtained. Obtained patterns of simple and partial correlations having nearly zero values will suggest causal relations among variables. Blalock presented a chart of 40 causal models of four-variable cases. His chart can be used to identify the causal model most suitable for an obtained pattern of correlations. In this process we must find a suitable corresponding empirical variable for each of the four variables in the theoretical models in which variables with smaller numbers as suffixes are always located earlier in causal chains.

Table 1 and Figure 6-4 summarize the 16 models in Series A of Blalock's chart in which X_1 and X_2 are mutually independent and in which X_1, X_2, X_3, and X_4 are represented by 1, 2, 3, and 4. Because two independent experimental variables are used and two dependent response variables are observed in many experiments of perception, these sources are especially useful. For more complicated experimental designs, in which experimental variables are mutually dependent, the readers should see Blalock's original chart. Table 1 shows the distributions of zero correlations for each of the 16 causal models in Series A. Zeros with asterisks indicate correlations that were described as zero in Blalock's chart, and simple zeros are those added by me according to Blalock's suggestion that some higher order partial correlations should become zero automatically from the lower order zero correlations described in his chart. Figure 6-4 marks causal relations in each model with arrows. It should be noted that causal relations that are too complex (all four variables are connected by six arrows) and too simple with only one or two arrows are omitted in Table 1 and Figure 6-4 as they are in the Blalock's original chart.

For causal inference of a given experimental result we must find the most suitable causal model whose pattern of zero correlations is similar to the obtained pattern. A comparison should be made between the obtained pattern of nearly zero correlations and expected patterns of zero correlations in various models. Table 1, or Blalock's original chart, will be useful for this process. Because exactly zero correlations can never be found in empirical data, we must use some rather arbitrary criteria to judge "nearly zero." As suggested in Namboodiri, Carter, and Blalock (1975), the statistical insignificance of the correlation might be suitable as the criterion for small samples, and it would be more appropriate to select some relatively small absolute magnitude, say .10, for large samples. As we shall see in the next section, obtained patterns of insignificant (or smaller than criterion) correlations may not be exactly the same on any of the theoretical causal models shown in Table 1 or in Blalock's original chart in spite of the fact that all possible causal relations are listed in the table and chart with the exceptions of the few described above. We can proceed only by eliminating models that do not fit the data rather than by establishing any specific model as the correct one. Statistical significance for any of the correlations that are expected to be zero in a particular causal model will be used as the criterion to eliminate that model. Elimination of simpler models by this criterion usually produces a more complicated or more general model that includes all causal arrows of the eliminated simpler models. More detail processes of selecting the most suitable model are discussed in the next section.

For interpretation of individual partial correlations Blalock (1962) presented a point of caution: "These vanishing partials involve controls for *all* variables which are either antecedent to, or intervening between, the particular variables being related, but they do *not* involve controls for variables taken to be dependent upon *both* of these variables" (p. 185); for example, if there were no arrow between variables 1 and 3, the value of $r_{13 \cdot 2}$ should be approximately zero, but $r_{13 \cdot 4}$ and $r_{13 \cdot 24}$ cannot always be expected to be zero.

APPLICATIONS OF CAUSAL INFERENCE TO EXPERIMENTAL DATA

In this section the procedure of causal inference by partial correlations is applied to experimental data. The data to be analyzed were obtained from published and unpublished studies on size, shape, and lightness constancies, and in every study reports of two perceptual properties were secured for each standard stimulus. In most of the studies the invariance hypotheses discussed earlier held exactly or approximately, but in a few they did not. It should be noted that an assessment of causal inference is important whether or not the invariance hypotheses hold.

Perceived Size and Perceived Distance

The analysis begins with two studies of size constancy in which reports of perceived size and perceived distance were obtained. One is Oyama's study in a stereoscopic situation; the other is Mori and Watase's study in an ordinary dark experimental room.

SIZE AND DISTANCE PERCEPTION IN STEREOSCOPIC DISPLAYS

Oyama (1974a) studied the effects of visual angle and convergence on the perceived size and distance of a familiar (playing card) and a nonrepresentational object (blank card). Pairs of positive colored photographic transparencies of the Queen of Spades and a blank white card were projected on a daylight screen in a polarizing stereoscope. The height of the projected images of the cards was varied in five steps from 1° 30′ to 3° 3′ in visual angle and the convergence angle was varied in six steps from 3′ to 3° 33′. Six subjects were asked to provide verbal estimates of the size (height) of the card and of its absolute distance in meters and centimeters. Thirty conditions (5 sizes x 6 convergences) were presented in random order for each of the playing card and white card series.

The results indicated that size estimates (S') increased almost proportionally to the visual angle (θ) and decreased with the convergence angle (α). Distance estimates (D') decreased almost linearly as the visual or convergence angle increased. No clear difference in results was observed between the playing card and white card series, although the familiar object might have been expected to show a stronger resistance to change in its apparent size and a stronger tendency to change apparent distance than a nonrepresentational object (Ittelson, 1951). The observed effect of visual angle on perceived distance may be attributed to the relative retinal sizes among the successively presented stimuli (Gogel, 1969) rather than to the assumed size of the stimulus object.

One of the most important findings was that the ratio of size estimate to distance estimate (S'/D') for a given visual angle was almost constant, regardless of convergence. In this sense the size-distance invariance hypothesis held. As discussed earlier in this chapter, the size-distance invariant relation found in this experiment can be explained in at least three ways. In Cases 1 and 2 S' or D' is first determined by convergence. The other perceptual variable (D' or S') is then determined by visual angle and the already determined perceptual variable. The invariance hypothesis holds because the second perceptual variable increases in proportion to the first perceptual variable. In Case 3 the two perceptual variables are determined by convergence directly and independently. If the functions relating the two perceptual variables to convergence have the same form, the variables will covary and the invariance hypothesis will hold. To identify the causal channel underlying the invariant relation found by Oyama (1974a) the method of causal inference described in the preceding section can be used.

To apply Blalock's procedure for four-variable cases to Oyama's results, six simple correlations, 12 first-order partial correlations, and six second-order partial correlations were computed for individual subjects (Oyama 1974b). (Oyama showed only the results of correlational analysis on the group means.) Table 2 contains all the insignificant correlations found for individual subjects, except for those between visual angle and convergence that are expected to be zero because the experimental design guarantees their independence.

The first-order partial correlation between size and distance estimates in which the visual angle was held constant, $r_{S'D'\cdot\theta}$, was highly significant, (.486~.915), except for Subject C in both series and Subject F in the white card series. This significant correlation means that the perceived size and distance of an object with a constant visual angle are positively correlated. This relation is concordant with the size-distance invariance hypothesis. On the other hand, significantly negative first-order partial correlations, $r_{S'D'\cdot\alpha}$, were found between the same two perceptual variables when convergence was held constant (−.543~−.851). The negative correlations indicate that the perceived size and distance of

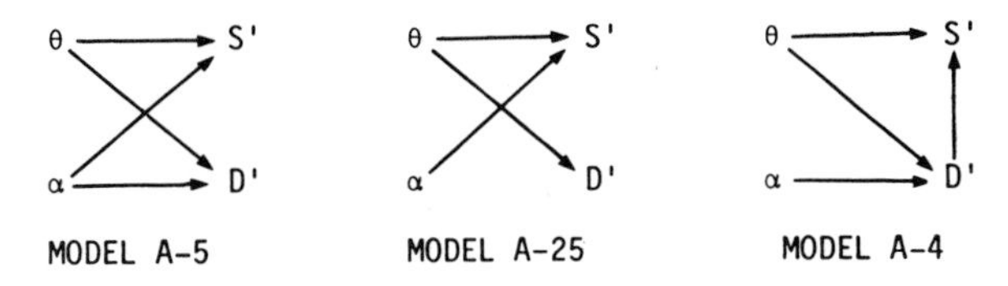

Figure 6-5 Suggested causal models in Oyama's (1974a,b) study on perceived size and perceived distance. θ, α, S' and D', represent visual angle, convergence, perceived size, and perceived distance, respectively.

the object with varying physical size presented at a constant distance are negatively correlated; the objects appear to be larger and closer as the visual angle increases. This relation has been reported by Epstein and Landauer (1969) and Landauer and Epstein (1969).

Except for Subject A in the white card series, the second-order partial correlation between the size and distance estimates when visual angle and convergence was held constant, $r_{S'D'\cdot\theta\alpha}$, was always insignificant. This means that no direct causal relation exists between perceived size and perceived distance. In four subjects (A, B, D, and E) in the playing card series and three (B,D, and E) in the white card series all other correlations except $r_{S'D'}$ were significant. Comparing this pattern of insignificant correlations with the patterns of zero correlations expected from Blalock's four-variable models shown in Table 1, we find that Model A-5 fits the obtained data best. Although this model does not require an insignificant $r_{S'D'}$, the existence of insignificant $r_{S'D'}$ in the obtained results does not contradict the model. Positive and negative correlational tendencies between S' and D', which are shown by the above mentioned positive and negative first-order partial correlations, might counteract each other and result in a nearly zero simple correlation between the two perceptual estimates. For Subject F in the playing card series two more simple correlations, $r_{\alpha S'}$ and $r_{\alpha D'}$, are insignificant, but partial correlations between α and S' and α and D' are all significant. Consequently the causal relations between α and S' and α and D' cannot be denied. With the exception of Model A-5, all other models are unsuitable for this pattern of insignificant correlations. This case should also be regarded as consistent with Model A-5. The causal relations indicated in Model A-5 are shown in Figure 6-5. Both perceived size and perceived distance are directly and independently determined by the two stimulus variables—visual angle and convergence. The observed correlations between S' and D' are spurious and are indirectly induced by their common stimulus determinants. This causal network is similar to that of Case 3 but a little more complicated.

For Subject C in the playing card series and Subjects C and F in the white card series $r_{\alpha D'}$, $r_{\alpha D'\cdot\theta}$, $r_{S'D'\cdot\theta}$, $r_{\alpha D'\cdot\theta S'}$ and $r_{S'D'\cdot\theta\alpha}$ are insignificant. Model A-25, shown in Table 1 and Figure 6-4, fits these three cases best if we regard θ,

α, D', and S' as variables 1,2,3,and 4. The causal relation shown in Figure 6-5 is applied to these cases. Perceived size is determined by the two stimulus variables, but perceived distance, by visual angle alone. It is similar to Case 3 in the earlier section, but in the present case the stimulus determinant of perceived distance is visual angle.

In the results of Subject A for the white card series a second-order partial correlation, $r_{\alpha S' \cdot \theta D'}$ instead of $r_{S'D' \cdot \theta \alpha}$, is significant. Model A-4, in Table 1 and Figure 6-4, fits this outcome best. As made apparent by Figure 6-5, a causal relation exists between the two perceptual variables S' and D'. Perceived distance has some influence on perceived size, which is determined by visual angle and perceived distance, whereas perceived distance is determined by the two stimulus variables. The effect of convergence on perceived size operates only through perceived distance. This causal relation is similar to that in Case 1 but a little more complicated. It is the only example of direct relation between the two perceptual properties in this study.

Size and Distance Perception in the Dark

Mori and Watase (1975) of Chukyo University have also studied reports of perceived size and distance. These authors presented a disk illuminated by a constant light on a dark background in a lightproof experimental room. Combinations of five visual angles ($42' \sim 3°28'$) and five physical distance ($1 \sim 5$ m) constituted 25 stimulus conditions. Five subjects served in two experimental series; a monocular and a binocular observation series, in each of which the 25 stimulus conditions were presented six time in random order. The subjects were asked to estimate the size and distance of each stimulus in meters and centimeters. In half of the stimulus presentations size estimation preceded distance estimation; in the other half the order was reversed.

The experimental results showed that both perceived size and perceived distance increased as power functions of physical distance, although the exponents for the two estimates were slightly different. Consequently a log-log linear relation (another power function) was also found between perceived size and perceived distance. In this sense size-distance invariance held in their experimental results.

To apply the causal inference procedure to Mori and Watase's results all variables, visual angle θ, physical distance D, size estimate S', and distance estimate D' were transformed into logarithms. Table 3 lists the insignificant simple and partial correlations found among these variables, with the exception of those between the two experimental variables θ and D which were independent according to the experimental design.

A comparison between Tables 1 and 3 indicates that Model A-4 is the best fit for the results of Subjects a, b, and e in the monocular series and a,c, and e in the binocular series, if we regard θ, D, D', and S' as variables 1,2,3, and 4. Some insignificant correlations were observed in addition to $r_{DS' \cdot \theta D'}$, which is expected to be nearly zero, but the significant correlations obtained reject the other models in Table 1. Consequently Model A-4 is considered the best model for these results. As shown in Figure 6-6, this model indicates that perceived size is determined by visual angle and perceived distance, whereas perceived distance is determined by the two stimulus variables. The effect of physical distance on perceived size operates only through perceived distance.

Model A-5 fits the results of Subjects b and d in the binocular series, although there are a few insignificant correlations unexpected in this model ($r_{S'D' \cdot \theta D}$ is the only zero correlation expected from Model A-5). In this model, as shown in Fig. 6-6, perceived size and distance are determined by the stimulus variables.

Table 3

INSIGNIFICANT SIMPLE AND PARTIAL CORRELATIONS FOUND IN MORI AND WATASE'S (1975) DATA

	Monocular					Binocular				
Subjects	A	B	C	D	E	A	B	C	D	E
$r_{\theta D'}$	−.22		−.37		−.10	−.21	−.32	−.24	−.09	−.14
$r_{DS'}$				−.00	.19		.35			
$r_{DD'}$				.04						
$r_{S'D'}$	.27	−.01	.25		.17		.02			
$r_{\theta D' \cdot D}$					−.16				−.37	
$r_{\theta D' \cdot S'}$				−.23						
$r_{DS' \cdot \theta}$				−.02						
$r_{DS' \cdot D'}$				−.06	.09					
$r_{DD' \cdot \theta}$				−.08						
$r_{DD' \cdot S'}$				−.08						
$r_{S'D' \cdot \theta}$				.12						
$r_{S'D' \cdot D}$	−.29				.03			−.29	−.29	−.33
$r_{\theta D' \cdot DS'}$				−.23			−.15			
$r_{DS' \cdot \theta D'}$	.15	.34		−.01	−.18	.20		.06		−.07
$r_{DD' \cdot \theta S'}$			−.20	−.08						
$r_{S'D' \cdot \theta D}$				.12			−.03		.35	
Fitted models	A-4	A-4	A-3	A-X	A-4	A-4	A-5	A-4	A-5	A-4

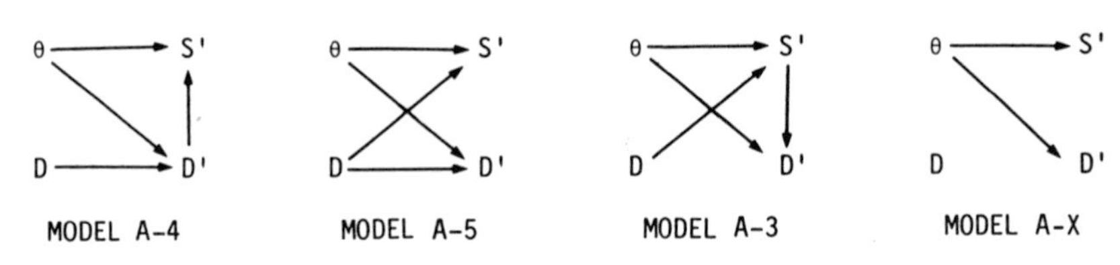

Figure 6-6 Causal models suggested from an analysis of Mori and Watase's unpublished data on perceived size and perceived distance; θ, D, S', and D' represent visual angle, physical distance, perceived size, and perceived distance, respectively.

There is no direct relation between the two perceptual variables.

Model A-3 fits the results of Subject c in the monocular series in which we regard the variables D, θ, S', and D' as variables 1,2,3, and 4. (The order was changed from that in the preceding cases.) As shown in Figure 6-6, S' is determined by the two stimulus variables θ and D, whereas D' is determined by θ and the thus determined S'. The effect of physical distance on perceived distance operates only through perceived size! This is an expected result but the fact that $r_{DD' \cdot \theta S'}$ approximates zero indicates the appropriateness of this model. Some subjects in Oyama's study indicated that convergence, which is usually considered as a "depth" cue was effective in determining perceived *size* but ineffective for perceived *distance*. Both results may show the same tendency.

A suitable model for the results of Subject d in the monocular series cannot be found in Table 1, in which simple models involving one or two causal arrows were omitted. The only significant second-order partial correlation was $r_{\theta S' \cdot DD'}$. We cannot deny, however, some causal relation between θ and D' because we cannot explain the obtained significant $r_{\theta D'}$, $r_{S'D'}$, $r_{\theta D' \cdot D}$, and $r_{S'D' \cdot D}$ without it. Consequently a suitable model may be Model A-X shown in Figure 6-6; both S' and D' are independently determined by visual angle. Physical distance has nothing to do with the two perceptual properties.

It should be noted that in Mori and Watase's study many subjects showed causal relations between perceived size and perceived distance; four subjects (a, b, c, and e) in the monocular series and three (a, c, and e) in the binocular series.

PERCEIVED SHAPE AND PERCEIVED SLANT

Kaiser (1967) presented three trapezoids as the standard stimuli, one at a time in random order in a viewing box. The sizes and shapes of the trapezoids were selected so that when placed at 15, 45, and 65° away from the frontoparallel plane they projected identical retinal images. Each trapezoid at its respective slant subtended the following visual angles: top, 8°, height, 5°, and base, 10°. Six groups of five subjects were assigned to six conditions which consisted of combinations of three stimulus-slant and two viewing-order conditions (monocular-first and binocular-first). The subject was asked to reproduce perceived shape and slant with a shape response and slant response apparatus. Both were placed on the inside back wall of the viewing box. The shape response apparatus was a vertical display board that could present a varying trapezoidal shape whose height and width at the base and bottom were independently adjustable. The slant response apparatus was a half black and half white disk. The subject represented perceived slant by rotating the black-white division in the frontal plane to show slant in depth. Two measures were used to represent perceived shape: one was the ratio of the adjusted height to the adjusted base (h/b), the other, the ratio of the adjusted top to the adjusted base (t/b). Kaiser's results showed that the perceived shape and slant generally increased as physical slant increased in

Table 4
CAUSAL ANALYSIS OF KAISER'S (1967) RESULTS

	Monocular		Binocular	
	Monocular First	Binocular First	Monocular First	Binocular First
Simple correlations between				
Physical slant and perceived slant	-.013	.747**	.904**	.931**
Physical slant and h'/b'	.038	.533*	.880**	.980**
Perceived slant and h'/b'	.664**	.649**	.835**	.884**
Physical slant and t'/b'	.030	.288	.618*	.734**
Perceived slant and t'/b'	.616*	.653**	.627*	.746**
Partial correlations between				
Physical slant and perceived slant with h'/b' held constant	-.051	.623*	.647*	.703**
Physical slant and h'/b' with perceived slant held constant	.063	.094	.533*	.921**
Perceived slant and h'/b' with physical slant held constant	.666**	.447	.193	-.404
Physical slant and perceived slant with t'/b' held constant	-.405	.771**	.844**	.849**
Physical slant and t'/b' with perceived slant held constant	.049	-.398	.154	.161
Perceived slant with t'/b' with physical slant held constant	.617*	.688**	.201	.251

* Significant beyond the 0.05 level.
**Significant beyond the 0.01 level.

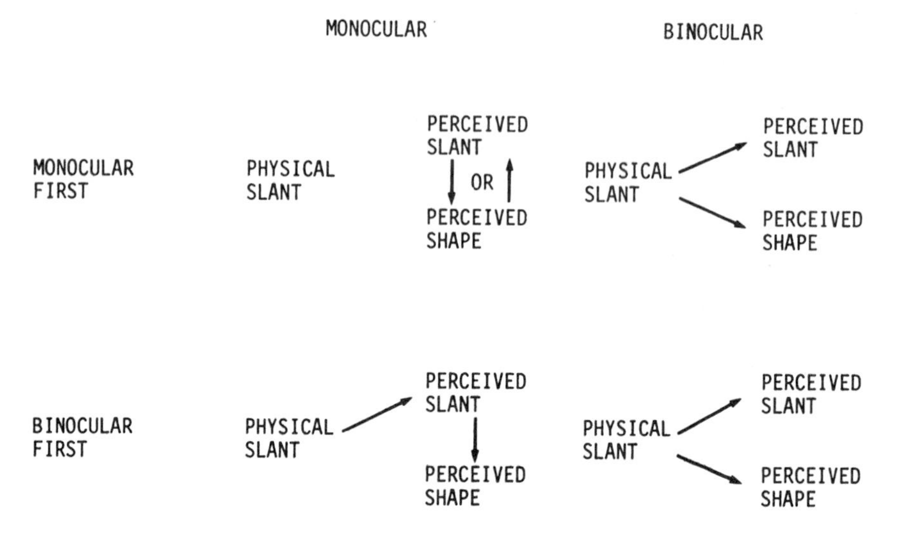

Figure 6-7 Causal relations suggested from an analysis of Kaiser's (1967) experimental results on perceived slant and perceived shape.

the binocular condition for both viewing-order groups and in the monocular condition for the binocular-first group, whereas they did not vary systematically with physical slant in the monocular condition for the monocular-first group.

Because of the nonlinear relations of the original measures to physical slant, the measures of perceived shape h/b and t/b were transformed into h'/b' and t'/b', respectively, according to Kaiser's (1966) Tables 6 and 7, before the causal analysis by partial correlations was applied to these results. Table 4 shows all the simple and partial correlations of physical slant, perceived slant, and the two transformed measures of perceived shape under the monocular and binocular conditions for the two viewing-order groups. Significant simple correlations between physical slant and perceived properties (slant and shape) reconfirmed the above mentioned trend which was found in all conditions except the monocular in the monocular-first group. The simple correlations between perceived slant and perceived shape were significant according to the shape-slant invariance hypothesis discussed earlier.*

Do these significant simple correlations indicate causal relations between perceived slant and perceived shape? The partial correlations answer this question. Simon's three-variable models, described in the preceding section, can be ap-

*Kaiser (1967) also performed correlational analyses of his data, but his were different from my analysis in that he used change scores (differences between the monocular and binocular conditions) and error scores (differences of the obtained measures from the physical values), whereas I used the absolute values of slant measures and transformed shape measures.

plied. In the binocular conditions for both viewing orders all partial correlations between perceived slant and perceived shape are small and insignificant. These facts indicate that the observed simple correlations between perceived slant and shape are spurious, not causal. They were observed because both perceived slant and perceived shape were independently affected by physical slant, as shown by significant partial correlations between physical slant and the perceived properties. Model 3 in Figure 6-2 fits these results well.

In the monocular condition partial correlations between perceived slant and perceived shape are all high and significant, except one between perceived slant and h'/b', which is slightly smaller than the significant level, which suggests that some causal relations existed between perceived slant and perceived shape. For the monocular-first condition there is no correlation (no causal relation) between physical slant and perceived slant or between physical slant and perceived shape and consequently the direction of causality cannot be decided. For the binocular-first group the partial correlations between physical slant and perceived slant with perceived shape held constant are significant, whereas the partial correlations between physical slant and perceived shape with perceived slant held constant are insignificant. It means that the effect of physical slant on perceived shape operated only indirectly through perceived slant. Model 1 in Figure 6-2 can be applied.

Figure 6-7 shows the suggested causal relations for these four conditions. In the binocular conditions the two perceptual properties (slant and shape) were independently determined by physical slant or, more directly, by binocular disparity induced by it. In the monocular condition for the monocular-first group both perceived slant and perceived shape were independent of physical slant but mutually related. Perhaps one of the perceptual properties was first determined by some inner information, which was independent of real slant, and this property determined the other. In the monocular condition for the binocular-first group perceived slant was determined by physical slant (probably through the memory of binocular observation of the same object), and the thus determined perceived slant determined perceived shape. It is interesting that causal relations between the two perceptual properties were found only in the monocular condition in which insufficient stimulus cues were available for the subjects to judge slant and shape.

PERCEIVED SURFACE LIGHTNESS AND PERCEIVED ILLUMINATION

Two investigations of perceived surface lightness and perceived illumination were analyzed by the Simon-Blalock procedure of causal inference. One was conducted by Oyama (1968) who studied perceived surface lightness and perceived illumination by the matching method. Another was performed by Kozaki and Noguchi (1976) who used the rating method for the same purpose.

Oyama's Study

Oyama (1968) studied perceived illumination and perceived surface lightness in a display box. Three boxes of the same size arranged side by side were used. The middle one was the *standard box* in which the standard stimulus (a gray disk) was presented. As is Hsia's (1943) experiment on lightness constancy, the back of the box was open. A black velvet curtain hung in the dark behind the box constituted the background of the standard disk, which was suspended by a thin black string in the back plane of the box. This set up kept the background very dark despite the illumination in the standard box. The inside of the box which was covered with a gray paper of 40% reflectance constituted the indirect surround of the standard disk and varied in luminance as the level of illumination changed. The right-hand box, covered inside with black paper, was used as *comparison box I*. Against the center of the black back wall of this box and on the same level as the front viewing aperture was a Maxwell top whose black-white ratio was adjustable during rotation. This top operated as the comparison disk. *Comparison box II*, the left-hand box, was used for to match illumination. Its inside was white and a white square on a black circle was placed on its back wall as a "cue" for judging illumination.

Combinations of the five standard disks (6 to 88% in reflectance) and five levels of illumination (10 to 160 lumens/m^2 on the standard disk) constituted 25 conditions. Five subjects were tested under all five conditions presented in random order. First the subject was asked to inspect the standard box and comparison box I alternately and to adjust the black-white ratio of the Maxwell top until it matched the standard disk in apparent lightness. After lightness matches were completed for each of the 25 conditions the subject was asked to compare the standard box and comparison box II and to make illumination matches by adjusting the illumination on comparison box II.

Log-log linear relations were found between standard reflectance (R) and matched reflectance (R'), between standard illuminance (E) and matched illuminance (E'), and between E and R'. The regression coefficients of log R' to log E were .31 to .44. This coefficient should be .00 for complete lightness constancy and 1.00 for complete lack of constancy. Consequently the obtained values indicated high (but not complete) constancies. Matched illuminance increased almost proportionally to E. These facts indicate that no invariant relation was found between perceived surface lightness and perceived illumination because the proportional increase of E' with E should be accompanied by complete lightness constancy, according to the invariance hypothesis. Beck (1959, 1961) had already found that the accuracy of illumination judgment and the completeness of lightness constancy were not necessarily related.

Two kinds of causal analysis were conducted on the results of this experiment. One was made among four variables, log R, log E, log R', and log E'. In another the logarithmic luminance (log I) of the standard disk replaced log E. Luminance is usually considered as a basic and more proximal stimulus variable than illuminance because retinal illuminance is directly proportional to luminance when pupil size is kept constant, whereas it is only indirectly related to illuminance when reflectance of surface is varied. However, illuminance had a direct retinal correlate in this experimental situation. This was the retinal illuminance of indirect surround because the inside reflectance of the standard box was always constant and the luminance of indirect surround was proportional to the standard illuminance. Consequently the luminance of the standard disk and the illuminance of the standard box had their retinal correlates. On this point there is no difference between the two analyses. The causal models, however, inferred from the first analysis, were a little simpler than those from the second analysis. For this reason mainly the results of the first analysis, in which illuminance was used as a stimulus variable, are emphasized.

Table 5 lists all insignificant correlations, except those between the two stimulus variables, found among log R, log E, log R', and log E' in the first analysis. The simple correlation r_{RE} was zero according to the experimental design. The second-order partial correlation between the two perceptual variables, $r_{R'E' \cdot RE}$, was insignificant for every subject, which indicates that there was no causal relation between perceived lightness and perceived illumination in this experiment. In all subjects $r_{RE' \cdot ER'}$ was insignificant, but causal relations, with the exception of Subject 4, between standard reflectance and perceived illumination cannot be denied because $r_{RE' \cdot E}$ was significant (.415~.885) for the other four subjects. A comparison between Tables 2 and 5 indicates that Model A-5

Table 5

INSIGNIFICANT SIMPLE AND PARTIAL CORRELATIONS FOUND IN OYAMA'S (1968) STUDY

Subjects	1	2	3	4	5
r_{RE}					
$r_{RE'}$	.150		.143	.023	.123
$r_{RE' \cdot E}$				.144	
$r_{ER' \cdot E'}$	−.381		−.311	.089	−.197
$r_{R'E' \cdot E}$				.034	.378
$r_{RE' \cdot ER'}$	.065	.208	.168	.216	.227
$r_{R'E' \cdot RE}$	.049	.258	−.015	−.188	−.132
Fitted models	A-5	A-5	A-5	A-15	A-5

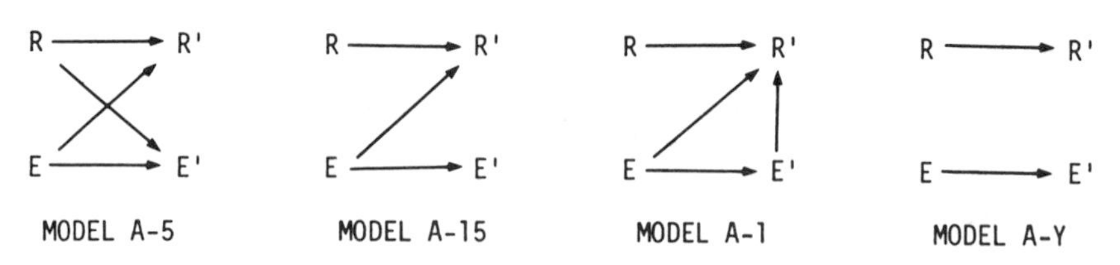

Figure 6-8 Causal models suggested by an analysis of the experimental results of studies by Oyama (1968) and Kozaki and Noguchi (1975) on perceived surface lightness and perceived illumination; *R, E, R'*, and *E'* represent reflectance, illuminance, perceived lightness, and perceived illumination, respectively.

fits the results of Subjects 1, 2, 3, and 5 and Model A-15, the results of Subject 4.

As shown by the left most diagram in Figure 6-8, the perceived lightness of the standard disk and the perceived illumination of the standard box in Subjects 1, 2, 3, and 5 were considered to be determined by the two stimulus variables, the reflectance of the standard disk and the illuminance on it. There was no causal relation between the two perceptual variables. In Subject 4 perceived lightness was determined by the two stimulus variables, but perceived illumination was determined only by illuminance, as shown by the second diagram in Figure 6-8.

Of course, the subject cannot experience reflectance and illuminance directly. More proximal stimuli mediate their effect. The retinal illuminance ratio between the image of the standard disk and the indirect surround varies proportionally with the reflectance of the standard disk and the retinal illuminance of the indirect surround is proportional to the illuminance of the standard box. In this sense the causal models we have discussed indicate that for most subjects perceived lightness and illumination were determined by the retinal illuminance ratio of the focal area to the peripheral field and the absolute retinal illuminance at the peripheral field. For one subject (Subject 4) perceived illumination was determined only by peripheral retinal illuminance.

The results of the second analysis, in which luminance instead of illuminance was used as one of the stimulus variables, showed a similar tendency. The second-order partial correlations between the two perceptual variables, $r_{R'E' \cdot RI}$, again were insignificant because they had exactly the same values as $r_{R'E' \cdot RE}$. There was no causal relation between the two perceptual variables. A smaller number of insignificant correlations was found than in the first analysis. In this case the results of Subject 4 were the same as those of the other subjects. Both perceived lightness and perceived illumination were considered to be determined by the reflectance and luminance of the standard disk or, in retinal terms, by the retinal illuminance ratio of the focal area to the peripheral field and the absolute retinal illuminance of the focal area, which was the retinal correlate of the luminance. A causal network similar to Model A-5 shown in Figure 6-8 can be applied to these results; however, it differs from Model A-5 in that two stimulus variables R and I are correlated ($r = 0.696$). It corresponds to Blalock's Model F in which a causal arrow intervenes between R and I.

Kozaki and Noguchi's Study

Kozaki and Noguchi (1976) presented various small patches of Munsell neutral papers (N 2.0 to N 9.0), one at a time, on three different backgrounds, black (N 3.0), gray (N 6.0), and white (N 9.0), under various illuminations (49~3690 lumen/m^2). For each background 29 to 34 combinations of patches and illuminations were used. In half of the sessions the subject was asked to rate the apparent lightness of patches in nine categories; in the other half he was required to rate the apparent overall illuminations in nine categories. Five subjects made both judgments under all conditions of patch, background, and illumination.

The average results showed that lightness and illumination judgment increased almost linearly with logarithmic reflectance and logarithmic illuminance. The regression coefficients of R' to log E ranged from .01 to 1.30 and decreased as the reflectance of background increased. Because smaller regression coefficients indicate greater tendencies to lightness constancy, a lighter background resulted in a greater tendency to lightness constancy. The regression coefficients of E' to log E ranged from .65 to .90. It was expected from the lightness-illumination invariance hypothesis that greater regression coefficients (greater sensitivities to the change of illumination) would be accompanied by greater tendencies to lightness constancy, and, consequently, greater regression coefficients would be found in lighter backgrounds in which tendencies to lightness constancy were greater. The results were not consistent with this expectation, however. No systematic difference in regression coefficients was observed on the three backgrounds.

An interesting relation was discovered in this experiment. When sums of R' and E' were plotted against sums of log R and log E for each background, a linear relation was found between these two kinds of sum. High correlations were obtained between them (.886 for the white background, .910 for the gray, and .967 for the black). This means that R' and E' were reciprocally related when the sum of log R and log E (or logarithmic luminance) was kept constant. In this sense the lightness-illumination invariance hypothesis held in this study.

The procedure of causal inference was applied to the relations among log R, log E, R', and E'. Table 6 shows all insignificant correlations of these four variables for individual subjects in the three backgrounds. In this analysis only the data obtained from the 24 combinations of patches and illuminations that were common to all three backgrounds were used. In this experiment the correlation between the two stimulus variables was not zero but insignificantly negative

Table
6

INSIGNIFICANT SIMPLE AND PARTIAL CORRELATIONS
FOUND IN KOZAKI AND NOGUCHI'S (1976) STUDY

Backgrounds	White					Gray					Black				
Subjects	AK	KM	MM	KN	KT	AK	KM	MM	KN	KT	AK	KM	MM	KN	KT
$r_{RE'}$	−.35	−.30	−.27	−.23	−.34	−.17	−.22	−.22	−.20	−.22	−.04	−.21	.02	−.21	−.25
$r_{ER'}$	−.16	−.20	−.16	−.15	−.23	−.02	−.16	−.09	−.11	−.03	−.24	−.01	.12	−.02	.03
$r_{E'R'}$	−.21	−.20	−.15	−.10	−.28	.08	−.09	−.02	−.02	.03		.11		.06	.07
$r_{RE' \cdot E}$	−.29	−.02	.10	.25	−.26		.39	.29	.34	.29		.34		.31	.27
$r_{RE' \cdot R'}$					−.31										
$r_{ER' \cdot R}$		.36			.23										
$r_{ER' \cdot E'}$	.23	−.00	−.05	−.23	.20			−.33	−.35	−.29				−.36	−.19
$r_{R'E' \cdot R}$		.36			.24										
$r_{R'E' \cdot E}$	−.27	−.02	.02	.20	−.25	−.27	.39	.32	.34	.29				.36	.20
$r_{RE' \cdot ER'}$	−.12	−.01		.22	−.11	.30	.04	−.11	.05	.02	.20		.19	−.16	.38
$r_{ER' \cdot RE'}$	.05	.04		.28	−.00	.24	.02	.02	.33	.22	.33		.25	.00	
$r_{R'E' \cdot ER}$	.04	.00		−.16	.05	−.12	.06	.18	−.01	.04	.09		.12	.25	−.33
Fitted models	A-15	A-Y	A-1	A-15	A-Y	A-5	A-5	A-15	A-15	A-15	A-5	A-1	A-5	A-15	A-15

(-.299). It can be regarded as neglegible.

As shown in Table 6, there are considerable individual and situational differences. For Subjects KM and KT on the white background almost all correlations except those between log R and R' and between log E and E' were insignificant. Simple causal relations shown as Model A-Y in Figure 6-8 can be applied to these results. This model is too simple to be found in Blalock's chart. Perceived lightness was determined only by reflectance (or the retinal illuminance ratio of the focal area to its surround) and perceived illumination was determined only by illuminance (or peripheral retinal illuminance). For the results of Subjects AK and KN on the white background Subjects MM, KN, and KT on the gray background, and Subjects KN and KT on the black background Model A-15 was most suitable. Their pattern of insignificant correlations was similar to that of Model A-Y, but significant $r_{ER' \cdot R}$ rejected Model A-Y. As in Subject 4 of Oyama's study, perceived lightness was determined by the two stimulus variables, whereas perceived illumination was determined by illuminance (or peripheral retinal illuminance) alone. For the results of Subject AK and KM on the gray background and Subjects AK and MM on the black background Model A-5 was most suitable, as in most subjects of Oyama's study. Perceived lightness and perceived illumination were independently determined by the two stimulus variables.

In all the results discussed so far no causal relation was found between the two perceptual variables, but in the results of Subject MM on the white background, and Subject KM on the black second-order partial correlations between the two perceptual variables were significant. The most suitable was Model A-1 in which perceived lightness was determined by three variables, the two stimulus variables plus perceived illumination; perceived illumination was determined only by illuminance (or peripheral retinal illuminance). In this experiment lightness judgments and illumination judgments were made in different sessions. Consequently any causal relation between these two judgments may sound curious. It can, however, be described as follows: perceived illumination was implicitly determined even when it was not required to report it explicitly; the implicitly perceived illumination in the lightness judgment session was approximately the same as the corresponding explicitly reported in the illumination judgment sessions; this implicit judgment determined explicit reports of perceived lightness.

It is to be noted that in as many as 13 of the 15 cases (five subjects times three conditions) in this study perceived illumination was determined by illuminance (or peripheral retinal illuminance) alone. In other words, stimulus patches had little influence on perception of illumination. This may point to an important difference between lightness constancy in such a situation and size and shape constancies in usual situations. Illumination can be judged only on a background without stimulus patches, whereas distance or slant cannot be judged without stimulus objects. The experimental situation in Oyama's (1968) study on lightness constancy, in which direct background (or surround) was kept constant in luminance regardless of illuminance on stimulus patches, may be more like the usual experimental situations of size and shape constancies.

CONCLUDING REMARKS

Perceptual constancies are frequently explained by the "taking-into-account" hypothesis; for example, lightness constancy is explained by the observer's hypothetical ability to take illumination into account. This interpretation presupposes a causal relation between illumination perception and lightness perception. Similarly, size constancy is often attributed to the observer's ability to take the distance of the object into account when judging its size, presupposing a causal relation between distance perception and size perception. The size-distance invariance, shape-slant invariance, and the lightness-illumination invariance hypotheses can be understood as expressions of these "taking-into-account" processes and the causal relations between the two perceptual properties. This is not the only explanation of perceptual constancies, however. The mutually correlated perceptual properties (e.g., size and distance) may be independently determined by some common stimulus variables. "Depth" cues determine not only perceived distance but perceived size. This hypothesis can explain the correlations between the perceptual properties without causal relations between them.

As a test of the validity of these two possible explanations of a given experimental result, the Simon-Blalock method of causal inference by partial correlations is effectively applicable. In this chapter this method was applied to the experimental data of five studies; two on size constancy, one on shape constancy, and two on lightness constancy. In all of them linear relations can be assumed between stimulus and perceptual variables or their logarithmic transformations, an important requirement for the application of this method. In four of the five studies some invariant relations were found between perceptual properties. The results of correlational analyses showed a number of individual differences, a fact that suggests that it is preferable to apply this procedure to individual data rather than group means.

Analytical results of individual data indicated that the second explanation is most often valid; two perceptual properties (e.g., perceived size and perceived distance) are independently determined by stimulus variables and no direct causal relation exists between them. In some cases, however, direct causal relations were apparent. In one of the 12 in Oyama's study on size constancy and in six of the 10 in Mori and Watase's study perceived distance determined perceived size. In one case in Mori and Watase's study the relation was reversed; perceived *size* determined perceived *distance*. In the monocular condition in Kaiser's study on shape constancy causal relations were found between perceived slant and perceived shape, whereas these perceptual properties were independently determined by stimulus variables in his binocular condition. In two of the 15 cases in Kozaki and Noguchi's study perceived illumination determined perceived lightness, whereas both were independently determined by stimulus variables in the other cases in their study as well as in all cases in Oyama's study on lightness constancy. As the results of correlational analysis in Kaiser's experiment suggested, the monocular observation, or curtailment of cue for depth, may favor the occurrence of a causal relation between the two perceptual properties. When outer information is reduced, inner information including judgment on some perceptual properties will be used for the determination of another perceptual property.

The causal relations found between perceptual properties in some experimental results analyzed in this chapter suggest a multistage model of perception (Attneave, 1962). When perceived distance determines perceived size, these two perceptual properties are produced in different stages, perception of distance occurring in an earlier stage than perception of distance. It is quite natural that a perceptual process produced in an earlier stage will influence a perceptual process occurring in a later stage. Our perception of the visual world may be a net result of many stages of perceptual processing, and we cannot ourselves identify which part of our perception is generated in which stage. The causal inference by means of partial correlation will be a useful means of identifying the processing stage for each perceptual property, but this processing stage may not be fixed; for example, perception of distance occurs in a stage earlier than the stage for perception of size in one situation but it may occur in a later stage in another situation, as suggested by the results of causal inference presented in this chapter.

REFERENCES

Attneave, F. Perception and related areas. In S. Koch (Ed.), *Psychology: a study of a science,* (Vol. 4). New York: McGraw-Hill, 1962, 619-659.

Beck, J. *Surface color perception.* Ithaca, New York: Cornell University Press, 1972.

Beck, J. Judgments of surface illumination and lightness. *Journal of Experimental Psychology*, 1961, **61**, 368-375.

Beck, J. Stimulus correlates for the judged illumination of a surface. *Journal of Experimental Psychology*, 1959, **58**, 267-274.

Beck, J. & Gibson, J. J. The relation of apparent shape to apparent slant in the perception of objects. *Journal of Experimental Psychology*, 1955, **50**, 125-133.

Blalock, H. M. Four-variable causal models and partial correlations *American Journal of Sociology*, 1962, **68**, 182-194.

Epstein, W. The process of 'taking-into-account' in visual perception. *Perception*, 1973, **2**, 267-285.

Epstein, W. *Varieties of perceptual learning*, New York: McGraw-Hill, 1967.

Epstein, W. & Landauer, A. A. Size and distance judgments under reduced conditions of viewing. *Perception & Psychophysics*, 1969, **6**, 269-272.

Epstein, W. & Park, J. Shape constancy: functional relationships and theoretical formulations. *Psychological Bulletin*, 1963, **60**, 265-288.

Epstein, W, Park, J., & Casey, A. The current status of the size-distance hypotheses. *Psychological Bulletin*, 1961, **58**, 491-514.

Gogel, W. C. The sensing of retinal size. *Vision Research*, 1969, **9**, 1079-1094.

Helmholtz, H. von. *Treatise on physiological optics* Vol. 3 (J. P. C. Southhall, Ed. and Trans.). Menasha, Wisconsin: Optical Society of America, 1925.

Hsia, Y. Whiteness constancy as a function of difference in illumination. *Archives of Psychology*, 1943, No. 284.

Ittelson, W. H. Size as a cue to distance: static localization. *American Journal of Psychology*, 1951, **64**, 54-67.

Kaiser, P. K. Perceived shape and its dependency on perceived slant. *Journal of Experimental Psychology*, 1967, **75**, 345-353.

Kaiser, P. K. An investigation of shape at a slant. Doctorial dissertation, University of California, Los Angeles, 1966.

Kilpatrick, F. P. & Ittelson, W. H. The size-distance invariance hypothesis, *Psychological Review*, 1953, **60**, 223-231.

Koffka, K. *Principles of Gestalt psychology*, New York: Harcourt Brace, 1935.

Kozaki, A. & Noguchi, K. The relationship between perceived surface-lightness and perceived illumination: a manifestation of perceptual scission. *Psychological Research*, 1976, **18**, in press.

Landauer, A. A. & Epstein, W. Does retinal size have a unique correlate in perceived size? *Perception & Psychophysics*, 1969, **6**, 273-275.

Mori, T. & Watase, T. Personal communication, 1975.

Namboodiri, N. K, Carter, L. E. & Blalock, H. M., Jr. *Applied multivariate analysis and experimental designs*. New York: McGraw-Hill, 1975.

Oyama, T. Perceived size and perceived distance in stereoscopic vision and an analysis of their causal relations. *Perception & Psychophysics*, 1974a, **16**, 175-181.

Oyama, T. Inference of causal relations in perception of space and motion. *Psychologia*, 1974b, **17**, 166-178.

Oyama, T. Stimulus determinants of brightness constancy and the perception of illumination. *Japanese Psychological Research*, 1968, **10**, 146-155.

Simon, H. A. Spurious correlation: A causal interpretation. *Journal of the American Statistical Association*, 1954, **49**, 467-479. (Chapter 3 in H. A. Simon, *Models of man*. New York: Wiley, 1957.)

Wallach, H. Brightness constancy and the nature of achromatic colors. *Journal of Experimental Psychology*, 1948, **38**, 310-324.

Woodworth, R. S. *Experimental Psychology*. New York: Holt, 1938.

CHAPTER 7

INSTRUCTIONS AND PERCEPTUAL CONSTANCY JUDGMENTS

V. R. CARLSON
National Institute of Mental Health

Leibowitz and Harvey (1969) have observed that the most effective experimental variable in the size-constancy experiment is the instruction given to the subject. This is a remarkable phenomenon, considering that this variable is verbal, auditory, and conceptual and not a visual variable at all. It is even more remarkable in that instructions deliberately devised to have no effect nevertheless have systematic and reliable effects. I know of no visual stimulus variable for which the same can be said.

It is not only that instructions affect the experimental results obtained. Instructions interact with the conditions of the experiment so that unless the instructional effect is understood it is difficult to know what effects are determined by the visual parameters and what effects are due to attitudinal influences. In order to specify a satisfactory theory of the perceptual process, one must understand the attitudinal process, but in order to understand the attitudinal process one must be able to specify something about the perceptual process. The only way out of this dilemma is to make some working assumptions about one and investigate hypotheses about the other.

Historically, theorists have made assumptions about the attitudinal process and have tested propositions about perception. In spite of almost a hundred years of research, this approach has not resulted in a satisfactory theory of perceptual constancy. I believe it is time to reconsider the assumptions that have been made about the observer's attitude and to do so with greater explicitness than has been done in the past. I shall concentrate primarily on size constancy rather than other constancies because what little systematic work has been done with instructions has been done in connection with judging size at varying distances.

THE PERSPECTIVE ATTITUDE

The size-constancy experiment consists essentially of asking the subject to vary the size of a test-object presented at a near distance so that it is the same size as a test-object presented at a greater distance. It makes a difference in the result just how the experimenter defines size to the subject, and the question is why that should be so. Generalized effects of set, motivation, and attention are relevant considerations (Haber, 1966; Goldner et al., 1971; Weber & Cook, 1972). In order for instructions to be as effective as they are, however, they must communicate with some attitude of the observer in a way that has a particular implication for how he should respond in the experimental situation.

A specific attitude that can provide a powerful vehicle for the effect of instructions is the *belief* that the apparent size of an object becomes smaller at a greater apparent distance—an object is said to "look" smaller if it "looks" farther away. Not only does this belief seem to be widely held but many observers insist that it is an undeniable fact of direct sensory experience. A doubter can usually be persuaded by asking him to hold his hand in the line of sight to a chair, say, across the room. He will agree that the image of the chair appears smaller than the image of his hand. Intuitive certainty notwithstanding, there is good reason for doubting that the verbal report that apparent size diminishes with increasing distance necessarily means that perceived object size does also.

A pervasive reason for supposing that a decrease in apparent size with increasing distance is perceptual rather than cognitive is the proposition that apparent size decreases because retinal image size decreases with increasing object distance. Small and large objects occur more or less randomly at near and far distances, yet object sizes are generally perceived correctly. If perceived object size varied with retinal image size, the veridicality of size perception would be zero. Logically there is little more reason to believe that perceived size should ordinarily vary with retinal image size than there is to believe that the world should appear upside down because the retinal image is inverted. We must search more carefully for the reason why observers say an object appears smaller when it appears farther away.

Looking out the window at the window of a house across the street, we may say that the frame of the near window appears much larger than the far window. It is not necessarily the near frame itself that appears to be larger. It may be the portion of the side of the house outlined by the projection of the lines of sight past the frame. That portion of the side of the house can be said to appear larger than the window it contains, or equivalently, the window appears small in relation to that portion of the side of the house. The appearance of smaller size in that sense does not in the least imply that the two windows are not perceived to be equal in size.

Sighting through a window, though a common experience, is not general. The apparent size of any nearby object is said to be greater than the apparent size of a more distant object of equal size, whether seen through a window or not. An automobile next to the observer is said to appear larger than an automobile parked across the street, even though there is no doubt that the two are identical in size. A frame is not present but something else is that can provide an analogous reference and that is the amount of combined head and eye movement needed to look from one end of the car to the other. The amount of lateral movement required to encompass the near car is enough to encompass several cars across the street. The perceived physical extent of one car is small in relation to the perceived physical extent of several cars, but that does not mean that the perceived size of one near car is not the same as the perceived size of one more-distant car.

The generalization to be made is that the magnitude m of the lateral physical extent that can be perceived within a given degree of head and eye movement increases proportionally with increasing distance from the observer; the proportion s/m of this magnitude associated with an object of fixed size decreases, not the perceived size s of the object itself. Gogel and Mertens (1968) have argued persuasively that perceived distance is proportional to perceived size per unit of visual angle s/θ, where θ is the angular size of the object. If actual and perceived distance are constant and actual and perceived size decrease (e.g., letting the air out of a balloon without letting it go), s/θ is constant (both s and θ decrease) and there is no ambiguity about reporting a decrease in apparent size in conjunction with a decrease in s/m (s decreases and m remains constant).

With an increase in the actual and perceived distance of an object an increase in lateral extent m is proportional to the decrease in angular size θ of the object; that is s/m varies inversely with s/θ. If the observer perceives object size to remain constant, he can report an increase in apparent distance, which is compatible with an increase in s/θ (θ decreases). He can also report a decrease in apparent size, which is compatible with a decrease in s/m (m increases). An "increase in apparent distance" and a "decrease in apparent size" are equivalent conceptual alternatives for conveying an increase in perceived distance with no change in perceived object size. Ordinarily behavior in the natural environment is a function of perceived sizes and perceived distances, and it does not matter that the observer associates apparent size cognitively with a change in s rather than a change in s/m. As long as the subjective change in apparent size is inversely correlated with the subjective change in apparent distance, behavior remains appropriate. If a person is said to look smaller and smaller as he walks away, there is no cause for alarm as long as the person is concomitantly seen to be getting farther and farther away. The observer may not be reporting the perceptual process correctly, but that has no consequence for his behavior with respect to the person walking away.

After Gibson (1952), I have called the belief that apparent size diminishes with apparent distance the *perspective attitude* (Carlson, 1960; Carlson & Tassone, 1962) because the observer seems to be interpreting the three-dimensional scene as though it were a two-dimensional perspective drawing. In perspective representation a nearer object is drawn larger than a farther object to portray equal object-size at different distances. How much larger in order to represent how much of a difference in distance has constraints but is not definitely given a priori. It is what the artist deems to be appropriate for the picture he is creating. I do not mean that the observer perceives the three-dimensional scene in two dimensions. The two-dimensional interpretation is a thought process not a perceptual one. I do not even propose that the subject necessarily deduces what he should do in the size-constancy experiment by explicit analogy to perspective drawing, although some subjects might. Rather the perspective attitude refers to the general expectation that if an object appears to be farther away it will appear smaller. That expectation is undoubtedly reinforced by experience with perspective representations, but it could arise without such experience. Because the decrease in apparent object size is a cognitive misnomer, how much smaller for how much farther away is completely subjective. It is what seems appropriate to the subject within the perceptual constraints of the particular stimulus situation, the particular instructions given by the experimenter, and the way the subject understands those instructions.

A situation that has been used to illustrate and debate the perspective-attitude hypothesis is the experience of looking down a pair of railroad tracks (Boring, 1952; Gibson, 1952). According to the hypothesis, the tracks are perceived to be parallel and never meet but are cognitively interpreted to converge. The situation only approximates an ideal demonstration but does capture the essence of the principle. As stated by Gibson in connection with an object on the ground (here taken to represent the separation between the tracks); "(A)n object can apparently be seen with approximately its true size *as long as it can be seen at all*"

(Gibson, 1950, p. 186). The eye is not a perfect optical instrument, however, and the tracks themselves disappear in a blur. Their angular sizes become too small for the eye to resolve and because of the blur actually become somewhat larger optically (Gubisch, 1966). A more precise statement of the principle would be this: if the tracks were on a flat roadbed and extended in straight lines to the horizon and the eye were optically good enough, the tracks would be perceived to be parallel and never meet but would be cognitively interpreted to converge.

It has often been assumed that the observer's report of apparent size is equivalent to a report of perceived size. According to the present viewpoint, perceived size is an inference on the part of the experimenter or the theorist. When we make a statement about perceived size, we hope that the statement will ultimately prove to be correct in terms of a satisfactory general theory of perception. It is for experiment and theory to determine under what circumstances a report of apparent size can be taken as a report of perceived size.

RESPONSE BIAS IN THE SIZE-CONSTANCY EXPERIMENT

Now, in the size-constancy experiment suppose that the experimenter instructs the subject to match the apparent size of a near object to the apparent size of a far object. The experimenter explains that the subject is to set the size of the near object so that it "looks" equal in size to the far object and he is not to be concerned about achieving equality of size in any other sense. According to the perspective attitude, if the two objects are equal in physical size, the nearer object will look larger; hence the perceived physical size of the near object must be set to some value smaller than the perceived physical size of the far object—just barely smaller, clearly smaller, or obviously smaller. We therefore cannot conclude from a smaller setting that perceived size decreases with distance. Neither can we conclude that a measure of the subject's attitude has been obtained because perceived size might, in fact, decrease with distance.

The situation is not much better if the experimenter instructs the subject to make the match in terms of actual physical ("objective") size. In this case there is no ambiguity about the result to be achieved, but operation of the perspective attitude is by no means precluded. When the two objects are equal in physical size, the near one should appear larger—just barely larger, clearly larger, or obviously larger. We therefore cannot conclude from a larger setting that perceived size increases with distance, nor, again, that the perspective attitude has been measured, because perceived size might actually increase with distance.

A comparison of the two settings, one for apparent size and one for objective size, provides some useful information. The hypothesis that the perspective attitude is affecting the subject's responses postulates that the setting representing equality of perceived size lies above the setting for apparent size and below the setting for objective size. This situation is not so bad; perceived size cannot be measured directly by any experimental operation and must always be inferred from the result of an experimental operation in terms of a theoretical interpretation of what the operation means. The specification of equal perceived size as lying between the two settings, however, does not allow for the possibility that one or the other might represent equal perceived size.

The instruction for objective size should logically include a statement to the effect that the two objects may or may not be equal in apparent size—equality or inequality in that respect is acceptable. The subject may still think that the near object must appear to be larger but not because the instruction itself implies that it should. In terms of the perspective attitude, the lower limit for the objective-size setting is shifted from the near object having to appear "at least barely larger" to "at least not smaller" than the far object. If it appeared smaller, the setting would contradict the perspective attitude. This modified objective-size instruction thus allows us to say that the setting for equal perceived size lies *at or below* the setting for objective size. Including a statement that the physical sizes may but need not be objectively equal in the apparent-size instruction similarly shifts the logically implied limit for that instruction from the near object having to be "at least barely smaller" to "at least not larger" in physical size and provides justification for saying that the setting for equal perceived size lies *at or above* the setting for apparent size. I refer to these modified instructions as the *neutral-objective-* and *neutral-apparent-size* instructions.

Although each of these neutral instructions allows for a setting of equal perceived size, they may not produce identical settings. Objective size provides a conceptually definite criterion that may be different from perceived size. Apparent size provides no tangible criterion. By eliminating accuracy in terms of objective size and not specifying any other tangible criterion the simplest criterion for the subject to adopt is equality in terms of perceived size. The neutral-apparent size instruction may therefore be more "neutral" than the neutral-objective-size instruction.

In order to indicate how even a minimal attitudinal bias can affect the subject's response as the far test-object is moved to more and more distant locations, I make the following analogy to perceptual noise. The perception of the size of an object depends primarily on the relationships between its angular size and the angular sizes of whatever other objects are in its immediate vicinity. The accuracy with which these relationships are perceived is not perfect and becomes less so at greater distances—that is, the perceptual noise distribution for the perceived size of the object in relation to its immediate surround becomes wider at more distant locations. The noise distribution can be thought of as fluctuations in perceived size due to random causes. The less the noise distributions overlap for the near and far objects, the more certain the subject is that the two objects are different in size. Coincidence of the means of the noise distributions corresponding to the near and far objects represents equal perceived size. This coincidence also represents the point of maximum subjective *uncertainty* that the near and far objects are *different* in apparent size or in objective size.

Consider that the subject has set the near object to the theoretical point of perceived equality with the far object. To achieve a match of objective equality he assumes that the near object must appear larger than the far object. To reduce his uncertainty that it appears larger he increases the size of the near object. As he increases it, he becomes more certain that it appears larger, but at the same time the mean of the near-object noise distribution moves away from the mean of the noise distribution for the far object and the subject becomes more uncertain that the two objects are equal in objective size. At some point the subject reconciles the two uncertainties sufficiently to be satisfied with the setting or at least not too dissatisfied. Just where that point will be depends on the width of the noise distributions (perceptual constraint) and the degree of subjective certainty that the near object should appear larger but the setting will be above the setting for equal perceived size. At a more distant location of the far object its noise distribution will be wider. Consequently for the same degree of attitudinal bias the setting will be more discrepant in the same direction. A similar argument leads to the expectation of settings that are increasingly farther below the point of perceived equality with increasing distance when the subject is attempting to match apparent size.

The foregoing considerations provide a basis for response by distance functions that increase or decrease minimally with increasing distance in relation to the perceived size by distance function. Also, the increase or decrease does not require that the same subject experience the different distances of the far object. Different subjects with the same degree of attitudinal bias (on the average) and the same widths of noise distributions at the various distances would produce the same result.

Other instuctions are more implicative of the perspective attitude. So-called "look" or "phenomenal" instructions need not but often have gone a step farther by including an attempted definition or illustrative instance of "apparent size." I know of no way to define this concept that would not be expected to invoke the perspective attitude. The more we try to define it, the more we imply that it is different from objective size, and the way it is different is that objectve size does not vary with distance but apparent size does. It may seem innocuous to point out to the subject that "the stars appear to be very small but we know they are very large." Why do they appear small? Because they are far away, of course.

By making a statement of that kind the experimenter introduces the notion, if the subject has not already thought of it, that the difference in the apparent sizes of the two test objects must bear some relation to the difference in their distances. The simplest possibility is a rank-order correspondence. At the first presented distance of the far object there is no basis for the subject to assume any particular degree of certainty that the apparent size of the near object must be smaller for an apparent-size match and larger for an objective-size match. Whatever that degree of certainty is, it is increased if the far object is moved to a more distant location and decreased if the far object is moved to a closer location. The result will be a rank-order relationship between the size settings and distance if the same subjects make judgments at the different distances of the far object.

If the subject understands the instructions to mean that there should be a proportional relationship between apparent size and distance, his responses may be expected to be a more precise function of the near/far distance-ratio for apparent size and of the far/near distance-ratio for objective size. One often used instruction is difficult to explain without implying that there is a proportional relationship. This instruction specifies apparent size as *projective size*. Geometrically it is the size at the distance of the near (or far) object subtending the visual angle of the far (or near) object—the "retinal image" size of the one projected to the distance of the other. It may be specified as the size that would appear on a photograph taken from the location of the eye. If the subject adopts this definition of apparent size, his settings must be proportional to the perceived near/far distance ratio for the apparent size instruction and proportional to the perceived far/near distance ratio for objective size. Otherwise he has not understood the instruction or is not conforming to what it logically requires.

Projective size, however, usually represents such an extreme criterion for equality of apparent size that the test objects are obviously different in perceived size. Hence the subject is not likely to use it as his criterion of apparent size when making an objective-size match. To determine whether an instruction would lead to a setting of the near object that is larger than that obtained with objective-size instructions a *perspective-size* instruction was devised (Carlson, 1962). The subject was told to adjust a near triangle so that imaginary lines extending from its corners through the corresponding corners of a far triangle would duplicate the apparent convergence of railroad tracks or the edges of a sidewalk or road. Logically this instruction specifies an equal match in terms of objective size but explicitly invites the perspective attitude. The subjects did what they might have been expected to do had they been asked to draw the triangles on a piece of paper. They set the near triangle to a size clearly larger than they did when asked to set it equal in objective size to the far triangle. Because the far object was overestimated with objective-size instructions, the perspective-size match was a gross overestimation. The overestimation was as much as the underestimation with projective-size instructions.

The perceived size by perceived distance function is assumed to be whatever it is determined to be by the particular stimulus situation. Instructions have no effect on the determination. The effect of instructions is hypothesized to be on the response bias associated with the perspective attitude and this effect is superimposed on the perceptual function. If perceived size does not change with distance, objective-size instructions will result in increasing size responses with distance ("overconstancy") and nonneutral apparent-size instructions will result in decreasing size responses with distance ("underconstancy"). Overconstancy and underconstancy represent the same bias operating in opposite directions in relation to perceptual size constancy.

PERCEPTUAL COMPROMISE

In the more common view of size constancy what I have referred to as the belief that apparent size decreases with apparent distance is considered to be a reflection of the fact that perceived size does decrease with distance. Either the observer learns the familiar sizes of objects or the perceptual system "corrects" for distance in some way. At near distances the cues to distance are good and perceived size is approximately veridical. At greater distances there are fewer cues to distance and the "correction" becomes insufficient to maintain perceptual constancy. Perceived size decreases with distance but not so much as retinal image size.

Brunswik (1956) developed a theory of perception that not only allowed for attitudinal effects on perceived size but for variability and inconsistencies as well. According to Brunswik's view, the perceptual system is continuously processing many cues to distance, some of which are reliable, some not, and some are in conflict with others. Perception is generally oriented toward the cues most reliably indicating distance but is also influenced by retinal image size. The result is a compromise in which perceived size is fairly accurate on the average but subject to error when the reliabilities of the cues to distance are not appropriate or the attitude of the observer does not result in an appropriate weighting of the cues.

Brunswik defined two perceptual attitudes in connection with size constancy, *naïve-realistic* and *analytical.* Instructions for the *naïve-realistic* attitude directed the subject to judge object size seen in the attitude of everyday perception with no special effort to discern objective or projective size. *Naïve-realistic* instructions have usually been referred to as *phenomenal* or *look* instructions because the subject is encouraged to judge size according to its "immediate impression," as it "looks," not as it might be "known to be." Because a test object of no known size is usually used in an experiment, the subject can logically interpret this instruction to mean that he is to respond in terms of objective or apparent size. Instructions for the analytical attitude, often referred to as *retinal* instructions, included statements that define projective size, but they also directed the subject to reorganize the cues in order to see the three-dimensional scene in two-dimensional perspective.

Two other attitudes were intellectual rather than perceptual, the *realistic betting* attitude and the *analytical betting* attitude, for which the subject was instructed to use whatever knowledge he could to make a "best bet" judgment of objective or projective size. In Brunswik's view, if the subject's attitude shifted from the naïve to the analytical, or vice-versa, perceived size also changed. The effect of a betting attitude, however, was an influence of judgment superimposed on perception and did not represent a perceptual change. Even under the realistic betting attitude perfect constancy would not be achieved or exceeded on the average. Gogel (1974) has proposed that the observer learns that perceived size and distance decrease with distance and in an experiment makes a compensatory correction. Perceived size and distance are more nearly veridical at near distances than the observer realizes, and the correction results in overconstancy at near distances. In the process of learning that a correction is necessary for greater distances, however, it is difficult to understand why the observer would fail to learn that a correction is not necessary at near distances or that a greater correction is needed at greater distances.

The principal differences between Brunswik's theory and the perspective attitude hypothesis are three. First, the perspective attitude hypothesis does not assume that the betting or the nonbetting attitudes affect perceived size or perceived distance. Instructions influence the subject's criterion as to what constitutes an appropriate response in terms of his assumption about the relationship between apparent size and distance. Second, perceived size is not assumed to be a compromise between a size dictated by distance cues and the retinal image size of the object. Perceived size can be accurate on the average in the natural environment and, if it is, the subject's response can be influenced as much in the direction of overestimation of physical size as in the direction of underestimation. Third, perceived size and perceived distance presumably conform to the size-distance invariance principle. Brunswik argued against any simple lawful relationship such as size-distance invariance as an adequate explanation of how perceived size and distance are determined. There is no disagreement with that view as such, but the supposition is that size-distance invariance is valid with respect to perceived size and perceived distance. As a result of the perspective attitude the subject's *responses* may not conform to size-distance invariance.

SIZE-DISTANCE INVARIANCE

The size-distance invariance hypothesis states that the ratio of perceived size to perceived distance is proportional to the visual-angle size of an object. If two objects of equal size are placed at equal distances but one is perceived to be farther away than the other, the one perceived to be farther away will be perceived to be larger. With objective-size instructions the subject may report the farther one as larger, but with apparent-size instructions he may report the two as equal. If one of the objects is twice as far away and is perceived to be twice as far, the objects will be perceived to be equal in size. The subject may report the nearer one as smaller or larger depending on the attitude he takes. If the subject is uncertain about distance, he may infer that an object perceived to be larger is nearer and an object perceived to be smaller is farther away. One problem therefore is that size-distance invariance may not seem valid when in fact it is.

A more complicated problem is that a combination of size-distance invariance and the perspective attitude may appear to account for a phenomenon when in fact that is not the correct explanation. The moon illusion may be used to illustrate this possibility. In conformance with the perspective attitude observers report that the horizon moon appears larger and closer than the zenith moon. Since the visual angle subtended is the same in both cases, the horizon moon should, according to size-distance invariance, be perceived to be farther away and larger or nearer and smaller. It can be argued, however, that the horizon moon is perceived to be farther away and larger, but because it is perceived to be larger it is reported as nearer (Dees, 1966; Gogel, 1974; Rock, 1975, pp. 39-47). This possibility may be correct and is not necessarily contradictory if we assume that what the observer reports is based partly on inference. Alternatively, the horizon moon may be perceived to be larger for reasons other than greater perceived distance (Restle, 1970); no difference in distance from that of the zenith moon is perceived, but the observer interprets the larger perceived size to mean closer apparent distance. The assessment of size-distance invariance can be quite uncertain without an assessment of the attitudinal influences that may also be operating.

EXPERIMENTAL MANIPULATIONS OF INSTRUCTIONS

Gilinsky (1955) placed a variable triangle at 100 ft and presented a standard triangle at distances of 100 to 4000 ft. Four sizes of a standard triangle, 42 to 78 in., were used at each distance on different trials. The average results for objective- and projective (retinal)-size instructions are shown in Figure 7-1. The

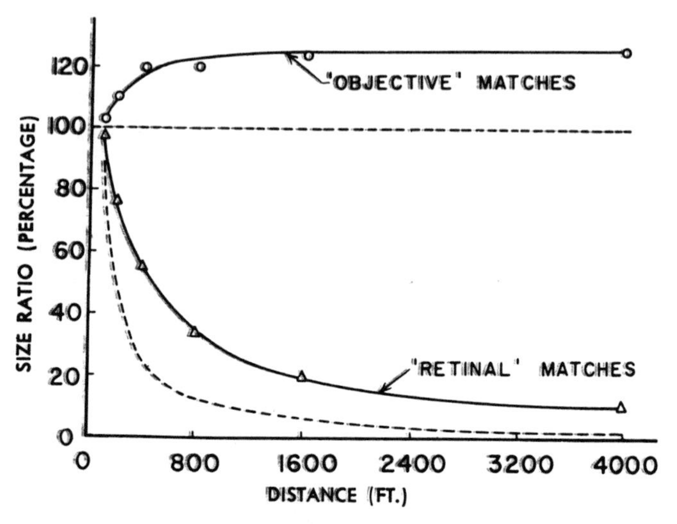

Figure 7-1 Matched size of near variable to far standard as a function of the distance of the standard. (From Gilinsky, 1955. Copyright 1955 by the University of Illinois Press. Reprinted by permission.)

average ratio of the subjects' settings of the variable triangle divided by the size of the standard is plotted as a function of the distance of the standard. The dashed straight line indicates perfect size constancy (ratio = 1), and the dashed curve indicates the ratios that would be expected if the responses conformed to retinal image size.

Gilinsky interpreted the function for the projective-size instruction as representing perceived size minimally influenced by perceived distance and the function for the objective-size instruction as representing a conscious inference of object size based on perceived distance. The objective-size matches tended to be proportional to a function of the far/near distance ratios, but the retinal matches were more precisely proportional to a function of the near/far distance ratios. The retinal instructions specified a concept of size that by definition is proportional to the near/far distance ratio. Hence the interpretation that the subjects were judging the distance ratios and reflecting those perceived ratios in their size judgments is more straightforward than the inference that the subjects were matching in terms of equal perceived size. The fact that the subjects were able to conform so well to the instructional requirement suggests that perceived size and perceived distance must have been reasonably veridical. According to the perspective attitude hypothesis, the retinal function illustrates the maximum response bias possible for an instructionally induced concept of apparent size, short of deliberately instructing the subjects to base their responses on judged distance.

Although the objective-size matches tended to be proportional to the far/near distance ratios, these matches do not necessarily represent the maximum possible bias in that direction. For the retinal matches the subjects' responses were not otherwise constrained and could become asymptotic to zero. For objective size the responses were limited by the variable apparatus, which had a maximum size only 8 in. larger than the largest standard. Although it is unlikely that the subjects would use projective size as the criterion for apparent size when making their objective-size matches, to allow for that possibility the maximum size of the variable triangle would have had to be *five times* as great as it was. The results for some of the subjects were discarded because the apparatus would not accommodate their responses, and Gilinsky did not include the values for the largest standard in the average results for that reason. Most of the subjects were willing to respond, however, even to the largest standard. For the other standards the subjects may have discovered that the maximum size of the variable was not adequate and may have scaled down their responses accordingly.

Because apparent size has no definite subjective quantitative relation to distance, subjects can accommodate their responses to the apparatus and circumstances of the experiment. This matter is important because most experimenters have probably designed size-constancy experiments on the supposition that maximum perceived size will not exceed maximum object size. That may be true, but the subjects' responses may exceed object size appreciably. From a theoretical viewpoint, Gilinsky should have also obtained size judgments with the variable and standard distances reversed, in which case objective-size settings would be expected to be smaller than the standard and projective-size settings larger. Gilinsky might have found that with this arrangement objective size becomes asymptotic to zero and projective size is limited by the maximim size of the variable apparatus. This criticism applies to other investigators of the problem, myself included. The reason for not using a far variable is that it is more difficult to construct an apparatus that can be controlled by the subject at different distances. The problem would have been especially difficult in Gilinsky's experiment because the variable triangle receded into a pit in the ground in order for the visible portion to be variable in size.

Gilinsky implied that overconstancy would occur only at great distances. Jenkin (1959) demonstrated overconstancy with objective-size instructions at distances of two to 30 feet. Moreover, Jenkin and Hyman (1959) found that the correlation between size and distance judgments was opposite in direction for objective-size and projective-size instructions. They interpreted the correlation with projective size as being consistent with size-distance invariance and the correlation with objective size as being inconsistent with that relationship. The distance judgments were estimates of the standard at a single distance (30 ft) and the correlations were in terms of individual differences. The differing directions of correlation are consistent with the hypothesis of a response bias operating in opposite directions for the two instructions.

Leibowitz and Harvey (1967, 1969) used relatively great distances in a series of experiments investigating the effects of instructions, a board versus a person as the far test object, and several test sites. The test sites were a university campus with many people in the field of view, an athletic field, and railroad tracks. The far test objects were located at distances of 100 to 500 or 1680 ft. The comparison object was located at approximately 50 ft and was a wood rod, an aluminum tube, or a strip of canvas, each of which could be varied in its visible length.

In addition to objective- and projective-size instructions, Leibowitz and Harvey used a naïve-realistic (phenomenal) instruction for apparent size. Generally, the objective-size matches were close to the correct physical values and did not change with distance. The projective-size matches were the smallest, the apparent-size matches were intermediate, usually closer to projective size, and both decreased with distance. These results are largely in agreement with the perspective attitude hypothesis, except for the veridicality of the objective-size matches. Assuming that Gilinsky's (1955) results indicate fairly good perceptual constancy, objective-size judgments should show overconstancy.

Several reasons are evident in these experiments for the lack of an increase in the objective-size matches with distance. At distances of the magnitude used, subjects cannot clearly discriminate among persons of more or less average height and are likely to assume that the person test object is about average in height (Carlson & Tassone, 1971). Hence an essential control was lacking. Subjects should have been asked to match the comparison test object to the height of an average person with no other particular standard designated. Possibly that is what the subjects were doing, and the distances were irrelevant as far as the objective-size matches with person test objects were concerned. Similarly, a judgment of the objective size of an object placed on railroad tracks is analogous to asking for a judgment of the objective separation between the tracks as a function of distance. An object judged to be a given proportion of the separation should be judged to be the same objective size, regardless of distance.

The only instance of appreciable overestimation of size occurred with objective-size instructions in the athletic field situation. In this experiment (Experiment II, Leibowitz & Harvey, 1969) the far test object was a 5-ft aluminum tube, the comparison was a similar tube with a maximal visible length of 12 ft, and the conditions compared were the athletic field and railroad tracks. The results were analyzed in terms of the Brunswik ratio, which is the difference between correct projective size and the subject's response, divided by the difference between the correct projective and objective sizes (see Brunswik, 1956, pp. 18, 21). The effect due to the tracks versus field conditions was significant for objective-size instructions but not for apparent- or projective-size instructions.

In Leibowitz and Harvey's Experiment II the maximum size of the comparison was 2.4 times the size of the standard so that this relationship was considerably greater than in Gilinsky's experiment and the objective-size ratios reached higher values. In Leibowitz and Harvey's Experiment III the maximum size of the comparison was 1.6 times as great as the standard. Again the Brunswik ratio for objective size was constant as a function of distance but was reduced in magnitude. In both Experiments II and III the apparatus containing the comparison was visible to the subjects and presumably indicated the maximum possible size of the comparison. It appears from these experiments that the asymptote for objective-size matches is related to the maximum size of the comparison in relation to the size of the standard. If there were no limitation in this respect, the size ratios for objective size might increase indefinitely with increasing distance.

Epstein (1963) employed much smaller test objects and shorter distances than Gilinsky and Leibowitz and Harvey but distance ratios up to 24:1. The variable test object was at 5 ft and the standard was presented at distances of 10 to 120 ft. Two different sizes of the standard were presented at each distance. Each test object (a white triangle) was seen against a relatively large immediate black background. The experiment was carried out in a normally illuminated corridor. The significance of this experimental arrangement is that the apparatus containing the test objects articulated with the floor; hence the sizes and distances of the apparatus were presumably perceived at least as accurately as the sizes and distances of the test objects in the Gilinsky and Leibowitz and Harvey experiments. The test objects themselves did not articulate with the floor or clearly with their surrounding frameworks, and there was probably less perceptual constraint on the size responses, especially at the greater distances at which the texture of the black background would not be so perceptible.

In the Gilinsky and Leibowitz and Harvey experiments the same subjects made judgments at the different distances and could make successive comparisons of them; the result was a more precise proportionality of the size responses to distance than might otherwise have been the case. Epstein used different subjects at the various distances, as well as different subjects for the various instructions, so that responses at one distance could not be influenced by responses at another. The percentage deviations of the size judgments from a correct match are shown as solid lines in Figure 7-3. This measure is proportional to the size ratios involved. The size ratios for the objective-size instruction, for example, were 1.10, 1.06, 1.14, 1.24, and 1.53 for distances 10 to 120.

The phenomenal- and objective-size instructions were not the same as those used by Gilinsky and Leibowitz and Harvey. Rather they were the neutral apparent- and objective-size instructions developed by Carlson (1960), in which the apparent-size instruction was expected to be more "neutral" than the objective-size instruction. This apparent-size instruction was the only one that did not result in a significant change in size response with distance. The perspective-size

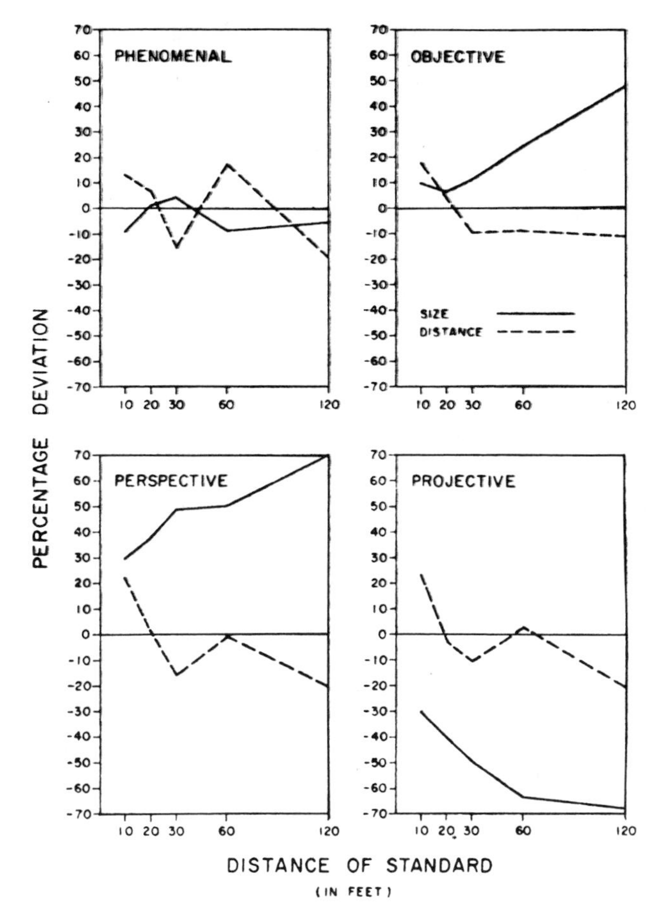

Figure 7-3 Deviations from correct size and distance judgments with different subjects making judgments at different distances. (From Epstein, 1963. Copyright 1963 by the American Psychological Association. Reprinted by permission.)

instruction was devised to enhance the bias assumed to be present with the objective-size instruction (Carlson, 1962). The difference in effect between these two instructions was consistent with that expectation.

There was a general increase with distance for objective and perspective size and a general decrease for projective size, consistent with the presumably lessened perceptual constraint for perceived size at the greater distances. A nearby comparison that is 1.5 times as large as a standard at 120 ft can be judged to be "just barely larger." With the standard at 10 ft a difference that great is obviously more than just barely larger and a smaller difference is required to satisfy the subject's response criterion. Except for the difference between 60 and 120 ft, none of the differences between adjacent distances was significant, including those for projective size. The size judgments therefore were only roughly protional to distance. From the general trend, however, it is clear that if Epstein had increased the distance beyond 120 ft, the objective and perspective responses would of necessity at some point have to become asymptotic to the upper limit of the variable apparatus and the projective responses would have to become asymptotic to zero.

Epstein also obtained distance judgments from his subjects. The average results are shown as dashed lines in Figure 7-3. There was nothing in them to indicate that subjects were basing their size judgments on anything but "perceived" distance—as opposed to "apparent," "objective," "projective," or "perspective," distance. The distance judgments did not differ significantly among the size-instructional groups.

Carlson and Tassone (1967) investigated the effect of providing subjects with the opportunity to compare the different distances of the standard test object. The circumstances of the experiment were similar to those of Epstein's experi-

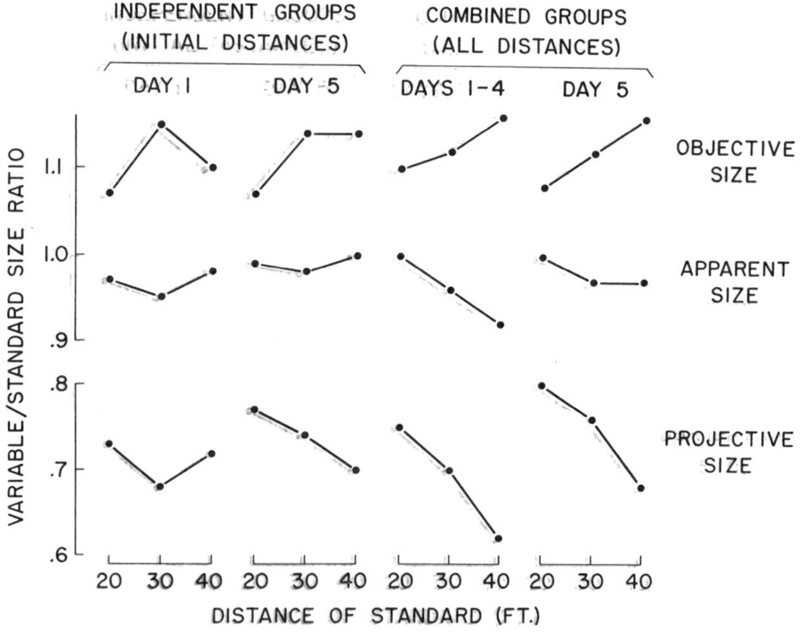

Figure 7-4 Trends in size by distance functions with judgments at different distances made on successive test days. (From Carlson & Tassone, 1967. Copyright 1967 by the American Psychological Association. Reprinted by permission.)

ment except that the variable triangle was at 10 ft and the standard triangle was presented at 10, 20, 30 or 40 ft. Ten different sizes of the standard were presented at each distance. On each testing day except the last each subject made size judgments at only one distance of the standard. All four distances were presented over four testing days in nonsystematic order. On the fifth day each subject made judgments at all four distances. The mean size-ratios for distance 10 (variable and standard at the same distance) were approximately equal to 1.0 for all three instructions. The question was whether an ordering of the size ratios for distances 20, 30, and 40 developed as a result of experience with the differing distances.

The nine points for Day 1 plotted in Figure 7-4 are the averages for nine different groups of subjects. These values are not statistically different from the corresponding values in Epstein's results for distance ratios of 4:1 or less. There was no systematic ordering of the size ratios with respect to distance on Day 1. On Day 5 the averages for these nine groups were more systematically ordered with respect to distance for objective and projective size. These averages are based on only one distance for each subject, and the trends were not statistically significant. Based on all distances for each subject on Days 1 to 4, the trend for each instruction was significant. Utilizing all distances for each subject on Day 5 resulted in significant trends for objective and projective size but not for apparent size. The trend in the apparent-size means for Days 1 to 4 is an anomalous result with respect to the data for Day 5 and with respect to Epstein's results with this instruction. But clearly the precision of the relationship between the size judgments and distance depended on the nonindependence of the judgments at different distances.

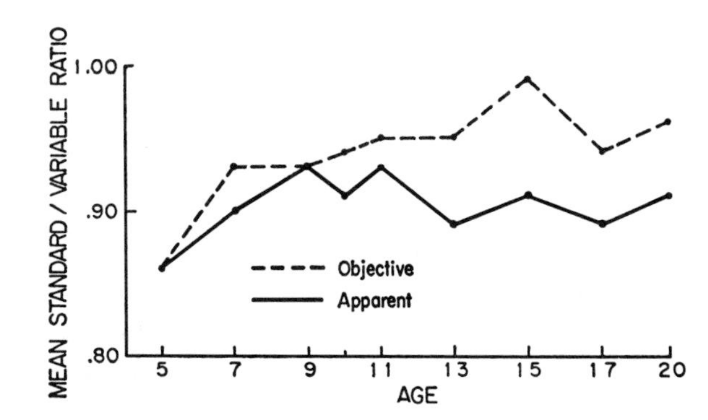

Figure 7-5 Differential effect of size instructions with increasing age. (From Rapoport, 1967. Copyright 1967 by the Society for Research in Child Development. Reprinted by permission.)

Most of the subjects in these experiments have been college students who might be disposed to interpret instructions in a special way. Rapoport (1967) tested subjects ranging in age from 5 to 20 years. The effect of instructions as a function af age is shown in Figure 7-5. In this case the standard was near and the variable far and the size ratio has accordingly been reversed. The general increasing trend with age for objective size was statistically significant, whereas the change for apparent size was not. Instructions had no differential effect up to age 9 but clearly did by age 13. In another experiment (Carlson & Tassone, 1963) subjects 61 to 80 years of age were tested with essentially the same instructions. These subjects did not show so great a difference as college students, but still their apparent-size judgments were not significantly different from the correct values and their objective-size judgments were overestimations. For both the children and the older subjects the instructions were neutral, not requiring a differentiation if the subjects were not inclined to make one. College students may be more sensitive to the specific instructions used, but the distinction between apparent size and objective size seems to have considerable generality.

I believe these experiments covering distances from a few feet to a few thousand feet provide a reasonable basis for supposing that with the right instructions it would be possible to demonstrate any degree of underconstancy or overconstancy within the limitations of the apparatus and the distances used in the experiment. On the basis of the argument that perceived size may be inferred to be within the limits indicated by responses to the neutral instructions, perceived size would have to be considered as remaining constant or increasing with distance. As I have indicated, there is reason to believe that the apparent-size instruction is neutral in effect but the objective-size instruction is not. Some further rationale and support for that notion can be given.

MINIMIZING THE BIAS

Joynson (1958a,b) explored in considerable detail his subjects' understanding of the concept "apparent size" or a phrase such as "look the same size." Some subjects use these expressions to mean "judged objective size" and some do not. Most subjects are willing to accept the notion that there is some such thing as "apparent" or "nonreal" size which is different from objective size, but just what it is is obscure. Experimenters have often resorted to stressing that apparent size *is* different from objective size without clearly specifying how it differs in that respect, yet is the same as perceived size. This strategy seems to lead to confusion in subjects' understanding of what is meant (Leibowitz & Harvey, 1969, p. 42).

In my own work on the problem (Carlson, 1960) the phrase "apparent *visual* size" seemed to be more effective than just "apparent size" for indicating something different from objective size. This phrase does not indicate what the concept means either, but subjects more readily understand it as not implying objective size. My supposition of the source of the difficulty was that "apparent size" or "apparent visual size" is in fact an illusory concept and cannot be defined satisfactorily in an instruction. Because the subject is likely to think there is some such entity as apparent size, it is necessary to tell him that apparent size may or may not be the same as objective size. He may make a distinction if he likes but the experiment does not require it. Because he really has no clear basis for making a distinction, he can forget about it and respond on the basis of perceived size without concern for what he should do to achieve a correct match or to achieve a match that is somehow different from a correct match. It is necessary to state which concept is indicated, apparent or objective size, but it is also necessary to indicate that these two concepts are not necessarily different.

It may be objected that this strategy does not prevent some subjects from adopting a criterion of objective size and other subjects from adopting some other criterion. This objection is certainly valid. There is no way to eliminate individual biases by instructions alone. The difference obtained with these two instructions on the average, however, is obtained whether the instructions are given to the same or different subjects and neither distribution is bimodal; that is, the average judgment is not an artifact due to some subjects taking one attitude, some another, and no one taking the neutral.

The neutral-objective-size instruction usually produces overconstancy and the neutral-apparent-size instruction results in more nearly accurate judgments. That the objective-size instruction will result in accurate judgments by some subjects is suggested in an experiment by Predebon, Wenderoth, and Curthoys (1974, Experiment I). Their task was to set the size of a near object equal to the size of a far object. The experimenter explained that it is possible to distinguish between apparent and objective size and for the purposes of the experiment it was important for the subject to state whether he could make this distinction. Subjects who professed inability to do so were assigned to the objective-size instruction group. Subjects who said they could make the distinction were assigned to the apparent-size instruction group. Under these circumstances the objective-size group produced an accurate size match and the apparent-size group produced underestimation and greater variability.

The results of the apparent-size group are understandable. Its subjects were, in effect, told that it was important to make the distinction between apparent and objective size and that they could choose either criterion. The instructions were not ambiguous for the objective-size group because its subjects acknowledged an understanding of only the criterion of objective size. It is evidently not the concept of objective size per se that introduces a bias with that instruction. It is the presumption that objective size and apparent size are different.

It would be valuable to have an assessment of the biasing effect of instructions by some measure other than effects on judged size. Epstein and Broota (1975) have compared reaction times for objective-size and neutral-apparent-size estimates of squares at distances of 1.2 to 5.5 m. Reaction times for the objective-size estimates increased proportionally with distance. Reaction times for the apparent-size estimates were shorter and did not change as a function of distance. This result is consistent with the veiw that a response bias related to distance is associated with objective-size judgments and the neutral-apparent-size instruction tends to eliminate that bias.

It seems to me that the best inference we can make from the data available is that perceived size is constant as a function of distance in the natural environment. That is the relationship indicated by the neutral-apparent-size instruction, and responses obtained under that instruction are the least likely to be biased by preconceptions of the nature of the relationship between apparent size and distance.

DISTANCE CONSTANCY

The supposition in the size-constancy experiment is that the subject is concerned with an assumed difference between apparent size and objective size as a function of distance but not with a similar distinction between apparent distance and objective ("real") distance. His responses are therefore assumed to be based on perceived distance, but when attention is directed toward judging distance the subject may wonder whether distance is what it appears to be. Cognitively a physical extent is the same size, whether it is oriented vertically, horizontally, or longitudinally and whether it is nearby or far away. The subject may assume that a nearer objective extent needs to be larger or not smaller in apparent extent in order to match a more distant extent, and, if so, his behavior may be expected to be analogous to what it is in the size-constancy situation. In additon, an inferential difference between apparent distance and objective distance is possible. An object may be inferred to be larger in objective size than it appears to be and farther away than it appears to be.

In the size-constancy experiment the direction of the apparent versus objective bias reverses when the variable or comparison and the standard are reversed. This reversal is a technical matter because the position of the variable is usually clear with respect to the standard when two sizes are being compared. Verbal estimates of "absolute" distance are often obtained in distance experiments, however, and the subject may assume a reference distance to which his judgments are relative. If no standard is designated by the experimenter in the method of magnitude estimation, the subject is free to assume whatever standard he wishes, in which case a reversal in the direction of the effect of instructions might be due to a difference in an assumed reference distance. The context in which the judgments are made can affect size judgments in different directions (Parducci, Perrett, & Marsh, 1969; Restle, 1971), and there is no reason to assume that context is any less important for distance judgments.

Rogers and Gogel (1975) investigated the effect of apparent and objective instructions in conjunction with several different procedures for obtaining distance judgments, including verbal estimates. The distances varied to about 7 m. In bisecting a depth interval, the subjects set the nearer half larger than the farther half for objective distance and equal for apparent distance. This overconstancy for relatively near distances with the objective attitude was established by Wohlwill (1963, 1964), who interpreted it as a judgmental bias but did not determine whether it was affected by instructions. The Rogers and Gogel result with apparent-distance instructions, perfect distance constancy on the average, suggests that the overconstancy is not perceptual. Accurate bisections have been obtained to much greater distances (Purdy & Gibson, 1955) as have bisections exhibiting underconstancy (Gilinsky, 1951). There is no way to know whether a difference in attitude accounts for these contrasting results because the experimenters did not investigate the effect of instructions.

In the method of interval reproduction the subject generates a series of depth intervals matched to a standard—the series progressing either away or toward himself. In the Rogers and Gogel experiment these intervals tended to be equal on the average for both instructions, although one group of subjects under a particular combination and order of conditions exhibited underconstancy with apparent-distance instructions. A complex interaction such as this should not be given much weight unless reliably established, but I would point out that this kind of effect suggests that distance underconstancy, no less than distance overconstancy, can be the result of a response bias. Harway (1963) found underconstancy by a similar method. His instructions may be characterized as "everyday objective", which from the present viewpoint do not indicate whether the subjects were judging objective or apparent distance.

In the Rogers and Gogel experiment the verbal apparent and objective estimates of a near depth-interval were the same. For a far interval the objective estimates remained the same (constancy), whereas the apparent-distance estimates decreased (underconstancy). Of particular interest was an interaction that involved the verbal estimates of the bisection interval, which was not changed in distance. For subjects who had not yet experienced the interval-reproduction task the apparent estimates of the bisection interval were greater than the objective estimates. For subjects who had experienced the interval-reproduction task the objective estimates were greater than the apparent estimates. The two were affected about equally but in opposite directions. Something to which these "absolute" judgments were relative must have been different.

Baird and Biersdorf (1967) made objective size and distance judgments according to the size-constancy paradigm. Rectangles were placed upright at distances to 18 ft and matched in height with a near and a far comparison. Overconstancy was obtained, as would be expected in this size-constancy condition with objective-size instructions. The rectangles were then laid flat on the table top so that the lengths of the rectangles became depth intervals. These lengths were judged in the same way as the heights had been. The result was underconstancy. By analogy to size constancy apparent-length instructions would not have produced greater constancy than objective-length instructions; hence we would infer that the perceived depth intervals decreased with increasing distance.

In addition, Baird and Biersdorf made magnitude estimations of distance with designated near and far standards. Either a near distance or almost the distance to the far end of the table was given the arbitrary value of 100; the distances of other points along the table top were rated in terms of multiples or fractions of 100. Essentially perfect constancy was obtained with the near-distance standard, underconstancy with the greater-distance standard. Again, we would infer perceptual distance underconstancy by analogy to size constancy, but the possibility of a different interaction between apparent-distance estimates and lesser and greater standard distances is not precluded.

Teghtsoonian and Teghtsoonian (1970a) obtained magnitude estimations of both sizes and distances in separate series of judgments. No standard and no reference number (modulus) were designated. The subjects assigned numbers of their own choosing proportionally to the sizes or distances of irregular shapes placed at distances of 5 to 45 ft in a large room. Apparent size and apparent distance were specified in the instructions. Some size overconstancy occurred in the first experiment, which the authors suggest may have been due to those subjects adopting a more objective attitude. In the second experiment the size estimates did not change as a function of distance. With this method the degree of distance constancy is reflected in the acceleration of the function relating the distance estimations to distance. Because the distances are discriminable, the function will increase. If it increases linearly, distance constancy may be said to occur. Positive acceleration in the function indicates overconstancy and negative acceleration indicates underconstancy. The function obtained was positively accelerated, which indicated overconstancy with apparent-distance instructions.

The same result—overconstancy with apparent-distance instructions—was obtained by these investigators in two other experiments, one also conducted in a large room and the other in a corridor (Teghtsoonian & Teghtsoonian, 1969). The latter condition involved a maximum of 80 ft. In an outdoor setting distance ranges of 37, 110, and 480 ft were used (Teghtsoonian & Teghtsoonian, 1970b). Perfect constancy was obtained with the 37-ft range but underconstancy with the two greater ranges. There was a change in instructions for the outdoor situation from apparent distance (indoors) to a neutral-apparent-distance instruction. The subjects were told that objective distances might be the same as apparent distances, and this instructional nuance may be as important for distance judgment as it is for size judgment.

Another factor that may be related to this indoor-outdoor difference in distance constancy is that the maximum perceptible distance available to the subjects is much greater outdoors. That the increase in the range of distances judged is not by itself responsible (see Teghtsoonian, 1973) is suggested by some experiments involving judged distances from several hundred yards to more than five miles (Galanter & Galanter, 1973). One experiment yielded distance constancy and one experiment, underconstancy. In general the functions exhibited overconstancy, similar to the Teghtsoonian's indoor results. Unfortunately the Galanters did not report what their instructions were with respect to apparent versus objective distance.

We thus see that distance constancy, underconstancy, and overconstancy can all occur at both near and far distances as can interactions among instructions, methods, and conditions. As with size constancy, it is possible that perceived distance increases veridically with distance under natural conditions of observation. Until the effects of instructions are systematically investigated, we cannot infer with any confidence how perceived distance varies with distance.

EQUAL DISTANCES

In size- and distance-constancy experiments variations in distance are usually obvious to the subject, and a response bias may be expected because the distances are perceived to be different. When test objects are at equal distances, effects of instructions must be reduced to that allowed by the perceptual constraint for perceived equality in distance or the subject's proclivity to assume that the objects are not at equal distances. If the distances are perceived to be equal and the subject has no reason to assume that the distances are not equal, apparent size should be the same as objective size as far as the cognitive association between size and distance is concerned. Similarly, if a single object is presented at a single given distance and is perceived to be at a definite distance, there is no reason for the subject to differentiate apparent from objective size. In order for instructions to be effective the subject would have to consider what the object would look like if it were at some distance other than where it is perceived to be. Instructional effects are not precluded, but instructions may be expected to be less effective than they are when distance is perceptibly varied.

The Gilinsky (1955) and Carlson and Tassone (1967) experiments, discussed above, were primarily concerned with size judgments at different distances, but a condition of equal distances was included. Although the average judgments were nearly the same for different instructions, a detectable effect was present (Carlson & Tassone, 1968). It was as though the standard were perceived or assumed to be a little farther away than the variable. An assumption seems more likely because the effect increased after experience with the standard at greater distances than the variable.

In another experiment (Carlson & Tassone, 1971) a person and a board of equal height and at equal distances (183 m) were compared. The person was judged to be smaller in apparent size and equal or larger in objective size but was nearly unanimously judged to be farther away. In this case the person was probably perceived to be farther away because there was no evident reason why the subjects should have assumed so.

With objective-size instructions Jenkin (1959) varied the distance of a comparison object from 20 to 160 in., with the standard sometimes within that range. Thus the subjects experienced the comparison at both lesser and greater distances than the standard. There was a displacement of the judgments which developed with practice, but this displacement occurred for different as well as equal distances. Jenkin argued that a concept of the standard as an object of fixed size had developed. Possibly this effect might not have occurred with apparent-size instructions, but experimenters usually present different sizes of the standard at each distance to prevent the subject from assuming that he is always judging an object of the same size.

Teghtsoonian (1965) presented figures one at a time at a constant distance and asked subjects to make magnitude estimations of apparent or objective size. There was no difference due to instructions in judging the lengths of straight lines or the areas of irregular polygons, but there was an effect for judging circles. In this case the subjects understood "objective area" to mean that they should estimate the linear extent of a circle and square it. With apparent-size instructions subjects may use several different strategies for estimating area (Macmillan et al., 1974). These effects, however, suggest no relation to the distance at which figures are presented.

In their Experiments II and III Predebon, Wenderoth, and Curthoys (1974) gave each subject a single object at one distance. There was no effect of apparent- versus objective-size instructions. One test object was an off-sized chair, which might have been more likely to produce effects on distance judgments than on size judgments (Gogel, 1969a; Carlson & Tassone, 1971).

That subjects will associate greater distance with smaller size even though the objects are perceived to be at the same distance was demonstrated in an experiment by Rapoport (1969). Two triangles of the same size were placed at a distance of 5 ft, and three groups of children were asked, "How can we change the size of this triangle (the variable) so that it looks twice as far away as the other one without really moving it?" Five-year-old subjects made the triangle larger or smaller equally often. In the seven-year-old group twice as many subjects made the triangle smaller. Nearly all the subjects in the nine-year-old group reduced the size of the triangle to make it "look" farther away.

A similar association between apparent size and apparent distance occurred when college undergraduates viewed random-letter stereograms (Tolin, 1969, 1970). When the two stereoscopic configurations were equal in size but different in depth (different in binocular disparity), the configuration appearing farther away was reported as larger and the configuration appearing nearer was reported as smaller. This result is in agreement with size-distance invariance and presumably the sizes and distances were perceived as reported. However, when the configurations were different in size but equal in binocular disparity, the larger configuration tended to be reported as appearing nearer and the smaller as appearing farther away. Telling the subjects that the configurations were the same or different in size or depth did not have a significant effect. Asking the subjects to report apparent depth was evidently sufficient to elicit an association of lesser depth with larger size and greater depth with smaller size.

These experiments suggest that subjects will interpret smaller preceived size as indicating greater apparent depth in the absence of a perceptually determined difference in depth. Size instructions may have some residual effects on size judgment or effects extraneous to the size-distance relationship. When a difference in size is perceived but not a difference in depth, instructions may be more important with respect to their effect on subjects' distance responses. Variations in the effects of specifying apparent versus objective distance and of indicating to the subject that apparent and objective distance may be the same have not been explored.

IMAGES IN THE DARK

The conditions considered so far have been those in which perceived sizes and perceived distances are presumably determined by the visual stimulus situation. In the darkroom experiment the visual stimulation is restricted to one or two test objects in an otherwise dark field with as little other visual or oculomotor information as possible. Under these circumstances it cannot be assumed with any certainty what is perceived (Foley, 1968, 1972). Two carefully controlled experiments indicate that visual angle is perceived as something that is neither size nor distance, as such, but can be interpreted by the subject as either size or distance (Epstein & Landauer, 1969; Landauer & Epstein, 1969). These investigators found an inverse relation between size and distance responses as a function of visual angle, in agreement with the perspective attitude expectation.

Gogel (1971) has argued that the Epstein and Landauer results are not in disagreement with the size-distance invariance hypothesis because the size/distance ratio of the judgments increased as a function of visual angle and this increase is in agreement with general size-distance invariance. This ratio, however, would also have increased if the size judgments had increased with visual angle, judged distance remaining constant, or if the distance judgments had decreased with increasing visual angle, judged size remaining constant; hence the size-distance invariance hypothesis can hardly be said to predict the results. In equally carefully controlled experiments Gogel (1969b) compared first presentations of test objects with successive presentations. On the first presentations size responses varied with visual angle; distance responses tended to remain constant. On successive presentations size responses tended to become independent of visual angle and distance responses to be inversely related to visual angle. Gogel postulates perceptual constraints on distance responses (equidistance and specific distance tendencies) and perceptual determination of size by visual angle and by information from preceding stimulus presentations. These tendencies probably exist, but they do not account for the Landauer and Epstein results without some additional assumptions about subjects' attitudes (Gogel & Sturm, 1971).

Coltheart provided subjects with verbal or haptic size information and asked for distance judgments (1969a, b); he also provided verbal distance information and asked for size judgments (1971). The implication of the results is that subjects will judge a given angular size to be larger if it is assumed to be more distant or will judge a given angular size to be more distant if it is assumed to be larger (positive relationship). If size is assumed to be constant and visual angle varies, judged distance will vary inversely with visual angle. The positive relationship between size and distance responses is not necessarily contradictory to the negative relationship obtained by Landauer and Epstein. If visual angle is constant, size and distance judgments will vary together. If visual angle varies, size and distance judgments can vary inversely. What is unknown is how perceived size and perceived distance vary or do not vary under both circumstances.

Complicating the problem are possible assumptions of equidistance or constant object size on the part of the subjects. These assumptions are not required by the perspective attitude hypothesis but would not contradict it either so long as they are compatible with the subject's general assumption that smaller apparent size is associated with greater distance. The subject could interpret an object of given apparent size at a given apparent distance as representing an object of larger objective size at a greater objective distance. This interpretation would be in agreement with the perspective attitude but would result in a positive correlation between objective size and distance responses. Rump (1961) found that subjects will make inferences that imply opposite directions of correlation, depending on whether equidistance or equal size is assumed. To compound the matter some of Rump's subjects responded in one direction and some in another either because different subjects made different assumptions or because different subjects made different inferences.

In the studies I have referred to Coltheart specified objective size and distance. Equally important is what the results would have been if he had provided objectively incorrect information and asked for apparent size and apparent distance judgments. Using off-sized familiar objects, Gogel (1969a) obtained data that indicated that if an object appears smaller than normal the subject will infer that it is farther away than it appears to be. Landauer and Epstein's subjects were asked to report the size and distance of the test object in terms of a common object and distance of the subject's choosing but not actually present. The authors report that several subjects gave responses such as "the distance of the moon," which presumably, though not necessarily, were interpretations. Gogel's instructions specified that the subjects judge the test object as it "appears to be," which is an intrinsically ambiguous specification with regard to apparent versus objective size. Excellent as these experiments are in other respects, they still leave open the question of what is assumed, what is inferred, and what is perceived.

Over (1960) devised instructions that differentiated projective and objective size, the instruction for the latter indicating that objective size-equality at different distances was meant. On successive trials pairs of differing angular sizes were presented in ascending or descending series approaching or diverging from equal angular size. For projective size the subject could assume equidistance and interpret the smaller angular object as smaller in both apparent and objective size and the larger angular object as larger in both apparent and objective size. For objective size the subject would have to assume unequal distances and interpret the smaller angular object as farther away and larger in its objective size than in its apparent size, reporting "equal" for a wider range of values from equal angular size. If the subject assumed equal objective sizes for the two objects, he would logically have to report "equal" on all trials. The obtained matches were in terms of visual angle equality for both the objective- and projective-size instructions.

Over did not indicate whether the subjects were told that a response of "equal" on all trials would be acceptable nor whether any subjects responded "equal" on all trials. The steps in the series were probably too large to detect a tendency toward a difference in the width of the "equal" category; and Over did not obtain distance judgments, which under the circumstances of this experiment would have been more likely to indicate an effect of instructions.

Baird (1963) presented a standard test object under restricted viewing conditions and a comparison under nonrestricted conditions. He found a positive correlation between distance estimates of the standard and objective size but no correlation with apparent size. This result would be expected if subjects assumed an object to be larger in objective size than in apparent size and correspondingly to be farther away in objective distance, different subjects making the assumption to differing degrees. The apparent-size instruction directed the subjects "not to take distance factors into consideration." There was a positive correlation between the distance estimates of the standard and the comparison with apparent-size instructions but none with objective-size instructions. This result is understandable if the subjects assumed equidistance as a result of the apparent-size instruction but differed in their estimates of the comparison distance.

Ono (1966) demonstrated that subjects learn more readily to associate unrestricted comparison size with the angular than with the physical size of an object seen under restricted conditions, but the subjects did learn to respond with the correct physical sizes. Levy (1967), in an experiment similar to Ono's, verbally reinforced subjects for correct ratings of physical-size equality between a standard and a comparison at different distances under restricted viewing conditions. Following this training, the subjects matched a comparison to the size of the standard and moved the farther object to the judged point of equidistance with the nearer object. These tasks were also performed under restricted viewing conditions. All subjects were told to make their judgments in terms of objective size, but some were told that apparent size can be different from objective size and that this distinction should be taken into account. The subjects who differentiated apparent and objective size more nearly matched physical size and distance than the other subjects. In these experiments it is not at all certain that the subjects were learning to perceive physical size. It is just as possible that they were learning to infer objective size on the basis of an assumed difference between apparent and objective size and between apparent and objective distance.

Given the uncertainty as to just what is perceived in the darkroom experiment and given the assumptions and inferences that subjects can make about sizes and distances, it would seem essential to explore the effects of instructions in this situation much more carefully than has been done. Especially lacking is an assessment of the effects of specifying apparent versus objective distance. It seems likely that there are attitude by methods by conditions interactions, perhaps even more so than in the distance-constancy experiment.

SHAPE CONSTANCY

The shape-constancy experiment is basically similar to the size-constancy experiment. A flat surface of some bounded shape is rotated in depth so that it is not perpendicular to the subject's line of sight. The subject's task is to vary or select the shape of a surface perpendicular to his line of sight to match the slanted shape. If a rectangle is rotated, its retinally projected shape is trapezoidal. If the subject selects a matching shape that is rectangular, shape constancy is said to occur. If he selects a shape that is trapezoidal, underconstancy is said to occur. Shape and slant are analogous, respectively, to size and distance in the size-constancy experiment. The similarity is evident when we consider that if the nearer and farther edges and the nearer and farther angles of a rotated rectangle are perceived to be equal in size then a rectangle rather than a trapezoid is perceived (see Mefferd, Redding, & Wieland, 1967). Most of the theoretical issues associated with size constancy have their counterparts in a consideration of shape constancy. So do the problems of the subject's attitude and the effects of instructions.

In natural unrestricted viewing the shapes and slants of objects are generally perceived veridically. When the subject tries to judge the projective relationships involved, he can do so only with inaccuracy and difficulty (Lappin & Preble, 1975). Yet much of the literature based on laboratory experiments indicates low degrees of shape constancy. Lichte and Borresen (1967) suggested an attitudinal reason for low constancy. If the experimenter presents a slanted circle and the subject perceives it as a slanted circle, the subject is likely to assume that the response "circle" is too simple. Something more must be required, especially if the experimenter asks for a judgment of "apparent" shape. Accordingly the subject attempts to base his judgment on some vague notion of retinal image relationships.

All the subject would need, however, to respond proportionally to projective shape is the assumption that lesser lateral occlusion of the background with rotation would produce a change in "apparent shape" but not in objective shape. His shape judgments could then be based on a proportionality to perceived slant. Joynson and Newson (1962) made a detailed study of subjective concepts of shape in relation to slant. They found that some subjects were aware only of objective shape. It was clear, however, that subjects can be aware of a conceptual difference between "real" and "nonreal" shape and can consider the difference to be related to slant and the narrowing in extent of a figure lateral to the axis of rotation. In any case, instructions have differential effects on shape-constancy judgments. These effects appear at about the same age in children as they do for size-constancy judgments (Vurpillot, 1964).

Lichte and Borresen (1967) found that objective-shape instructions produced nearly perfect shape constancy, apparent-shape instructions produced underconstancy, and projective-shape instructions produced the least constancy. Their apparent-shape instruction defined apparent shape as a "compromise" between real and projective shape and yielded a bimodal distribution, some subjects adopting an objective criterion, others, a projective criterion. Landauer (1969) also used these three kinds of instruction but defined apparent shape as the "immediate impression" of object shape without mentioning projective shape. On the average, overconstancy occurred with the objective-shape instruction, perfect constancy with the apparent-shape instruction, and underconstancy with the projective-shape instruction. The variation among subjects was somewhat greater for the apparent-shape group, but the distribution was not bimodal.

Landauer had found earlier that this apparent-shape instruction was not bimodal and did not differ in effect from a nondirective instruction (Landauer, 1964a, b). Judgments under both instructions varied somewhat with slant, more so than objective judgments and less so than projective judgments. There was less differential effect of instructions under more restricted viewing conditions. Epstein, Bontrager, and Park (1962) also found instructional effects with less restricted viewing but none with restricted viewing. In an experiment similar to Levy's (1967) on size judgment, however, Gregg and Pasnak (1971) obtained a significant effect of instructions for subjects "learning" to judge objective or projective shape under restricted viewing conditions.

Subjects in Epstein, Bontrager, and Park's (1962) experiment made slant judgments in addition to shape judgments. These authors report no effect of shape instructions on the slant judgments and their average results appear to indicate none. Kraft and Winnick (1967) obtained a main effect of shape instructions on shape judgments but not on slant judgments, although there were signigicant interactions involving instructions and every other variable in the experiment. Their "projective" instruction seems more clearly to imply an objective-shape judgment than does their "constancy" instruction, and the subjects seem to have understood the instructions in this implied way. No manipulation of slant instructions on slant judgments appears to be available.

The kinds of effect that objective, apparent, and projective instructions have on shape judgments at different slants thus seem generally to resemble the kinds of effect instructions have on size-constancy judgments. It is possible that a similar response bias is involved—a judgment proportional to perceived slant in one direction for a projective-shape response and in the other for an objective-shape response. Although overconstancy does occur (Winnick & Rogoff, 1965), it is less often reported than size overconstancy. This difference may be due to restrictions in the apparatus or in the viewing conditions. In any case the expectation based on the perspective attitude hypothesis is that a response indicating perceived shape will be more likely to be elicited by a neutral-apparent-shape instruction than by any other kind. Such an instruction would direct the subject to judge apparent shape but would also inform him that apparent shape need be no different from objective shape.

IMPLICATIONS

The thesis I wish to advance is that in order to understand what is perceived in a given situation it is necessary to understand how the subject's attitude interacts with the particular stimulus-conditions. If the interpretation of an experimental result is to have generality, it is as important to vary instructions systematically as it is to vary the stimulus conditions. The Müller-Lyer illusion offers an appropriate specific illustration.

Whether this illusion involves a depth component has been the subject of a great deal of controversy and research (see Fisher, 1970). If there is any differential depth in this figure, perceived or assumed, the segment with the ingoing arrowheads is the "near" segment and the segment with the outgoing arrowheads is the "far" segment. By analogy to size constancy the subject would set the near segment larger for objective-size instructions (greater illusion) and smaller for apparent-size instructions (lesser illusion). If the two segments are perceived and assumed to be equal in depth, apparent- and objective-size instructions would produce no difference. If apparent-size instructions result in greater illusion, the effect is more likely to be related to some aspect of the phenomenon other than depth.

J. A. Carlson (1966) tested the effects of apparent- and objective-size instructions on size and shape constancy as well as on the Müller-Lyer illusion with the same subjects. Her rationale was somewhat different from the present one, but the conclusion is the same. Because apparent-size instructions resulted in greater illusion, the simplest interpretation is that depth was not involved in the judgments. The lack of correlation of illusion magnitude with size- and shape-constancy performance is additionally consistent with that interpretation. Over (1968) also found greater illusion with apparent-size instructions.

Eaglen and Kirkwood (1970), however, obtained just the opposite result—greater illusion with objective-size instructions—and this result would be consistent with a depth interpretation. Obviously an interaction exists between instructions and conditions that requires explication. Such explication by itself will almost certainly not be sufficient to explain the illusion, but without it no explanation can be said to be satisfactory.

The apparent versus objective dichotomy is potentially applicable to any dimension of stimulation (Braine & Shanks, 1965). No instance of perceptual constancy, at least, seems immune. Interactions between instructions and experimental conditions occur for brightness constancy (Landauer & Rodger, 1964; Lindauer & Baust, 1970) and for orientation constancy (Wade, 1970; Ebenholtz & Shebilske, 1973). It is probable that if no effect of instructions has been reported for a constancy phenomenon it is either because the matter has not been investigated or has been investigated under a limited set of circumstances.

I propose a neutral instruction that, if it works as well generally as it seems to in the size-constancy situation, could go a long way toward eliminating discrepant results from one experiment to another. Whether it is size and distance or some other combination of variables between which the subject has a cognitive association, the neutral instruction may serve to minimize the bias resulting from that association. It is difficult enough to reconcile findings when methods and conditions differ. When methods, conditions, and instructions differ, reconciliation can be impossible.

I do not propose that the goal should be to eliminate the effects of instructions or to render those effects constant. What is "response bias" for one theorist is "percept" for another. If theories are to be reconciled, it will be necessary to understand the effects of instructions, whether they are assumed to be cognitive or perceptual.

REFERENCES

Baird, J. C. Retinal and assumed size cues as determinants of size and distance perception. *Journal of Experimental Psychology*, 1963, **66**, 155-162.

Baird, J. C. & Biersdorf, W. R. Quantitative functions for size and distance judgments. *Perception & Psychophysics*, 1967, **2**, 161-166.

Boring, E. G. Visual perception as invariance. *Psychological Review*, 1952, **59**, 141-148.

Braine, M. D. S. & Shanks, B. L. The development of the conservation of size. *Journal of Verbal Learning and Verbal Behavior*, 1965, **4**, 227-242.

Brunswik, E. *Perception and the representative design of psychological experiments.* Berkeley: University of California Press, 1956.

Carlson, J. A. Effect of instructions and perspective-drawing ability on perceptual constancies and geometrical illusions. *Journal of Experimental Psychology*, 1966, **72**, 874-879.

Carlson, V. R. Size-constancy judgments and perceptual compromise. *Journal of Experimental Psychology*, 1962, **63**, 68-73.

Carlson, V. R. Overestimation in size-constancy judgments. *American Journal of Psychology*, 1960, **73**, 199-213.

Carlson, V. R. & Tassone, E. P. Familiar versus unfamiliar size: A theoretical derivation and test. *Journal of Experimental Psychology*, 1971, **87**, 109-115.

Carlson, V. R. & Tassone, E. P. Size-constancy judgments at equal distances. *Perceptual & Motor Skills*, 1968, **27**, 193-194.

Carlson, V. R. & Tassone, E. P. Independent size judgments at different distances. *Journal of Experimental Psychology*, 1967, **73**, 491-497.

Carlson, V. R. & Tassone, E. P. Size-constancy and visual acuity. *Perceptual & Motor Skills*, 1963, **16**, 223-228.

Carlson, V. R. & Tassone, E. P. A verbal measure of the perspective attitude. *American Journal of Psychology*, 1962, **75**, 644-647.

Coltheart, M. The effect of verbal size information upon visual judgments of absolute distance. *Perception & Psychophysics*, 1971, **9**, 222-223.

Coltheart, M. Effects of two kinds of distance information on visual judgments of absolute size. *Nature*, 1969a, **221**, 388.

Coltheart, M. The influence of haptic size information upon visual judgments of absolute distance. *Perception & Psychophysics*, 1969b, **5**, 143-144.

Dees, J. W. Moon illusion and size distance invariance: An explanation based upon an experimental artifact. *Perceptual & Motor Skills*, 1966, **23**, 629-630.

Eaglen, J. & Kirkwood, B. The effect of instructions on judgment of the Müller-Lyer illusion with normal and haptically mediated visual inspection. *Perception & Psychophysics*, 1970, **8**, 35-36.

Ebenholtz, S. M. & Shebilske, W. Differential effects of instructions on A and E phenomena in judgments of the vertical. *American Journal of Psychology*, 1973, **86**, 601-612.

Epstein, W. Attitudes of judgment and the size-distance invariance hypothesis. *Journal of Experimental Psychology*, 1963, **66**, 78-83.

Epstein, W., Bontrager, H., & Park, J. The induction of nonveridical slant and the perception of shape. *Journal of Experimental Psychology*, 1962, **63**, 472-479.

Epstein, W. & Broota, K. D. Attitude of judgment and reaction time in estimation of size at a distance. *Perception & Psychophysics*, 1975, 18, 201-204.

Epstein, W. & Landauer, A. A. Size and distance judgments under reduced conditions of viewing. *Perception & Psychophysics*, 1969, 6, 269-272.

Fisher, G. H. An experimental and theoretical appraisal of the perspective and size constancy theories of illusions. *Quarterly Journal of Experimental Psychology*, 1970, 22, 631-652.

Foley, J. M. The size-distance relation and intrinsic geometry of visual space: Implications for processing. *Vision Research*, 1972, 12, 323-332.

Foley, J. M. Depth, size and distance in stereoscopic vision. *Perception & Psychophysics*, 1968, 3, 265-274.

Galanter, E., & Galanter, P. Range estimates of distant visual stimuli. *Perception & Psychophysics*, 1973, 14, 301-306.

Gibson, J. J. The visual field and the visual world: A reply to Professor Boring. *Psychological Review*, 1952, 59, 149-151.

Gibson, J. J. *The perception of the visual world.* Boston: Houghton Mifflin, 1950.

Gilinsky, A. S. The effect of attitude upon the perception of size. *American Journal of Psychology*, 1955, 68, 173-192.

Gilinsky, A. S. Perceived size and distance in visual space. *Psychological Review*, 1951, 58, 460-482.

Gogel, W. C. Cognitive factors in spatial responses. *Psychologia*, 1974, 17, 213-225.

Gogel, W. C. The validity of the size-distance invariance hypothesis with cue reduction. *Perception & Psychophysics*, 1971, 9, 92-94.

Gogel, W. C. The effect of object familiarity on the perception of size and distance. *Quarterly Journal of Experimental Psychology*, 1969a, 21, 239-247.

Gogel, W. C. The sensing of retinal size. *Vision Research*, 1969b, 9, 1079-1094.

Gogel, W. C. & Mertens, H. W. Perceived depth between familiar objects. *Journal of Experimental Psychology*, 1968, 77, 206-211.

Gogel, W. C. & Sturm, R. D. Directional separation and the size cue to distance. *Psychologische Forschung*, 1971, 35, 57-80.

Goldner, J., Reuder, M. E., Riba, B., & Jarmon, D. Neutral vs. ego-orienting instructions: Effects on judgments of magnitude estimation. *Perception & Psychophysics*, 1971, 9, 84-88.

Gregg, C. L. & Pasnak, R. Effects of instructions and training on shape constancy. *Perceptual & Motor Skills*, 1971, 32, 485-486.

Gubisch, R. W. Over-constancy and visual acuity. *Quarterly Journal of Experimental Psychology*, 1966, 18, 366-368.

Haber, R. N. Nature of the effect of set on perception. *Psychological Review*, 1966, 73, 335-351.

Harway, N. I. Judgment of distance in children and adults. *Journal of Experimental Psychology*, 1963, 65, 385-390.

Jenkin, N. A relationship between increments of distance and estimates of objective size. *American Journal of Psychology*, 1959, 72, 345-363.

Jenkin, N. & Hyman, R. Attitude and distance-estimation as variables in size-matching. *American Journal of Psychology*, 1959, 72, 68-76.

Joynson, R. B. An experimental synthesis of the Associationist and Gestalt accounts of the perception of size. Part I. *Quarterly Journal of Experimental Psychology*, 1958a, **10**, 65-76.

Joynson, R. B. An experimental synthesis of the Associationist and Gestalt accounts of the perception of size. Part II. *Quarterly Journal of Experimental Psychology*, 1958b, **10**, 142-154.

Joynson, R. B. & Newson, L. J. The perception of shape as a function of inclination. *British Journal of Psychology*, 1962, **53**, 1-15.

Kraft, A. L. & Winnick, W. A. The effect of pattern and texture gradient on slant and shape judgments. *Perception & Psychophysics*, 1967, **2**, 141-147.

Landauer, A. A. Influence of instructions on judgments of unfamiliar shapes. *Journal of Experimental Psychology*, 1969, **79**, 129-132.

Landauer, A. A. The nature of "apparent" shape judgments. *Australian Journal of Psychology*, 1964a, **16**, 209-213.

Landauer, A. A. The effect of viewing conditions and instructions on shape judgments. *British Journal of Psychology*, 1964b, **55**, 49-57.

Landauer, A. A. & Epstein, W. Does retinal size have a unique correlate in perceived size? *Perception & Psychophysics*, 1969, **6**, 273-275.

Landauer, A. A. & Rodger, R. S. Effect of "apparent" instructions on brightness judgments. *Journal of Experimental Psychology*, 1964, **68**, 80-84.

Lappin, J. S. & Preble, L. D. A demonstration of shape constancy. *Perception & Psychophysics*, 1975, **17**, 439-444.

Leibowitz, H. W. & Harvey, L. O., Jr. Effect of instructions, environment, and type of test object on matched size. *Journal of Experimental Psychology*, 1969, **81**, 36-43.

Leibowitz, H. W. & Harvey, L. O., Jr. Size matching as a function of instructions in a naturalistic environment. *Journal of Experimental Psychology*, 1967, **74**, 378-382.

Levy, L. H. The effects of verbal reinforcement and instructions on the attainment of size constancy. *Canadian Journal of Psychology*, 1967, **21**, 81-91.

Lichte, W. H. & Borresen, C. R. Influence of instructions on degree of shape contancy. *Journal of Experimental Psychology*, 1967, **74**, 538-542.

Lindauer, M. S. & Baust, R. F. Instructions and knowledge of the situation in brightness perception. *American Journal of Psychology*, 1970, **83**, 130-135.

Macmillan, N. A., Moschetto, C. F., Bialostozky, F. M., & Engel, L. Size judgment: The presence of a standard increases the exponent of the power law. *Perception & Psychophysics*, 1974, **16**, 340-346.

Mefferd, R. B., Jr., Redding, G. M., & Wieland, B. A. Depth perception and its special case, slant in depth, as independent of apparent orientation (perspective) in depth. *Perceptual & Motor Skills*, 1967, **24**, 679-690.

Ono, H. Distal and proximal size under reduced and non-reduced viewing conditions. *American Journal of Psychology*, 1966, **79**, 234-241.

Over, R. The effect of instructions on visual and haptic judgment of the Müller-Lyer illusion. *Australian Journal of Psychology*, 1968, **20**, 161-164.

Over, R. The effect of instructions on size-judgments under reduction-conditions. *American Journal of Psychology*, 1960, **73**, 599-602.

Parducci, A., Perrett, D. S., & Marsh, H. W. Assimilation and contrast as range-frequency effects of anchors. *Journal of Experimental Psychology*, 1969, 81, 281-288.

Predebon, G. M., Wenderoth, P. M., & Curthoys, I. A. The effects of instructions and distance on judgments of off-size familiar objects under natural viewing conditions. *American Journal of Psychology*, 1974, 87, 425-439.

Purdy, J. & Gibson, E. J. Distance judgment by the method of fractionation. *Journal of Experimental Psychology*, 1955, 50, 384-380.

Rapoport, J. L. Size-constancy in children measured by a functional size-discrimination task. *Journal of Experimental Child Psychology*, 1969, 7, 366-373.

Rapoport, J. L. Attitude and size judgment in school age children. *Child Development*, 1967, 38, 1187-1192.

Restle, F. Instructions and the magnitude of an illusion: Cognitive factors in the frame of reference. *Perception & Psychophysics*, 1971, 9, 31-32.

Restle, F. Moon illusion explained on the basis of relative size. *Science*, 1970, 167, 1092-1096.

Rock, I. *An introduction to perception*. New York: Macmillan, 1975.

Rogers, S. P. & Gogel, W. C. Relation between judged and physical distance in multicue conditions as a function of instructions and tasks. *Perceptual & Motor Skills*, 1975, 41, 171-178.

Rump, E. E. The relationship between perceived size and perceived distance. *British Journal of Psychology*, 1961, 52, 111-124.

Teghtsoonian, M. The judgment of size. *American Journal of Psychology*, 1965, 78, 392-402.

Teghtsoonian, M. & Teghtsoonian, R. Scaling apparent distance in natural indoor settings. *Psychonomic Science*, 1969, 16, 281-283.

Teghtsoonian, R. Range effects in psychophysical scaling and a revision of Stevens' Law. *American Journal of Psychology*, 1973, 86, 3-27.

Teghtsoonian, R. & Teghtsoonian, M. The effects of size and distance on magnitude estimations of apparent size. *American Journal of Psychology*, 1970a, 83, 601-612.

Teghtsoonian, R. & Teghtsoonian, M. Scaling apparent distance in a natural outdoor setting. *Psychonomic Science*, 1970b, 21, 215-216.

Tolin, P. Relative size as a cue to the relative distances of random stereogram images. *Psychonomic Science*, 1970, 20, 212-213.

Tolin, P. Size-distance judgments with random letter stereograms. *Perception & Psychophysics*, 1969, 6, 340-342.

Vurpillot, É. Perception et représentation dans la constance de la forme. *L'Année Psychologique*, 1964, 64, 61-82.

Wade, N. J. Effect of instructions on visual orientation. *Journal of Experimental Psychology*, 1970, 83, 331-332.

Weber, S. J. & Cook, T. D. Subject effects in laboratory research: An examination of subject roles, demand characteristics, and valid inference. *Psychological Bulletin*, 1972, 77, 273-295.

Winnick, W. A. & Rogoff, I. Role of apparent slant in shape judgments. *Journal of Experimental Psychology*, 1965, 69, 554-563.

Wohlwill, J. F. Changes in distance judgments as a function of corrected and noncorrected practice. *Perceptual & Motor Skills,* 1964, 19, 403-413.

Wohlwill, J. F. The development of "overconstancy" in space perception. In L. P. Lipsett and C. C. Spiker (Eds.), *Advances in child development and behavior,* Vol. 1. New York: Academic, 1963, pp. 265-312.

CHAPTER

ILLUSIONS AND CONSTANCIES

STANLEY COREN
University of British Columbia

JOAN STERN GIRGUS
The City College of the City
University of New York

A number of simple two-dimensional stimuli composed of lines drawn on paper produce percepts that are at variance with the physical reality. When an observer is asked to assess stimulus relations, including the size, shape, or direction of pattern components, his report systematically differs from what might be expected on the basis of direct physical measurements of the stimulus array. Such patterns have been known for more than a century. The first serious psychological treatment of such distortions was presented by Oppel in 1854. He coined the phrase "geometrischoptishe Tauschung," which may be translated as "geometrical optical illusion," to describe such effects. Since Oppel's original publication several hundred different illusion-producing configurations have been cataloged.

This research was supported in part by grants from the National Research Council of Canada (A9783) and the National Science Foundation (74-18599).

The number, variety, and strength of these distortions is quite striking. Consider, for instance, the set of configurations shown in Figure 8-1. A contains the well-known Mueller-Lyer illusion, in which the left half of the horizontal shaft appears to be longer than the right, despite the fact that they are equal in length. The remarkably large magnitude of this distortion results in over and underestimations of shaft length as much as 30%. B contains the horizontal-vertical-illusion in which the vertical line appears longer than the horizontal. C contains the Zoellner illusion, in which the vertical lines are actually parallel but appear tilted. D is a shape distortion, first introduced by Hering in 1861, in which the horizontal lines are, in fact, straight but appear to be bowed outwards. E is the Ponzo illusion, in which the upper horizontal line appears to be longer than the lower. F contains the Poggendorf illusion, which is a distortion of direction. In this configuration the segments of the transversal are colinear, although the segment on the right appears to be too high in relation to the segment of the left. G shows the Oppel-Kundt illusion, which was the focus of Oppel's original paper. Here the upper divided extent looks longer than the lower undivided extent. H is the Ebbinghaus illusion in which the two central

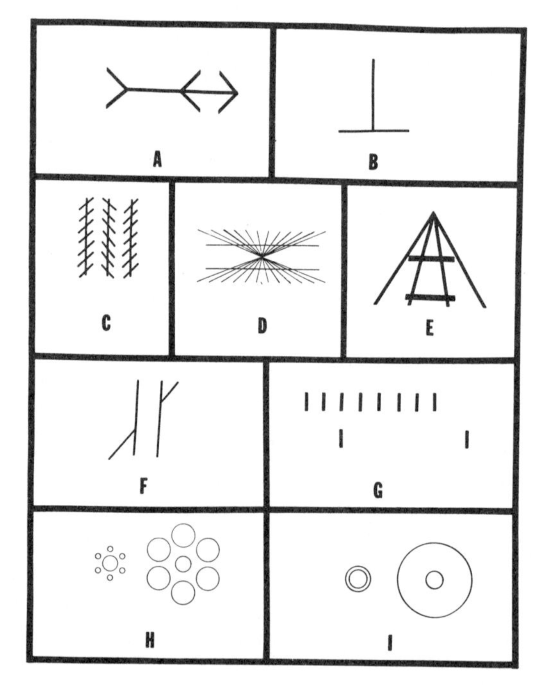

Figure 8-1 A, the Mueller-Lyer illusion; B, The Horizontal-Vertical illusion; C, The Zollner illusion; D, The Wundt illusion; E, The Ponzo illusion; F, The Poggendorf; G, The Oppel-Kundt illusion; H, The Ebbinghaus illusion; I, The Delboeuf.

circles are physically the same size, although the one surrounded by the larger circles is reliably reported as appearing smaller than the one surrounded by smaller circles. I shows a similar size distortion, known as the Delboeuf illusion, in which the circle surrounded by the smaller concentric circle appears larger than the circle surrounded by a concentric circle with a larger diameter.

Historically, there have been two major motivations for studying visual illusions. Probably the first is their curiosity value as a form of "visual magic." The second, more serious, is based on the firm belief that an understanding of these distortions will provide important clues about the nature of normal perceptual processing. In this vein Helmholtz (1883) argued as follows:

> The study of what are called illusions of the senses is, however, a very prominent part of the psychology of the senses; for just those cases which are not in accordance with reality are particularly instructive for discovering the laws of those means and processes by which normal perception originates.

This is a viewpoint that has been reiterated many times; for example, Baldwin (1890) contended that the study of pathological perception or, more simply illusions, is just as important to the understanding of normal veridical perception as the study of pathological states of the body is to the understanding of normal bodily functioning.

The general consensus has been that illusions represent the action of normal mechanisms that, because of the unusual circumstances in the stimulus array or viewing situation, lead to a distorted percept. Helmholtz (1867) summarized this position a hundred years ago when he said,

> The explanation of the possibility of illusions lies in the fact that we transfer the notion of external objects, which would be correct under normal conditions, to cases in which unusual circumstances have altered the retinal picture.

This type of reasoning suggests that we should view illusion configurations with an eye toward isolating perceptual mechanisms that are normally adaptive but inappropriately applied to the particular stimulus. It is the misapplication of a normal perceptual process that results in an erroneous percept.

Visual illusions may be understood more clearly if we interpret the process of perceiving as a search for meaning in the stimulus array. If we view the organism as an information-processing computer, the stimulus represents the input data, the cognitive judgmental strategies represent the program, and the solution or output is displayed as the conscious representation, which contains the organism's current hypothesis about the meaning of the stimulus input. For any living organism attempting to negotiate its way around a complex visual environment the word *meaning* in this context refers to a *referential meaning.* The organism is attempting to ascertain the identity and location of the object, or set of objects, that is represented by any particular pattern of stimulation. Thus the behaviorally relevant meaning of any stimulus rests in what Ames (1946) has called its *thatness* and *thereness.*

In this context we may now consider an observer who is looking at a simple illusion configuration. It is composed of a pattern of lines drawn on the two-dimensional surface of a piece of paper. How does an observer go about assigning a referential meaning to this array? It seems likely that the answer to this question lies in the area of picture perception, for we are posing a problem in which the lines drawn on the surface of a paper must be interpreted as representing something more than mere lines. As Gibson (1951) has shown, even line drawings of random shapes tend to be seen in terms of objects. Thus a pair of curved lines converging toward the top of the page was not reported as two curved lines but rather as a "horn" or a "road receding into the distance." The predisposition to assign a representational status to arrays of two-dimensional contours is apparently quite automatic. Helmholtz (1867) described this active process of search for referential meaning in such simple configurations when he wrote,

> . . . such objects are always imagined as being presented in the field of vision as would have to be there in order to produce the same impression on the nervous system, the eyes being used under normal conditions.

Because normal conditions usually entail viewing a three-dimensional world, filled with solid objects, it may be that observers are looking for the three-dimensional analog to the two-dimensional array presented to them. It is this predisposition that ties constancies to illusion formation.

PICTURE PERCEPTION AND ILLUSIONS

Let us begin by considering an extremely primitive pictorial representation. This "picture" consists of one vertical line, as shown in Figure 8-2A. What does this line represent? The very fact that we are asking this question indicates the almost automatic nature by which the perceptual system attempts to assign an object referent to any stimulus configuration presented to it. It is certainly possible that the conscious percept could simply represent the physical situation—here a thin layer of ink resting on a flat plane of paper. Yet, if the observer occludes one eye and views this primitive pattern for a few moments, a number of alternative meanings or conscious representations begin to suggest themselves. Thus the line may apparently take on the aspect of an unsharpened pencil lying on a flat surface or a flagpole standing vertically, imaged against a field covered with snow. Not only does the line contain the possibility of assuming a number of object identities but it may also evoke a number of alternative depth interpretations. The single vertical line we have been looking at could just as well be seen as a line lying on a horizontal plane receding into the distance, with the lowest point closest to the observer. We can easily demonstrate this alternate depth representation by integrating this vertical line into a more complex figure as we have done in Figure 8-2B. The same line could also represent an overhead line or contour in which the highest point in the line is closest to the observer, as shown schematically in C. Thus a number of alternative referential meanings are possible in any simple stimulus array. Whenever the organism selects a conscious representation of the stimulus that implies a specific orientation of an object in three-dimensional space, it has, at the same time, set in motion the constancy scaling mechanisms appropriate to that orientation. These mechanisms then interact with and shape the final conscious percept such that the resultant interpretation differs considerably from the actual image on the retinal surface. Let us see how such predispositions may, in fact, result in the formation of visual illusions and distortions.

Let us consider a sketch of a real world scene (see Figure 8-3A). Basically it represents a road receding into the distance. To most observers the log marked x appears to be the same size as the log marked y. This is certainly not a surprising percept, for, despite the fact that a ruler will immediately show that x is considerably shorter than y, a number of cues indicate that x is farther away than y. These cues include the perspective lines represented by the edge of the road, the decreasing retinal size of objects in the array, and the increasing density of textural elements in the scene.

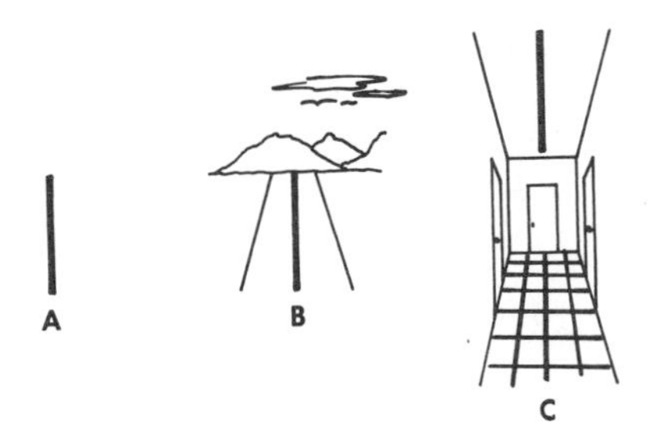

Figure 8-2 A, a minimal picture; B, vertical line with bottom end near; C, Vertical line with bottom end far.

In the real three-dimensional world, when two objects of the same physical size lie at different distances from the observer, the more distant object projects a smaller retinal image. By the operation of size constancy, however, the apparent distance of an object is taken into account; thus the size represented in consciousness remains constant despite the changes in retinal angle due to distance variations. In a two-dimensional representation of such a scene all the objects, by definition, must lie at the same distance from the observer, namely at the picture plane. The artist must indicate the appropriate depth cues so that they appear to lie at different distances from the observer and must draw objects of different sizes so that they will project images whose sizes in relation to one another mimic those obtained from viewing a three-dimensional scene. Under these circumstances it is clear that there is sufficient information contained in the two-dimensional array to trigger the size constancy scaling mechanism. This mechanism seems to operate on the picture or line drawing just as it would operate had the observer been viewing the comparable real-world scene. For the average viewer there is no question of an illusion in this configuration. It is merely an "accurate" representation of two equal-sized objects lying at different distances from him. Nontheless, if we consider the final percept, we have indeed created an illusion. Figure 8-3A presents two different physical extents at the same distance from the observer, yet the resulting percept is on two equal extents.

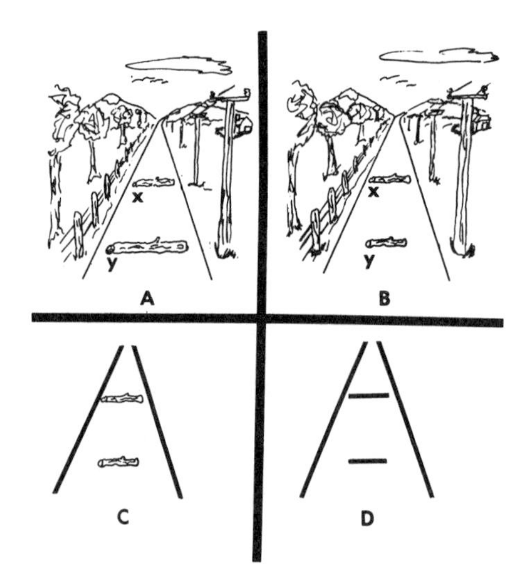

Figure 8-3 A, the logs x and y appear at different distances but the same size; B, log x appears larger than log y although they are drawn to be the same size; C, a reduced version of the preceding figure; D, The Ponzo illusion.

Let us now manipulate this drawing slightly, as shown in B. Here, again, we have two logs lying on a road at apparently different distances. However, because the two logs are drawn so that they are physically equal on paper, they now project equal-sized retinal images. In this configuration the two equal physical extents are seen as unequal, with the apparently more distant log (x) appearing to be larger than the apparently nearer log (y). This percept is again a reasonable application of the size constancy scaling mechanism. Certainly, if this were a real-world scene, the only way that two objects at different distances could subtend the same visual angle would be that the more distant object were indeed physically larger than the nearer object. It is interesting to note that although few observers would maintain the presence of any illusory distortion in A, a reasonably high percentage of observers looking at B would be somewhat puzzled by the resultant percept and be inclined to attribute the inequality of the two physically equal stimuli to an illusory distortion. To the extent that we are viewing a picture, the resultant percept in B, in which two equal physical objects are seen as if they were unequal in size, is as accurate as that evoked by A, in which two unequally sized physical objects appear to be equal in size. Most observers therefore could be convinced that B is an accurate pictorial representation of some possible real-world array.

Let us now reduce the configuration in B to the much more minimal form shown in C. As we can readily see, we have removed all the depth cues except the height in the picture plane and the perspective lines that converge along the sides. Notice that in this configuration the upper log still looks longer. It is easy to see that we are gradually beginning to approximate a variant on the Ponzo illusion which we saw as Figure 8-1E, while maintaining a schematic representation of the same real-world scene we presented in Figure 8-3B. It is only a small step from C to D, which is clearly a classical variant on the Ponzo illusion.

What makes D an illusion configuration and B a picture? Perhaps it has something to do with the fact that few observers would describe D as a drawing of objects at different distances (Worrall & Firth, 1971), for we do not consciously encode the array as a two-dimensional representation of a three-dimensional scene. We may be somewhat taken aback that two objects which are physically the same size on the page and which project the same sized retinal images are phenomenally represented as different in size. Despite the fact that the depth cues in D are not sufficiently powerful to evoke an overt perception of tridimensionality, they are apparently strong enough to trigger the constancy scaling mechanism, thus producing an apparent inequality in the two targets.

Using a pattern of interpretation similar to that offered above, a number of investigators have suggested that the inappropriate application of size constancy mechanisms to two-dimensional stimuli which are not intended to represent three-dimensional depth is responsible for a large number of the classical illusions. These investigators include Thiery (1896), Tausch (1954), Kristopf (1961), Gregory (1963, 1966, 1968a,b, 1970), Day (1972), and Girgus and Coren (1975). The basic line of reasoning employed by these theorists was initially laid down by Thiery (1896) who addressed himself to the Mueller-Lyer illusion (Figure 8-4A). In analyzing the standard configuration, he noted that the converging lines could be interpreted as pictorial perspective cues, which indicated that components of the figure are actually at different distances from the observer. He suggested that the apparently shorter half of the illusion may be seen as a sawhorse or trestle viewed from above, with the legs (represented by the wings) extending away from the observer. The apparently shorter half of the illusion could be viewed as the same sawhorse viewed from below, with the ends of the legs extended toward the observer.

A more familiar example of a three-dimensional configuration suggested by the Mueller-Lyer illusion is shown in B and C (Gregory, 1968a). B is the outer edge of a building that may be seen as the apparently shorter half of the Mueller-Lyer illusion, with the receding perspective lines of the roof and the base of the building making up the inward pointing wings and its protruding corner providing the shaft. C shows the corner of a room, which is apparently the longer half of the Mueller-Lyer illusion with the lines of the ceiling and floor making up the outward pointing wings and the corner making up the shaft. If the observer assumes that the closest point in the figure is roughly at the plane of the paper, the vertical line that represents the extent to be judged must be farther away from the observer in C than it is in B.

Because both lines are physically the same length and at physically the same distance, they subtend the same visual angle in the retinal image. Under these conditions the application of size constancy would correct for the *apparent* difference in distance and the phenomenally more distant line would be seen as larger than the phenomenally closer line. According to this mode of analysis, the Mueller-Lyer illusion in A may be interpreted as a schematic representation

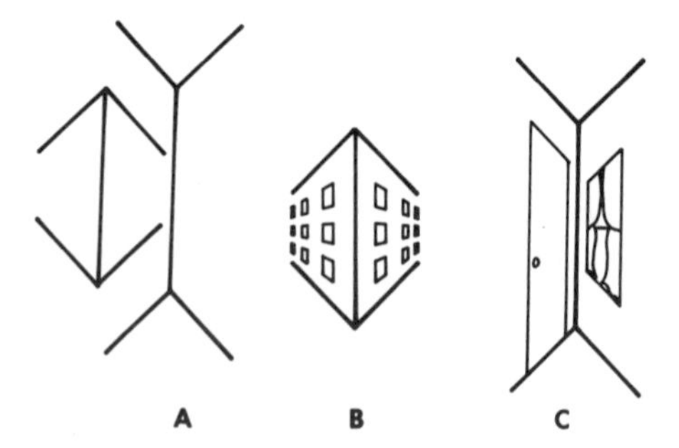

Figure 8-4 A, the Mueller-Lyer illusion; B, suggested depth in the apparently shorter portion; C, Suggested depth in the apparently longer portion.

of the three-dimensional depth relations shown in B and C, much as the Ponzo illusion shown in Figure 8-3D may be seen as a schematic representation of a roadway with logs lying across it. However, because the simple line configurations which make up the illusion are not intended to be representational, the evocation of the constancy mechanism is inappropriate. The resultant percept surprises the observer, for it produces an apparently inexplicable distortion in the size relations of a series of two-dimensional lines.

PICTORIAL DEPTH AND ILLUSIONS

A number of questions must be answered before we can accept the misapplication of size constancy as the explanation for many classical visual geometric illusion. First, we must ask whether observers interpret the depth cues in a pictorial array as if they were depth cues in the real world, despite the fact that they are quite simplified and improverished, for in the absence of the registration of depth constancy scaling cannot be evoked.

Much evidence indicates that depth relations may be extracted from two-dimensional representations. Smith and Gruber (1958) for example, presented subjects with a black and white photo mural of a 360 ft corridor. Subjects were asked to estimate the number of paces from their viewpoints to specific parts of the pictured scene. They were as accurate in their estimates based on the flat representation as they were when presented with the actual corridor. Furthermore, varying the magnification of the two-dimensional scene changed their estimates of the number of paces in a manner predictable from the changed projection of the optic array. However, large wall-sized photographs might be considered a rather special two-dimensional array. Despite the fact that it lacks color and distorts such depth cues as binocular disparity and motion parralax, this array is a cue-laden configuration and sizes are faithfully reproduced.

Smith, Smith, and Hubbard (1958) attempted to assess the accuracy of distance judgments in a variety of more simplified arrays. To begin with, they had a black and white photograph of a corridor, as in the Smith and Gruber (1958) experiment. In addition, they used a line drawing of the corridor that contained a great deal of detail, a line drawing that contained only the corners of the corridor and the junctions of the walls with the floors and ceilings, and in which the more distant parts of the corridor were slightly darker, and finally an extremely reduced version in which even the shading cue was eliminated. The observers were required to compare the apparent distance to the end of the corridor in the photograph with the apparent distance to the end of the corridor in the line drawings, using a magnitude estimation technique. They were asked also to estimate the size and distance of various parts of the corridor. Smith, Smith, and Hubbard report that the photographs and the line drawings produced equivalent perceptions of distance. They also found that changes in magnification of the display systematically altered the distance estimates. They concluded that "there is no evidence that depth perceived in perspective line drawings differs from that of photographs" (p. 674).

Although these data seem to indicate that adults can extract information regarding three-dimentional relationships from two-dimentional linear arrays, they do not tell us whether cues in any of the standard visual geometric illusions will elicit this depth encoding. Several investigators have tried to assess the presence or absence of depth cues in illusion configurations by asking subjects to report what they see when they look at these figures; for example, Worrall and Firth (1971) and Porac, Ward, Coren, and Girgus (1976) asked observers to report what they saw when they viewed the converging lines that provide the inducers in the Ponzo illusion (Figure 8-3D). Few observers report that the vertex of the angle is seen as more distant than the lower open end, although small variations in the inducing lines, such as leaving the top of the angle open (Worrall & Firth, 1971; Worrall, 1974) or increasing the number of lines that converge to a single point (Porac, Ward, Coren, & Girgus, 1976) increases the likelihood of a report of appropriate depth interpretations. Such data seem to imply that the standard Ponzo configuration does not usually elicit a conscious experience of depth, although some variants of the configuration may tend to do so.

The data based on phenomenal reports of depth in the Mueller-Lyer illusion is equally ambiguous. Even when possible conflicting depth cues that indicate the flatness of the array, such as the paper texture or specular reflection, are removed, most subjects report no depth for the Mueller-Lyer configuration (Hotopf, 1965). Pike and Stacey (1968) find that judgments of the depths of the fins in relation to the shaft are rather erratic and frequently may even be the opposite of the depth relations necessary to produce the illusion. Porac, Ward, Coren, and Girgus (1975) presented the apparently shorter and longer version of the Mueller-Lyer illusion separately and asked observers to describe an object or scene suggested to them by the array. Their results are consistent with those of Hotopf (1965) and Pike and Stacey (1968) in that less than half of their observers reported configurations that incorporated tridimensionality. Among those who did report tridimensional configurations, however, there is a significantly greater tendency to report the shaft as being closer in the apparently longer configuration, which is consistent with the implicit depth cue analysis we have presented.

In addition, some direct measurements seem to indicate that depth cues are implicit in some of the classical illusion figures manifested under certain measurement conditions. Gregory (1966, 1970) has used a device in which a luminous Mueller-Lyer figure is presented against a dark background. The figure is

presented to only one eye, thus eliminating cues from binocular disparity in regard to the locus of parts of the figure. In addition, it was only dimly luminous and thus eliminated textural cues for flatness as well. The observer was asked to adjust a binocularly seen point of light so that it appeared to be at the same distance as various points on the illusion configuration. When presented with the apparently longer half of the Mueller-Lyer illusion, which supposedly mimics the depth relationships shown in C, observers set the matching point as if the shaft were seen as more distant than the wing tips. When observers were presented with the apparently shorter half of the illusion, which supposedly evokes the depth relations shown as B, the shaft was indicated as being closer than the ends of the wings. Gregory (1966) has also presented data that seem to show that the variations in the angle of the wings lead to changes in the depth settings that correspond to the changes in the magnitude of the Mueller-Lyer that occur when wing angle is varied. Further support for such registered depth in simple illusion figures comes from Coren and Festinger (1967) who utilized Gregory's technique on a curved variant of the Ponzo illusion. They report depth differences that are in accord with the presumption that implicit depth cues occur in such configurations. Thus, when all cues to flatness are removed from the Mueller-Lyer and Ponzo illusions, the data seem to indicate that observers perceive the predicted depth relationships, which suggests that at least some implicit cues to depth are in these illusion configurations.
figurations.

When we consider all these studies, however, the weight of the evidence seems to indicate that most observers do not glance at the usual illusion configuration drawn on a sheet of paper and immediately interpret it as a schematic representation of a set of tridimensional relations. Their phenomenology is clearly more consistent with an interpretation in which lines are seen as resting on a flat surface of paper with no implied depth variations. Although at first glance such results may seem to run contrary to a constancy scaling in picture perception explanation of visual illusions, this is not necessarily the case. It may simply be another instance of the rather knotty problem of *phenomenal* versus *registered* depth, which has long puzzled researchers in the areas of constancies (cf Epstein, 1973; Rock, 1975). Although alterations in the cues to distance seem to lead to predictable changes in apparent size, most investigators have failed to find the concomitant changes in phenomenal depth judgments that would be expected under these circumstances (Wheatstone, 1852; Judd, 1897; Gruber, 1954; Hermans, 1954; Heinemann, Tulving, & Nachmias, 1959; Rock & McDermott, 1964; Gogel, Wist and Harker, 1963). Thus the failure of the stimulus configuration to evoke apparent depth differences does not necessarily mean that there are not sufficient cues to evoke the constancy scaling mechanism at some level of registration.

CUES FOR PICTORIAL CONSTANCY SCALING

To the extent that constancy effects contribute to the final magnitude of some visual geometric illusions, there may be theoretical and methodological benefits to be gained by inverting the argument that we used above. To the extent that illusion-producing configurations are minimal two-dimensional representations of three-dimensional scenes, which trigger the constancy scaling mechanism, they could provide us with a powerful tool for the investigation of factors that affect constancy scaling. Thus, rather than ask what constancy scaling mechanisms can tell us about illusion formation, we can ask what illusions can tell us about how constancy mechanisms operate on pictorial stimuli.

To do this let us first consider a most concrete question associated with constancy scaling, namely the isolation of depth cues that are powerful enough to evoke the constancy mechanism. Clearly, because most classical geometric illusions constitute lines drawn on a flat surface, we must limit our analysis to monocular depth cues commonly associated with pictorial representations.

Let us first see which of the most common depth cues are sufficiently strong to elicit size variations in simplified arrays. In both the Ponzo and Mueller-Lyer configurations, we have been dealing with the cue of linear perspective. This cue consists of converging or slanting lines that tend to be seen as representing variations in distance. In general, regions in which the lines are closer together tend to be seen as farther away than regions in which the lines are farther apart. An extreme example of how perspective can affect size estimates is shown in Figure 8-5A, in which the linear perspective cues indicate the presence of a plane receding into the distance. Note that the two horizontal bars, which are physically drawn as the same size, appear to differ in size. It is hard to decide whether we should classify this perception as a variant of the Ponzo illusion or as a correct interpretation of a pictorial array. As we have already noted, the distinction between pictures and illusions is often quite indistinct.

Another common cue to depth is interposition, in which an object that partly blocks another object from view is seen as closer to the observer than the partially obscured object. Figure 8-5B shows a size illusion based on an interposition cue in a minimal figural array (Coren & Girgus, 1975).

The upper figure in Figure 8-5B is drawn so that the line can be viewed as standing in front of the circle, whereas the lower figure is drawn so that the circle can be viewed as standing in front of the line. Note that these are possible interpretations of the array on the basis of an implied interposition between the line and circle rather than a phenomenal impression that suggests itself spontaneously. If the interposition cue in this array triggers the constancy scaling mechanism, the apparently more distant circle (the one with the line interposed in front of it) should appear to be larger than the apparently nearer circle, which is interposed in front of the line. The data confirm this prediction when compared with appropriate control configurations.

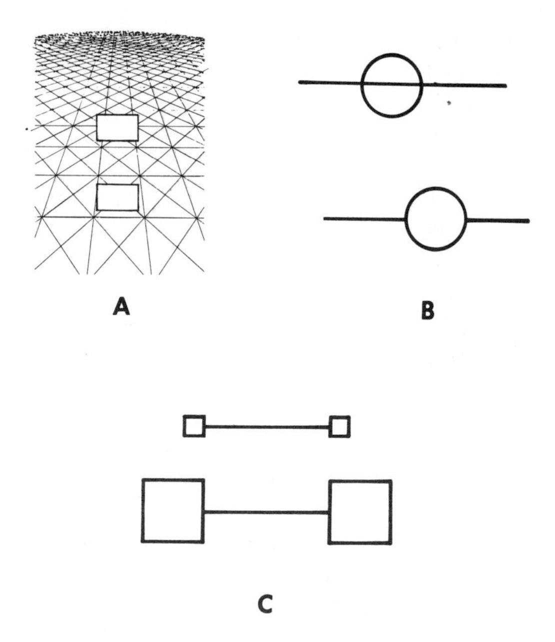

Figure 8-5 A, size differences induced by drawn perspective; B, a minimal interposition illusion; C, the Baldwin illusion.

The relative size of stimuli can also serve as a cue to depth because stimuli that are more distant tend to produce smaller retinal images. Day (1972) has suggested that this cue may be enough to trigger the size constancy mechanism in such simple configurations as the Baldwin illusion, a variant of which is shown as C. Note that in this figure the upper horizontal line flanked by the smaller squares appears to be larger than the lower horizontal line flanked by larger squares. Day suggests that the observer assumes that the squares represent objects that are all the same size, hence interprets differences in the retinal image as depth differences, thus eliciting the size constancy mechanism which leads to a distortion of the apparent line lengths. Direct evidence for this size identity assumption has been presented by Epstein and Baratz (1964). If the reader will study Figure 8-5C for a few minutes, this apparent depth difference may begin to manifest itself in consciousness. Note that the same explanation may be used to explain the apparent-size differences observed in the Ebbinghaus illusion (Figure 8-1H) and may even be extended to the Delboeuf illusion (Figure 8-11).

Another common monocular cue is the relative height of objects in the picture plane. In a configuration in which most objects are shown below the horizon, objects that are higher on the picture plane tend to be seen as farther away (cf. Epstein, 1966). This relationship is depicted in Figure 8-6A, in which line A is seen as closer than line B. Girgus and Coren (1975) have suggested that this pictorial cue is powerful enough to trigger the constancy scaling mechanism and lead to the horizontal-vertical illusion (Figure 8-1B), in which a vertical line is overestimated in length in relation to a horizontal line. These investigators suggest that although a horizontal line in a two-dimensional array offers little in the way of suggested depth cues a vertical line may frequently depict a line receding into the distance on the basis of the height in the plane cue. This may be more clearly seen if we consider B. In this figure an eye is viewing a line lying horizontally along a surface. Note that point a is farther away from the observer than point b. C shows a two-dimensional projection of this line as it would appear in the retinal image or in a photograph taken from the vantage point of the eye in the preceding panel. Notice that here point a is represented by a point higher in the picture plane than point b. This is merely another example of the height in the picture plane cue. If relative height is used to represent depth differences, the geometry of the situation is such that a constant retinal unit of length should represent a greater change in distance for the upper part of a vertical line than for the lower part of the same line. Thus in the scene in B the three points, a, b, and c, are all equally spaced, whereas the nearer points b and c are more widely in the two-dimensional projection of the same configuration. We might therefore expect that if an observer is asked to bisect a vertical line and that vertical line is registered as receding in depth, the size constancy correction should cause him to overestimate the upper half of the line in relation to the lower half. D demonstrates that such an illusion does indeed exist. The hatchmark actually bisects the vertical line precisely, but it appears to be placed a bit too low for most observers.

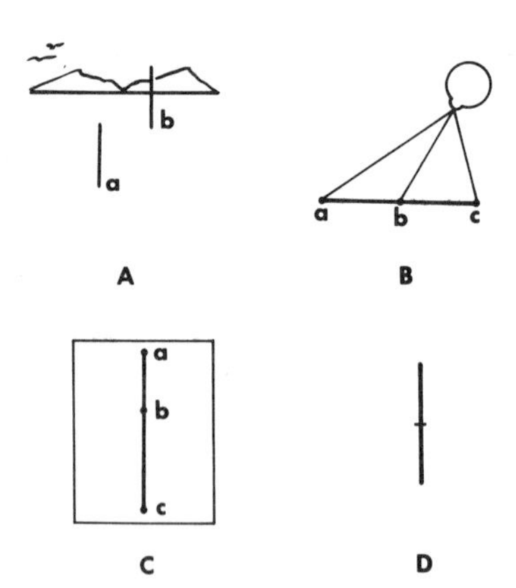

Figure 8-6 The line b appears further away than the line a, because it is higher in the picture plane; B, view of other horizontal line extending away from an observer; C, a frontal projection of the image of the line seen in B; D, the bisection illusion.

Gibson (1950) has analyzed the pictorial cue of texture gradient in some detail. This cue combines some of the aspects of relative size and perspective. Basically it may be described by noting that regions in the field in which objects or visual elements are more densely packed together seem to be farther away. In Figures 8-7A and B the texture gradient cue is strong enough to produce the phenomenal perception of fields or plains receding into the distance. Notice that objects superimposed on these texture gradients appear to vary in size, thus suggesting that constancy scaling has been triggered.

This list of monocular depth cues could be continued to include relative brightness, aereal perspective, familiar size, shadowing, and other pictorial depth cues to show that they also have, or have not, sufficient salience in a two-dimensional representation to evoke the constancy scaling mechanism. These examples, however, are sufficient to show that such analyses can be conducted. Let us now see how the utilization of the minimal depth cues found in some illusory configurations may be used to specify the field characteristics of constancy corrections in pictures.

NONUNIFORMITY OF CONSTANCY SCALING IN PICTURES

It has been generally accepted that constancy scaling is fairly uniform across the visual field under normal three-dimensional viewing conditions. Thus in an experiment on size constancy in which the target is a circle placed at varying distances, most experimenters would be content to have observers estimate the size of the circle by reproducing its horizontal diameter, simply presuming that the vertical diameter would covary in the same manner. Illusion research suggests that, at least for constancy scaling in pictures, this presumption may be unwar-

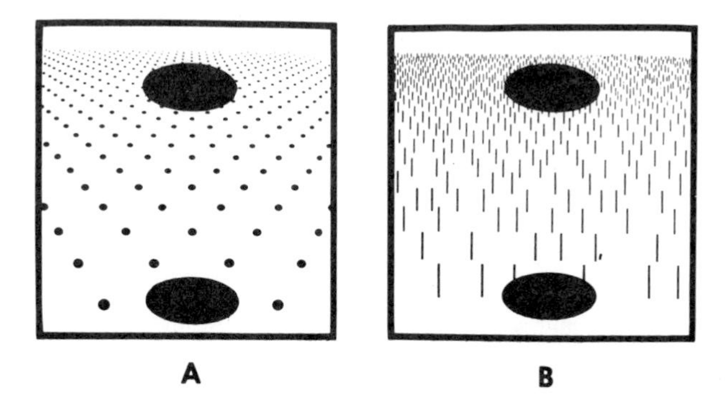

Figure 8-7 Objects vary in apparent size due to depth employed by texture gradients.

ranted, and that constancy scaling may operate nonuniformly over visual space, depending on the nature of the cues present in the array.

As an example, let us again turn to the Ponzo illusion, which contains, according to the constancy scaling analysis, a minimal perspective cue sufficient to elicit size constancy. In viewing A, it is clear that the upper portion of the angle, at which the lines converge more closely, will be registered as more distant. Thus the size constancy correction should lead us to perceive the upper line as larger. Let us now modify this configuration so that the two lines are oriented vertically rather than horizontally, as shown in B. This configuration was introduced by Humphrey and Morgan (1965) and later studied in some detail by Gillam (1973). The majority of observers note a massive reduction in the illusory distortion when the lines are oriented vertically. This is somewhat surprising because a constancy scaling analysis would predict that the upper vertical line should be seen as more distant than the lower vertical line and thus should be seen as larger. The fact that the size difference is so much smaller with vertical lines than it is with horizontal lines suggests a hithertofore unnoticed aspect of the interaction between pictorial depth cues and constancies in that it suggests that constancy corrections may be nonuniformly applied to various dimensions.

Let us note that the major change in the Ponzo configuration is in the horizontal spacing of the inducing lines. As we move up the angle, the inducing lines tend to compress around the test elements. The importance of this tendency can be seen in a varient of the Ponzo, introduced by Fisher (1968, 1970) and redrawn in C, in which the converging lines have been removed and replaced

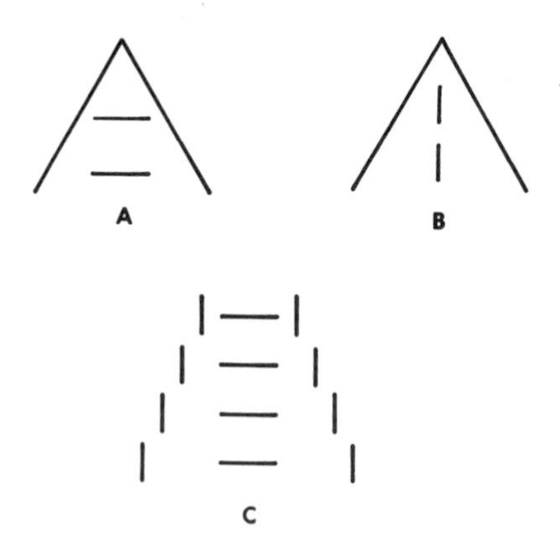

Figure 8-8 A, The Ponzo illusion showing apparent size differences in the horizontal lines; B, a Ponzo variant showing little distortion in the enclosed lines; C, A variant of the Ponzo illusion without converging line elements.

with vertical line segments that vary in distance from the ends of the test elements. It is clear that the expected distortion occurs in this configuration and suggests that constancy scaling may primarily be evoked along the axes of maximal gradation or change in depth cues. Gillam (1973) tested this hypothesis by using a texture gradient similar to that shown in Figure 8-9, in which the maximal rate of change is found in the vertical direction, where the line elements compress together. Gillam notes that despite the fact that one is left with the impression of a plane receding away from the observer, little size change is found for horizontal lines imposed on the gradient. Maximal change is found for vertical lines that are oriented in the direction of maximal cue change.

Other data collected from modified illusory distortions seem to indicate that constancy scaling does not occur uniformly across elements but rather selectively responds to orientations or directions of change. Consider the modified Mueller-Lyer figures shown as Figures 8-10A and B. These configurations were investigated by Waite and Massaro (1970), Dengler (1972), Massaro (1973) and Griggs (1974), all of whom have argued that if the operation of size constancy is homogeneous across the figure the entire rectangle in A should apparently be more distant from the observer than the rectangle in B. Thus, in addition to the length distortion, there should also be a width distortion, with an overestimation of the width in A in relation to that of B. Actually the distortion of width is in the opposite direction to that predicted by a uniform constancy

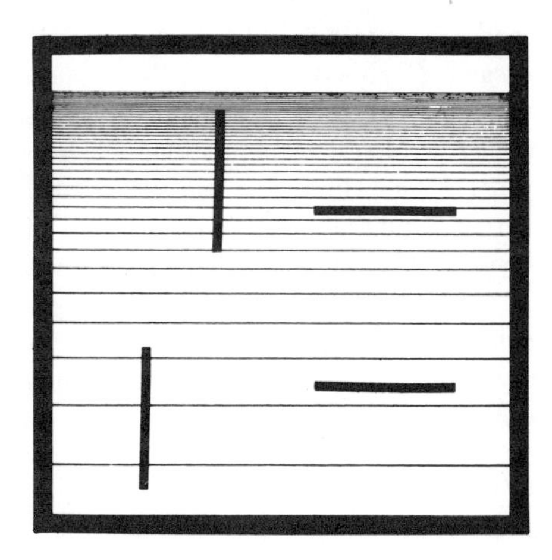

Figure 8-9 The vertical lines which are orthogonal to the direction of change of the depth view vary in size more than the horizontal lines.

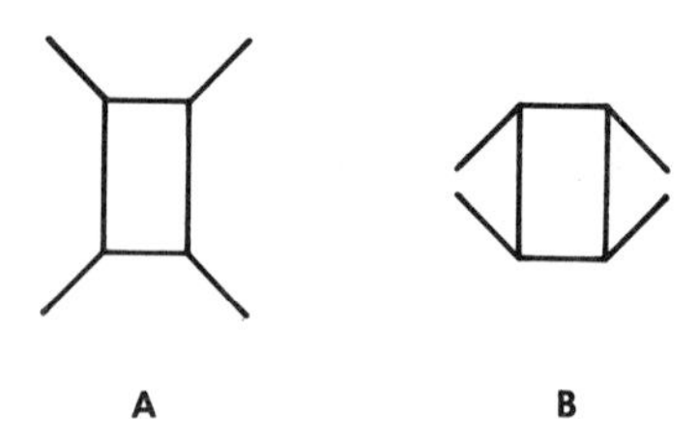

Figure 8-10 A modified Mueller-Lyer illusion.

scaling analysis and thus provides further evidence for the anisotropy of constancy scaling as a function of orientation of cue action.

There is another possible source of ambiguity in constancy scaling of pictorial arrays. When looking at pictures, particularly the kind of minimal pictorial arrays that illusions represent, it is possible for observers to engage in size adjustment centered around several alternate zero points. The observer might first assess the overall amount of depth in the array and then select some midpoint value. Items seen as more distant than this midvalue would be relatively overestimated, whereas items seen as closer would be underestimated. Alternatively, the observer could select the most distant item as being of zero value and systematically underestimate all targets in relation to its size or the closest item as the baseline and overestimate other targets in relation to this referent. In general, the data on visual illusions seem to indicate that the latter strategy is adopted. Thus in the Coren and Girgus (1975) configuration which utilized interposition to evoke constancy scaling (Figure 8-5B) the circle interposed in front of the line (i.e., the nearest target) is not seen as signigicantly different in size from control configurations involving no depth manipulation. The apparently more distant targets, which are interposed behind the line, are overestimated in relation to control configurations. A similiar pattern emerges from Gillam's (1973) work on the Ponzo illusion and its textural analogue (Figure 8-9). Here, again, the test element indicated as being closer by the presence of perspective or textural cues is not judged as being significantly different in size from control elements, whereas the apparently more distant targets are overestimated.

ADDITIVITY OF CONSTANCY SCALING IN PICTURES

Data based on visual illusions seem to indicate that constancy scaling in pictures acts in an additive fashion. Thus Leibowitz, Brislin, Perlmutter, and Hennessy (1969), working with the Ponzo illusion, showed that the strength of the illusion is considerably increased when it is displayed as part of a photograph rather than in its usual impoverished form. This may be due to the fact that the photograph provides many more depth cues that can serve to evoke constancy scaling. This effect may be seen in Figure 8-3A versus D, in which the size difference between the test elements is stronger for the version with added pictorial depth cues.

Similar results have been reported by Newman and Newman (1974) who found that increasing the depth cues in a pictorial array increased the magnitude of the illusions. Interestingly it seems to be possible to increase the constancy scaling correction through the addition of depth cues which point in a similar direction but it does not seem to be possible to decrease the illusion through the addition of conflicting depth cues. Newman and Newman (1974) presented the converging line elements of the Ponzo illusion in an apparently flat pictorial array and found no reduction in the magnitude of the illusion. Fisher (1968, 1970) altered some of the classical illusion figures by adding pictorial cues to achieve a depth impression the opposite of that necessary to explain the illusion. Figures 8-11A and B are examples of this kind of manipulation. In A we have drawn the legs or fins to be consistent with the depth relations suggested by the perspective cues, whereas in B we have reversed this apparent depth relation and

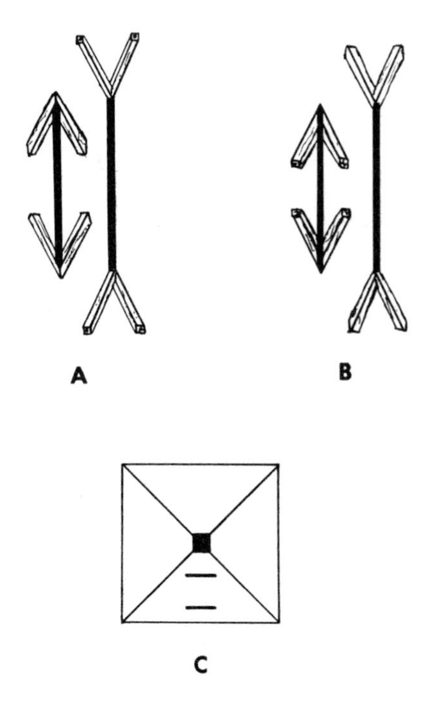

Figure 8-11 A, A modified Mueller-Lyer illusion with drawn depth cues consistant with employed perspective cues; B, A modified Mueller-Lyer illusion with inconsistent suggested pictorial depth; C, a Ponzo configuration with reversible apparent depth.

it now runs contrary to that necessary for the explanation of the illusory distortion. Despite this reversal, the usual illusory effect is still present.

These data may simply provide additional evidence for our earlier suggestion that the cues used to elicit the phenomenal impression of depth are not necessarily those that are registered and used to evoke the constancy scaling mechanism. This principle is demonstrated in C, which is patterned after Fisher (1968, 1970). At first glance this figure is a flat symmetrical pattern of lines resting on the page. Notice that in the lowest quadrant there are two horizontal lines that are flanked by converging elements, hence forming a varient of the Ponzo illusion. If this configuration is studied for a few moments, it will begin to take on some depth characteristics. Sometimes it may look like a long receding hallway in which the upper line is farther away than the lower; at other times it may appear to be a pyramid viewed from above in which the lower line is farther away than the upper. Notice that the apparent length of the two line segments does not change as the phenomenal depth varies from flat, to pointed inward, to pointed outward. Thus the constancy computation that presumably leads to the Ponzo illusion is clearly using a data base quite different from that used in the computation of apparent depth. Observers looking at pictorial stimuli, like observers looking at three-dimensional stimuli (Wheatstone, 1852; Gruber, 1954; Hermans, 1954; Heinemann, Tulving, & Nachmias, 1959; Rock & McDermott, 1962; Gogel, Wist & Harker, 1963), seem to utilize registered depth rather than phenomenal depth information when computing size corrections via constancies. Furthermore, the depth registration used in constancy scaling seems to attend to some cues (e.g., converging perspective lines) in preference to others.

EXPERIENCE AND CONSTANCY SCALING IN PICTURES

There is one area in which visual illusions have been used extensively as a research tool to explore the nature of constancy scaling. To the extent that visual illusion configurations serve as minimal pictorial stimuli capable of evoking constancy scaling mechanisms, they seem to provide an interesting set of tools to explore the effects of experience on constancy scaling. Unfortunately, the results have not always been clear cut or consistent.

Leibowitz and Heisel (1958), Hanely and Zerbolio (1965), Leibowitz and Judisch (1967), and Farquar and Leibowitz (1971) have consistently shown that size constancy for distant objects increases as a function of chronological age, hence presumably with experience in the environment. On the basis of such results it would seem reasonable to suppose that illusions based on the application of size constancy would also increase with age. Unfortunately, the developmental trends are actually quite mixed. For the Ponzo illusion, which is quite clearly reducible to a minimal pictorial array as we demonstrated earlier, there seems to be an increase in illusion magnitude with increasing chronological age (Zeigler & Leibowitz, 1957; Leibowitz, Pollard, & Dickson, 1967; Jenkin & Feallock, 1960; Leibowitz, 1961).

Unfortunately, the Ponzo illusion is the only configuration that consistently shows such trends. Most other illusions seem to show decreases in magnitude as a function of increasing chronological age. This trend has often been confirmed in Mueller-Lyer illusion (Binet, 1895; Girgus, Coren, & Fraenkel, 1975; Piaget & von Albertini, 1950; Pollack, 1964, 1969). In much the same fashion the horizontal-vertical illusion shows a gradual decrease in illusion magnitude as a function of increasing chronological age (Doyle, 1967; Piaget, Matalon, & Bang, 1961; Walters, 1942), with only a few reports of inconsistent changes or increases (Segall, Campbell, & Herskovits, 1966; Hanley & Zerbolio, 1965). It is somewhat difficult to reconcile the developmental increase and leveling off of the magnitude of size constancy with the developmental decrease and stabilization in the illusion magnitudes if we believe that misapplied constancy scaling is the major causal mechanism in the formation of illusions. Data such as these make it seem likely that illusory distortions are compounded of several mechanisms all of which need not operate in the same direction.

On the other hand, there is only one case history report in which a genuinely naïve subject was exposed to visual illusion. Gregory and Wallace (1963) had the unusual opportunity of testing a 52-year-old man who, virtually from birth, had had no useful vision as the result of a corneal opacity. If visual experience is necessary to develop reliance on the cues evoking constancy scaling in pictorial arrays and if such constancy scaling is responsible for many visual distortions, we would expect that an individual with this degree of restricted experience would not show the classical distortions associated with visual geometric illusion. Two months after vision was restored via a corneal transplant Gregory and Wallace administered a number of visual tests to this subject. Included in the battery of tests were the Mueller-Lyer, Poggendorff, Zoellner, and Hering illusions. It was interesting to note that for this observer, who was visually naïve and free from experience with visual depth cues, the usually observed illusory distortions were either totally absent or greatly reduced in magnitude.

Another procedure for studying the effects of experience with depth cues, hence their saliency for the evocation of constancies in illusion configurations, involves cross-cultural manipulations. Here individuals who have grown up in environments in which certain depth cues are more or less available are tested. In general, we may hypothesize that individuals should show greater illusions when presented with configurations that contain implicit cues with which they are familiar.

Cross-cultural investigation of visual illusions seems to have begun around the turn of the century with the work of Rivers (1901, 1905). Rivers tested individuals from two non-western and one western society on the Mueller-Lyer and the horizontal-vertical illusions. In general, the non-western groups showed considerably less susceptibility to the Mueller-Lyer illusion but greater susceptibility to the horizontal-vertical illusion than the western group. On the basis of these data Rivers suggested that the two illusions come from different classes. He proposed a physiological basis for the horizontal-vertical illusion, but suggested that, because the Mueller-Lyer illusion is associated with cues similar to those found in square carpentered rooms or buildings (as analyzed in Figure 8-5), the differences in susceptibility to this distortion may result from differential experience. These general findings have been replicated many times (Heuse, 1957; Jahoda, 1966; Morgan, 1959; Mundy-Castle & Nelson, 1962).

In the most extensive cross-cultural study conducted to date Segall, Campbell and Herskovits (1966) prepared a field packet of stimuli to test the magnitude of a number of illusion variants, including the Meuller-Lyer, the horizontal-vertical, the Sander parallelogram, the Ponzo, and the Poggendorff. The actual testing was done by anthropologists in 28 different societies located in Africa, the Philippines, and the United States. To test the differential experience hypothesis characteristics of the environment were rated to assess the presence or absence of certain types of depth cues. The results indicate that susceptibility to illusions such as the Mueller-Lyer or the Ponzo is increased by experience in an industrialized or carpentered society in which rooms are square, buildings are rectangular, and roads are relatively straight, presumably because such environmental features render the observers in these societies more familiar with examples of tridimensional configuration that incorporate the cues implicit in these illusion configurations. On the other hand, subjects who live in environments in which there are long uninterrupted expanses and in which height in the plane would be a more salient or frequently encountered depth cue were more susceptible to the horizontal-vertical illusion.

Some of the data have failed to confirm the "carpentered world" hypothesis. For example, Jahoda (1966) tested two groups of Ghanaian subjects and a group of Europeans on the same stimuli used by Segall, Campbell, and Herskovits. One of the Ghanaian groups lived in an urban environment, hence came from a more westernized or carpentered world, whereas the other Ghanaian group came from a rural environment. It might be expected that the urban Ghanaian group and the European group would manifest larger Mueller-Lyer and Ponzo illusions than the rural Ghanaian group. Unfortunately both Ghanaian samples showed reduced effects on these configurations, regardless of their degree of urbanization. Similar results have been reported for other studies in which carpentered environment and availability of open vistas have been varied within single societies (Berry, 1966; Gregor & McPherson, 1965). Thus it is clear that simple exposure to square buildings and a carpentered environment is not sufficient to augment the magnitude of illusory distortions elicited by constancy scaling based on certain forms of perspective cues.

These studies, of course, assume an equivalence between constancy scaling in the three-dimensional evnironment and constancy scaling in pictorial stimuli. Indeed, there is some evidence, cited earlier, that this might be the case (Smith & Gruber, 1958; Smith, Smith & Hubbard, 1958). However, the observers in the Smith and Gruber and Smith, Smith, and Hubbard studies were adults in a western society who presumably had had much experience with two-dimensional representations of three-dimensional scenes. There is considerable evidence that the interpretation of depth in two-dimensional displays is augmented by schooling, which presumably would provide specific experience on such displays (Dawson, 1963; Kilbride & Robbins, 1968).

There is, of course, one major difference between a real-world scene and a photograph (or a line drawing) of that same scene. The two-dimensional representation always contains two conflicting sets of depth cues. On the one hand, there are the cues indicating tridimensionality which are common to the real-world scene and its two-dimensional representation. On the other, there are the cues that indicate that the two-dimensional representation is flat. Perhaps by exposure to two-dimensional representations, such as television and books, most observers in western societies learn to ignore the "flatness" cues in two-dimensional stimuli and to respond only to the depth cues indicating tridimensionality. If this were the case, we might expect that schooling, particularly in non-western societies, would augment the encoding of depth cues that indicate tridimensionality in illusion stimuli.

A similar hypothesis has been suggested by Leibowitz (1974) to explain data from an experiment in which the depth cues in pictorial stimuli were directly manipulated in a cross-cultural paradigm. Following Holway and Boring's (1941) classic experiment which indicated that the addition of cues for depth for objects viewed in tridimensional space results in improved constancy, Leibowitz and Pick (1972) varied the cues in the Ponzo configuration by presenting stimuli that ranged from a schematic representation composed of four lines to an enriched photograph of a railroad track. Ugandan college students responded in the expected manner by showing greater size constancy for the more cue-laden displays. Ugandan villagers, however, showed no Ponzo illusion in the basic configuration. Furthermore, adding cues to the pictorial array did not result in the elicitation of size-constancy responses. Thus it seems likely that these subjects, although demonstrating normal constancy in the environment, are not capable of displaying it for graphic representations. This lack manifests itself not only in the assessment of the size of photographed objects but also in the assessment of illusions that owe their existance to constancy scaling. Kilbride and Leibowitz (1973) have provided data that converge on the same conclusion.

These investigators separated samples of Ugandans on the basis of their ability to see three-dimensional relations in a photograph of a rural road. The mean educational level of individuals who did not perceive depth in this picture was less than that of those who did. In addition, the individuals who did not perceive depth in the photgraph of the road also failed to manifest the Ponzo illusion or to respond to depth cue variations in graphic displays. All these data suggest that subjects must learn to ignore the cues that indicate the flatness of two-dimensional pictorial stimuli before they can respond to the cues to tridimensionality in such stimuli.

LIMITATIONS AND IMPLICATIONS

It is important to note that even if we accept a constancy scaling component in some illusions no presumption can be made that all visual geometric illusions are caused by constancy mechanisms. It would be unwise, even for those configurations in which implicit depth cues have been implicated, to attribute the total illusory effect to any one mechanism. Coren and Girgus (1973, 1974) and Girgus and Coren (1973) have indicated that most illusory distortions are actually a compound of many different mechanisms, many of which operate at the structural level, involving optical and neural interactions. For instance Coren (1969) has presented evidence to show that the optical degrading of the image in its passage through the crystalline lens and optic media may contribute to some illusions, as suggested theoretically by Einthoven (1898) and Chiang (1968). It has also been suggested that lateral neural interactions on the retina may augment many of the classical distortions (Bekesy, 1967; Ganz, 1966). These positions have been experimentally supported by Coren (1970) and Girgus, Coren and Horowitz (1973) who showed that the magnitude of the Mueller-Lyer illusion could be varied by altering the opportunity for optical and neural interactions to occur. These investigations, however, indicate that even when all such opportunities are removed significant illusory effects still remain. Once peripheral structural effects have been eliminated it is most likely that the residual illusion is due to various cognitive strategies of information processing, as typified by constancy scaling effects (Girgus & Coren, 1975). With these limitations in mind let us see where our analysis has taken us.

We began this discussion by noting the fine line between pictorial representations and visual illusions. To the extent that the commonplace distortions known as visual geometric illusions represent constancy scaling processes, we have found that there is a large body of data which may shed light on constancy scaling in pictures. Some of the more suggestive findings revealed by these data include the following:

1. Constancy scaling in pictures and illusions is evoked by many of the same cues that operate in normal tridimensional viewing.

2. Constancy scaling in pictorial representations and illusions does not act uniformly in all dimensions but responds selectively to direction of cue change. It would be interesting to test this suggestion under tridimensional conditions.

3. Experiential components do exist in the evocation of constancy scaling in pictorial arrays.

One of the reasons why we have chosen to reverse usual question about the effect of constancy scaling on illusions to ask what the data on illusions can tell us about constancy, is that, after all, it is no more of an illusion to judge two lines unequal in length, despite the fact that they are physically equal in size, than it is to judge two targets differing in distance as equal in size despite the fact that they cast retinal images of unequal size. The first percept is erroneous because it leads the final percept away from the veridical physical situation, whereas the second is adaptive because it leads toward an appropriate conscious representation of the physical environment. Both are misrepresentations of the retinal image, hence illusions.

REFERENCES

Ames, A., Jr., Some demonstrations concerned with the origin and nature of our sensations (What we experience). Hanover, 1946.

Baldwin, J. M. *Handbook of psychology* (2nd ed.). New York: Holt, 1890.

Bekesey, G. von. Sensory inhibition, Princeton, New Jersey: Princton University Press, 1967.

Berry, J. W. Temne and Eskimo perceptual skills. *International Journal of Psychology*, 1966, 1, 119-128.

Binet, A. La mesure des illusions visuelles chez les enfants. *Revue Philosophique*, 1895, 40, 11-25.

Chiang, C. A new theory to explain geometrical illusions produced by crossing lines. *Perception & Psychophysics*, 1968, 3, 174-176.

Coren, S. Lateral inhibition and geometric illusions. *Quarterly Journal of Experimental Psychology*, 1970, 22, 274-278.

Coren, S. The influence of optical aberrations on the magnitude of the Poggendorff illusion. *Perception & Psychophysics*, 1969, 6, 185-186.

Coren, S. & Festinger, L. An alternate view of the "Gibson normalization effect." *Perception & Psychophysics*, 1967, 2, 621-626.

Coren, S. & Girgus, J. S. Visual spatial illusions: many explantions. *Science*, 1973, 179, 5034.

Coren, S. & Girgus, J. S. A size illusion based upon a minimal interposition cue. *Perception*, 1975, 4, 251-254.

Coren, S. & Girgus, J. S. Transfer of illusion decrement as a function of perceived similarity. *Journal of Experimental Psychology*, 1974, 102, 881-887.

Dawson, J. L. M. Psychological effects of social change in a West African community, Unpublished Ph.D. Thesis, Oxford University, 1963.

Day, R. H. Visual spatial illusions: A general explanation. *Science*, 1972, **175**, 1335-1340.

Dengler, M. A test of constancy scaling theory in a modified Mueller-Lyer illusion. *Perception & Psycholophysics*, 1972, **12**, 339-341.

Doyle, M. perceptual skill development: A possible resource for the intellectually handicapped. *American Journal of Mental Deficiency*, 1967, **71**, 776-782.

Einthoven, W. Eine einfache physiologische Erklarung fur verschiedene geometrisch-optische Tauschung. *Pfluger's Archiv für Physiologie*, 1898, **71**, 1-43.

Epstein, W. Perceived depth as a function of relative height under three background conditions. *Journal of Experimental Psychology*, 1966, **72**, 335-338.

Epstein, W. The process of 'taking-into-account' in visual perception. *Perception*, 1973, **2**, 267-285.

Epstein, W. & Baratz, S. S. Relative size in isolation as a stumulus for relative perceived distance. *Journal of Experimental Psychology*, 1964, **67**, 503-513.

Farquar, M. & Leibowitz, H. W. The magnitude of the Ponzo illusion as a function of age for large and small stimulus configurations. *Psychonomic Science*, 1971, **25**, 97-99.

Fisher, G. H. An experimental and theoretical appraisal of the perspective and size-constancy theories of illusions, *Quarterly Journal of Experimental Psychology*, 1970, **22**, 631-652.

Fisher, G. H. Gradients of distortion seen in the context of the Ponzo illusion and other contours. *Quarterly Journal of Experimental Psychology*, 1968, **20**, 212-217.

Ganz, L. Is the figural after-effect an after-affect? *Psychological Bulletin*, 1966, **66**, 151-165.

Gibson, J. J. What is form? *Psychological Review*, 1951, **58**, 403-412.

Gibson, J. J. The perception of the visual world. Boston: Houghton Mifflin, 1950.

Gillam, B. The nature of size scaling in the Ponzo and related illusions. *Perception & Psychophysics*, 1973, **14**, 353-357.

Girgus, J. S. & Coren, S. Depth cues and constancy scaling in the horizontal-vertical illusion: The bisection error. *Canadian Journal of Psychology*, 1975, **29**, 59-65.

Gingus, J. S. & Coren, S. Peripheral and central components in the formation of visual illusions. *American Journal of Optometry and Archives of the American Optometric Society*, 1973, **50**, 533-580.

Girgus, J. S., Coren , S., Durant, M., & Porac, C. The assessment of components involved in illusion formation using a long-term decrement procedure. *Perception & Psychophysics*, 1975, **18**, 144-148.

Girgus, J. S., Coren, S., & Fraenkel, R. Levels of perceptual processing in the development of visual illusions. *Developmental Psychology*, 1975, **11**, 268-273.

Girgus, J. S., Coren, S., & Horowitz, L. Peripheral and central components in variants of the Mueller-Lyer illusion. *Perception & Psychophysics*, 1973, **13**,157-160.

Gogel, W. C., Wist, E. R., & Harker, G. S. A test of the invariance of the ratio of perceived size to perceived distance, *American Journal of Psychology*, 1963, **76**, 537-553.

Gregor, A. J. & McPherson, D. A. A study of susceptibility to geometric illusion among cultural subgroups of Australian aborigines. *Psychol. Afr.*, 1965, **11**, 1-13.

Gregory, R. L. *The intelligent eye.* London: Weidenfeld and Nicolson, 1970.

Gregory, R. L. Perceptual illusions and brain models. *Proceedings of the Royal Society*, (London), 1968a, Section B, **171**, 279-296.

Gregory, R. L. Visual illusions. *Scientific American,* 1968b, **219,** 66-76.

Gregory, R. L. Visual illusions. In B. Foss, (Ed.), *New Horizons in psychology.* Baltimore: Penguin, 1966.

Gregory, R. L. Distortion of visual space as inappropriate constancy scaling. *Nature,* 1963, **199,** 678-680.

Gregory, R. L. & Wallace, J. G. Recovery from early blindness: A case study. Experimental Psychology Society Monographs (Cambridge), 1963, No. 2.

Griggs, R. Constancy scaling theory and the Mueller-Lyer illusion: more disconfirming evidence. *Bulletin of the Psychonomic Society,* 1974, **4,** 168-170.

Hanley, C. & Zerbolio, D. J. Developmental changes in five illusions measured by the up-and-down method, *Child Development,* 1965, **36,** 437-452.

Heinemann, E. G., Tulving, E., & Nachmias, J. The effect of ocular motor adjustments on apparent size. *American Journal of Psychology,* 1959, **72,** 32-45.

Helmholtz, H. Wissenschaftliche Abhandlugen von Hermann Helmholtz, (J. A. Barth Ed.), Leipzig, 1883.

Helmholtz, H. *Handbuch der physolgischen Optik,* J. P. C. Southall, Hamburg, Leipzig, 1867, (reprinted, New York: Dover, 1962.)

Hermans, T. G. The relationship of convergence on elevation changes to judgments of size. *Journal of Experimental Psychology,* 1954, **48,** 204-208.

Heuse, J. A. Etudes Psychologiques sur les noirs soudanais et guineens. *Revue Psychologique des Peuples,* 1957, **12,** 35-68.

Holway, A. H. & Boring, E. G. Determinants of apparent visual size with distance variant. *American Journal of Psychology,* 1941, **54,** 21-37.

Hotopf, W. H. N. The size constancy theory of visual illusions. *British Journal of Psychology,* 1966, **57,** 307-318.

Humphrey, N. K. & Morgan, M. J. Constancy in the geometric illusions, *Nature,* 1965, **206,** 744-746.

Jahoda, G. Geometrical illusions in environment: A study in Ghana. *British Journal of Psychology,* 1966, **57,** 193-199.

Jenkin, N. & Feallock, S. M. Developmental and intellectual processes in size-distance judgment. *American Journal of Psychology,* 1960, **73,** 268-273.

Killbride, P. L. & Leibowitz, H. W. Factors affecting the magnitude of the ponzo illusion among the Baganda. *Perception & Psychophysics,* 1975, **17,** 543-548.

Killbride, P. L. & Robbins, M. C. Linear perspective, pictoral depth perception and education among the Baganda. *Perceptual & Motor Skills,* 1968, **27,** 601-602.

Kristopf, W. Ueber die einordnung geometrisch-optischer Tauschungen in die Gesetzmaessigkeit der viseullen wahrnehmung. *Teil. I. Arch. Ges. Psychol.,* 1961, **113,** 1-48.

Leibowitz, H. W. Multiple mechanisms of size perception and size constancy. *Hiroshima Forum for Psychology,* 1974, **1,** 47-53.

Leibowitz, H. W. Apparent Visual size as a function of distance for mentally deficient subjects. *American Journal of Psychology,* 1961, **74,** 98-100.

Leibowitz, H. W., Brislin, R., Perlmutter, L., & Hennessy, R. Ponzo perspective illusion as a manifestation of space perception. *Science,* 1969, **166,** 1174-1176.

Leibowitz, H. W. & Heisel, M. A. L'evolution de L'illusion de Ponzo en fonction de l'age. *Archives Psychologique, Geneve,* 1958, **36,** 328-331.

Leibowitz, H. W. & Judisch, J. M. The relationship between age and the magnitude of the ponzo illusion. *American Journal of Psychology,* 1967, **80,** 105-109.

Leibowitz, H. W. & Pick, H. A. Cross-cultural and educational aspects of the ponzo perspective illusion. *Perception & Psychophysics*, 1972, 12, 430-432.

Leibowitz, H. W., Pollard, S. W., & Dickson, D. Monocular and binocular size matching as a function of distance at various age levels. *American Journal of Psychology*, 1967, 80, 263-268.

Massaro, D. W. Constancy scaling revisited. *Psychological Review*, 1973, 80, 303.

Morgan, P. A study in perceptual differences among cultural groups in South Africa using tests of geometric illusions. *Journal of the National Institute of Personnel*, Johannesburg, 1959, 8 (9), 39-43.

Mundy-Castle, A. C. & Nelson, G. K. A neuropsychological study of the Knyssa forest workers. *Psychol. Afr.*, 1962, 9, 240-272.

Newman, C. V. & Newman, B. M. The ponzo illusions in pictures with and without suggested depth. *American Journal of Psychology*, 1974, 87, 511-516.

Oppel, J. J. Ueber geometrisch optische Tauschungen, *Jber. Phys. Ver Frankfurt*, 1854, 37-47.

Piaget, J., Matalon, B., & Bang, V. Research on the development of perceptions: VLII. The evolution of the horizontal-vertical illusion from its constituent elements and the Delhoef illusion in tachistoscopic presentation. *Archives Psychologique, Geneve*, 1961, 38, 23-68.

Piaget, J. & von Albertini, B. Recherches sur le developpement des perceptions: XI. L'illusion de Mueller-Lyer. *Archives Psychologique, Geneve*, 1950, 33, 1-48.

Pike, A. R. & Stacey, B. G. The perception of luminous Mueller-Lyer figures and its implications for the misapplied constancy theory. *Life Sciences*, 1968, 7, 355-362.

Pollack, R. H. Simultaneous and successive presentation of elements of the Mueller-Lyer figure and chronological age. *Perceptual & Motor Skills*, 1964, 19, 305-310.

Pollack, R. H. Some implications of ontogentic changes in perception. In J. Flavell & D. Elkind, (Eds.), *Studies in cognitive development. Essays in honor of Jean Piaget*, New York: Oxford University Press, 1969.

Porac, C., Ward, L. M. , Coren, S., & Girgus, J. S. Depth impressions evoked by illusion configurations: Misapplied constancy scaling revisited. Unpublished manuscript, University of Victoria, 1976.

Rivers, W. H. R. Observations on the senses of the Todas, *The British Journal of Psychology*, 1905, 1, 321-396.

Rivers, W. H. R. Vision. In A. C. Haddon, (Ed.), *Reports of the Cambridge Anthropological Expedition to the Torres Straits*, Vol. II, Part I Cambridge: Cambridge University Press, 1901.

Rock, I. *An Introduction to perception.* New York: Macmillan, 1975.

Rock, I. & McDermott, W. The perception of visual angle. *Acta Psychologica*, 1964, 22, 119-134.

Segall, M. H., Campbell, D. T., & Herskovits, M. J. *The influence of cultural and visual perception.* Indianapolis, Indiana: Bobbs-Merrill, 1966.

Smith, O. W. & Gruber, H. Perception of depth in photographs. *Perceptual & Motor Skills*, 1958, 8, 307-313.

Smith O. W., Smith T. C., & Hubbard, D. Perceived distance as a function of the method of presenting perspective. *American Journal of Psychology*, 1958, 71, 662-675.

Tausch, R. Optische Taeuschungen als artifizelle effekts der Gestaltungsprozesse von Groessen-und Formenkonstanz in der natuerlichen Raumwahrnehmung. *Psychologique Forschung*, 1954, 24, 299-348.

Thiery, A. Uber, Geometrisch-optische Tauschungen. *Philosophische Studiern*, 1896, 12, 67-126.

Waite, H. & Massaro, D. W. Test of Gregory's constancy scale and explanation of the Mueller-Lyer illusion. *Nature*, 1970, 227, 733-734.

Walters, A. A genetic study of geometric optical illusions. *Genetic Psychology Monographs*, 1, 1942, 25, 101-155.

Wheatstone, C. On some remarkable and hitherto unobserved phenomena of binocular vision: Part II, *Philosophical Magazine*, 1852, Series IV, 504-523.

Worrall, N. A test of Gregory's theory of primary constancy scaling, *American Journal of Psychology*, 1974, 84, 505-510.

Worrall, N. & Firth, D. Extension cues in open and closed figures. *Quarterly Journal of Experimental Psychology*, 1971, 23, 311-315.

Zeigler, H. P. & Leibowitz, H. W. Apparent visual size as a function of distance for children and adults, *American Journal of Psychology*, 1957, 70, 106-109.

CHAPTER

CONSTANCIES IN THE PERCEPTUAL WORLD OF THE INFANT

R. H. DAY
Monash University

B. E. MCKENZIE
La Trobe University

There are few more vivid descriptions of the perceptual constancies in every-day situations than that given by Woodworth in the first edition of his *Experimental Psychology*.

> . . . the retinal image continually changes without much changing the appearance of objects. The apparent size of a person does not change as he moves away from you. A ring turned at various angles to the line of sight, and therefore projected as a varying ellipse on the retina, continues to appear circular. Part of a wall, standing in shadow, is seen as the

> same in color as the well-lighted portion. Still more radical are the changes in the retinal image that occur when we move about a room and examine its contents from various angles. In spite of the visual flux the objects seem to remain in the same place. In short, what we perceive is the objective situation. (Woodworth, 1938, p. 595).

For the adult observer such extraordinary stability of the perceived environment in the face of often-marked fluctuations in the sensory image of external objects and events is commonplace. The question can be asked whether the same stability obtains in infancy. In Woodworth's terms, do infants also perceive the objective situation as its sensory representation varies? Do they, for example, perceive an approaching or receding person as the same size, a feeding bottle as the same shape as it is tilted in depth, and their hands as the same lightness when parts of them are in shadow? It is with these and similar questions that this chapter is mainly concerned. In particular, it is concerned with whether some of the perceptual constancies occur during the first year. Before attempting to answer these questions, however, one point needs to be made clear. Despite a considerable resurgence of interest in infant behavior and development over the last couple of decades, research on perceptual constancy is by no means extensive. Furthermore, firm conclusions are often difficult to reach because some early studies are methodologically questionable and more recent studies are sometimes in conflict. The likelihood that at least some perceptual stability with variable sensory representation is present in the first year of life is increased by observations that indicate a marked degree of perceptual competence during this period.

Before turning to the experimental observations we raise a number of basic issues that bear closely on the question of constancy in infancy. First, the range and classes of constancy at maturity; second, the discriminability of object features and qualities, and, third, the bearing of the occurrence of perceptual constancies in infancy on perceptual theory. In regard to the first it is worthwhile to point out that most early and recent discussions of the perceptual constancies have been confined largely to those of size, shape, brightness (or lightness), and ocasionally color, the *classical constancies.* Moreover, these constancies are generally treated as belonging to a single class. It is shown that there appear to be at least three classes. The second issue, that of feature discriminability, is fundamental. Because an object property can be discriminated in adulthood and exhibits perceptual constancy it cannot be taken to mean that it can be discriminated in infancy. Third, to consider the question of the occurrence or nonoccurrence of perceptual constancies in infancy without some reference to the major theoretical positions—as distinct from specific theories—would be to remove the issue from its context and thereby lose the main point of a review of this kind. The question is of long standing and the answer or answers to it bear closely on some of the main theoretical standpoints. These three issues are taken up in more detail below.

In general, interest here is centered mainly, but not exclusiverly, on evidence for the occurrence of perceptual constancies during the first year of human infancy. Two issues already mentioned briefly—the range and classes of perceptual constancies at maturity and the implications of early constancies for theory—are discussed first. Apart from their obvious relevance to the main question, they have not been comprehensively or systematically dealt with before.

Evidence for the presence of perceptual constancies in the first year is considered under the three headings: visual egocentric constancies, visual object constancies, and visual identity-existence constancies, also called object permanence. This arrangement follows the classification of the constancies arrived at after considering the range of constancy phenomena known to occur for adult observers; it is discussed in the following section.

RANGE AND CLASSES OF PERCEPTUAL CONSTANCY

The treatment accorded perceptual constancies in most texts and reviews is almost entirely confined to three, occasionally four, instances—visual size constancy with object distance variant, visual shape constancy with slant in depth variant, lightness constancy with luminance variant, and sometimes color constancy with the composition of reflected light variant. Notable exceptions to this limitation are the recent texts by Rock (1975) and Kaufman (1974). Restriction of interest to these four reflects the history of philosophical concern and empirical enquiry. Berkeley (1709) in his *An Essay Towards a New Theory of Vision* drew attention to the problem of size constancy (Section LX) which much later was the first to be tackled experimentally, initially by Martius (1889) and then by Hillebrand (1902) and Beryl (1926). The later and well-known contributions by Brunswik (1929, 1933, 1934) and Thouless (1931, 1932) were variously concerned with visual size, shape, lightness, and color. It is worth noting also that Brunswik's collaborators demonstrated the occurrence of perceptual "thing" constancies in the auditory and tactile-kinesthetic modes (Mohrmann, 1938. Schreiber, see Brunswik, 1938).

Consideration of experimental data indicates that the constancies can be conveniently grouped into three major classes, the egocentric, the object, and the identity-existence constancies. Before discussing these classes and giving examples of each it is worth noting that investigations of their occurrence in infancy have been concerned so far only with visual object and identity-existence constancies. As far as we know, no systematic studies have been done of the egocentric constancies in vision or hearing in infancy.

Egocentric, Object, and Identity-Existence Constancy

Perception not only includes response to object features such as size, shape, orientation, lightness, and hue and, in auditory sources, loudness and quality of sounds, but to position in relation to the observer, which includes responses to direction, distance, and movement. Sensory representation of direction and distance in vision and hearing vary widely but perceived location of object and source remain relatively unchanged. Such effects are called egocentric constancies. They refer to stability in the perception of the position of things in relation to the observer as the sensory representations of position and change in position vary with observer posture, locomotion, and head and eye movements.

The best known and by far the most extensively investigated constancies are those of object features and properties with variation in their sensory representations. Thus upright objects continue to appear upright as the observer tilts his head or body (orientation constancy), lightness is more or less constant with change in ambient light intensity, and apparent shape is constant as the object is tilted in depth. Object constancies refer to stability in the perception of numerous properties or features of an object as their sensory projections change. Visual size and shape constancy are probably the best-known examples.

The same object can be present in different places and at different times. It can disappear, often occluded by intervening objects, and then reappear in a different place at a different time. Likewise, it can disappear and reappear later in the same place. It has long been known that when one object moves behind another or is covered by another so that the retinal representation of the first is progressively reduced, eventually being eliminated completely, the object is "perceived" as being "still there," although it is only partly visible or not visible at all (Michotte, 1946, 1955; Piaget, 1954). Similarly, an object first presented in one location and later in another may be perceived as the same object, although time and place of occurrence have changed. This apparent continuity of identity as time and place of object occurrence varies has been variously referred to as object permanence and existence constancy. Here the term identity-existence constancy is used in order to emphasize two of its aspects: the constancy of an object's identity with change in place and time and its continuity of existence when it is out of sight.

The difference between the object constancies and the identity-existence constancies is by no means clear-cut; both can involve recognition or identification of an object when its sensory representation varies. The main difference lies in the extent to which sensory information is immediately available and can be "taken into account" or "discounted." When an object is rotated in depth, thus changing its retinal image, constancy of shape is mediated by information for its slant which is available in the stimulus array. In the identity constancies the object may occur in one place at one time and then in another place at another time. There must be some memory for what the object was like before it changed its place. The difference between the first two classes and the third seems to lie in the availability in the stimulus array of the information necessary to preserve the constancy of the object or the retention in memory of that information.

Egocentric Constancies

Visual direction, movement (including stationariness), and distance constancy have been demonstrated respectively by Hill (1972), Wallach and Kravitz (1965), and Purdy and Gibson (1955). Hill (1972) showed that if the head is in a fixed position and the eyes deflected to one side so that the locus of an object image on the retina changes the apparent position of the object hardly changes at all. Somewhat similarly, Wallach and Kravitz (1965) showed that as the head is moved from side to side so that the image of a stationary object on the retina moves also the object is nevertheless perceived as nearly stationary. If if can be assumed that the apparent velocity of a moving object would also be unchanged, stationariness constancy can be regarded as a limiting case of constancy of velocity of moving objects.

Distance constancy can be regarded as a special example of size constancy along the median axis. The retinal projection of a fixed median distance between two points, that is, the size or extent of the space between the points, is greater when they are near than when they are far (Over, 1961). Perceptual constancy of near and far extents which are physically equal but project unequal images has been demonstrated by Purdy and Gibson (1955) and Smith (1958). The former concluded: "observers can divide stretches of distance (up to 300 yards) into halves or thirds with very good accuracy. Perceived magnitudes of distance appear to correspond well with physical magnitudes of distance" (Purdy & Gibson, 1955, p. 380).

Auditory egocentric constancies have also been demonstrated. Day (1968) showed that with marked changes in the binaural sound stimulus consequent on head rotation with a fixed source the apparent direction of the source relative to the observer was nearly constant. Even with the head rotated through 40° an intermittent noise immediately in front of the blindfolded observer was heard in that direction. Some evidence for auditory distance constancy was adduced by Engel and Dougherty (1971). However, the interpretation of these data in terms of distance has been questioned (Day, 1972).

Object Constancies

Visual size, shape, lightness, and color constancy with variation in the relevant properties of the retinal image are sufficiently well documented and reviewed not to require detailed description here. There are full discussions in Woodworth (1938), Woodworth and Schlosberg (1954). Vernon (1962), and Hochberg (1971).

Other object constancies are those of visual speed and orientation, auditory loudness, and tactile-kinesthetic weight. Visual speed constancy may be regarded as a variant of size constancy. The object moves through a fixed frontal extent at different distances from the observer. The time taken to travel through this extent is fixed but the retinal representation of the extent varies with its distance so that retinal speed also varies. Nevertheless, apparent object speed is more or less constant with distance (Rock, Hill, & Fineman, 1968).

If the head is tilted laterally, the orientation of the retinal image in relation to the normally vertical meridian of the eye is altered. For small tilts the apparent orientation of an upright object changes slightly in the direction of head tilt and for larger tilts—about 60°—in the opposite direction. The complete function is called the Aubert-Müller effect but represents instances of over- and underconstancy of object orientation (Day, 1969; Day & Wade, 1969; Wade, 1968). A limiting case of orientation constancy, recently referred to as uprightness constancy (Gajzago & Day, 1972), occurs when the observer views the object while looking between his legs, the Matanozoki posture (Howard & Templeton, 1966), so that the image is rotated through 180° at the retina. Some adult observers but not children perceive the object as upright under these conditions (Gajzago & Day, 1972).

Constancy in the loudness of an auditory sound with object-observer distance variant and in the weight of an object with force variant have also been reported. Mohrmann (1938) showed that the loudness of sounds such as speech, music, metronome clicks, and tones located .75, 2.37 and 7.50 m from the observer varied less than would be expected in terms of intensity variation with distance. Constancy of loudness occurred with and without visual information for distance. Schreiber, also working in Brunswik's laboratory (Brunswik, 1938) showed that when balls were dropped from different heights, thus varying the force with which they struck the hand, variation in apparent weight varied less than would be predicted from change in the force of impact. This phenomenon is referred to as weight constancy (Postman & Tolman, 1959).

Identity-Existence Constancy

Objects can appear in different parts of the visual field in different orientations at different times. They can also pass behind other objects or the latter can pass in front of them so that the object is occluded from view. Nevertheless, the object in different positions and orientations viewed at different times retains its identity as that particular object. Moreover, when it disappears from view it is conceived of as being "still there," although its sensory representation has ceased. Here the perception of an object's identity with variation in its position, orientation, and time of occurrence is referred to as identity constancy and that of its continuity of existence with temporary disappearance as existence constancy. In most experiments the two are examined together so that the term identity-existence constancy is appropriate. Retention of identity with variation in place, time, and visibility has also been called object permanency and extensively studied developmentally (Michotte, 1946; Piaget, 1954). Experiments with infants have recently been reviewed in detail by Harris (1975).

Typically, studies of identity constancy involve an initial view of the object, its progressive occlusion as it passes behind a screen (or a screen moves in front of it), and its progressive reemergence into view from behind the screen. However, experiments by Rock and his colleagues (Rock, 1956, 1973, 1975; Rock & Heimer, 1957) involving the tilting of an object in a frontoparallel plane can also be regarded as studies of identity constancy. The observer is required to indicate whether the object is recognized when it is presented in various orientations. The difference between this situation and that involving tilt in depth—the classical shape constancy condition—is that the observer is required to recognize the object, not simply its shape. Of course, object recognition could also be required with tilt in depth. Rock's experiments have been concerned *inter alia* with the conditions that produce constancy of identity as the object is rotated in a frontoparallel plane.

Summary of Classes of Constancy

Egocentric constancies refer to constancy of an object's apparent position in space in relation to the observer as the sensory representation of that position varies. Object constancies refer mainly to constancy of object properties as their sensory images vary. Identity-existence constancies refer to constancy of an object's identity as the object's position in space or time of presentation varies. This class also incorporates the "perceived" continuity of an object's existence—its permanence—even though it passes from view and no representation of it occurs at the sensory surface.

The question whether these three classes are related in terms of the processes that underlie them or whether they in fact represent different phenomena that are only superficially connected is an important one. It is reasonable to assume that the egocentric and object constancies are essentially similar phenomena. In each case the sensory information, which must be "taken into account" or alternatively "discounted" if constancy is to occur, is available. Thus in visual size constancy there is concurrent information about the object itself and its distance. In visual orientation constancy information for the object and for head and body tilt is present. When an object retains its stationariness in relation to the observer as his head is moved from side to side, presumably kinesthetic and labyrinthine indications of head movement can likewise be integrated with those from the retina. For both classes of perceptual constancy the sensory information necessary for the maintenance of constancy is present in the total stimulus array. The process of constancy can be presumed to involve integration of information from two or more sources.

The situation for the identity-existence constancies is somewhat different. When the object occurs first in one position in the visual field and then in another, analogous concurrent information to that "taken into account" in egocentric and object constancies is not available. Information about the object and its earlier position must be carried in memory. Likewise, when an object disappears from view and is perceived as the same object on its reappearance, its characteristics such as size, shape, lightness, and color must be stored for it to be identified. Similarly, if an object is "perceived" as continuing to exist when it is partly or fully hidden from view there must be stored information about its features. Thus identity-existence constancies must rely more heavily on memory. The egocentric and object constancies presumably can occur on the basis of immediately given sensory information. However, to categorize identity-existence constancies as a cognitive phenomena and the other two as purely perceptual is probably overrigid and misleading. Rather it would seem closer to the actual events to characterize the involvement of both the memorial and perceptual processes as constituting gradations. Some constancies may well be purely perceptual events involving here and now stimulus information and others may be entirely cognitive, depending entirely on stored information. Between these extremes the extent of perceptual and cognitive contributions varies.

Constancies in Infancy and Theoretical Standpoints

Without implying commitment to any particular theoretical view the perceptual constancies can be regarded as "distal achievements" (Brunswik, 1943). The organism perceives the environmental state of affairs rather than that represented in the pattern of stimulation at the receptors. Such distal reference in perception presumably serves the interest of adaptation and survival. Three main theoretical standpoints and a less systematic fourth viewpoint from which this distal achievement can be viewed are identifiable: organizational, functionalist, invariance detection, and resolution of ambiguity views. The organizational standpoint is, of course, that of Gestalt psychology which has stressed the innate organization of the nervous system and allows for interaction between parts of the centrally represented external world. Perceptual constancies are conceived of as relationally determined states that result from central neural interactions. In opposition, the functionalist standpoint exemplified in Brunswik's (1940, 1943) views emphasizes the role of the observer's experience in perceptually "achieving" the distal object. Experience and learning are regarded as essential to the occurrence of perceptual constancies. Gibson (1966) has argued that the constancies in perception depend on the ability of the individual to detect invariants in the environment. This ability may involve some learning, but in the main the process is one of "registering" invariants, something that nervous systems are held to be geared to do. A fourth view of the origins of perceptual constancy is that of perceptual resolution, which has not been argued through in detail (Day, 1969). The sensory projection is equivocal in that it can represent a variety of external situations; for example, a retinal projection of a particular size may signal a small object that is near or a large object that is far. Perception can be regarded essentially as a process that resolves this ambiguity. Thus equivocalities of size are resolved on the basis of information for distance. Misleading information for distance would, of course, resolve the ambiguity in the direction of illusory rather than veridical perception of the external environment (Day, 1972). Whether the use of the information on which perceptual resolution is based is innate or an outcome of experience is an open question and not critical for the theory.

The relevance of the occurrence of perceptual constancies in infancy for theoretical standpoints is fairly clear-cut. Those standpoints that stress organization and the role of neural interaction and the automatic registration of invariances would be supported by the occurrence early in development of some or all of the constancies in perception. With the maturation of neural structures and functions the constancies, by and large, would be expected to occur. Contrariwise, those standpoints that emphasize the role of learning in perception would be called into question by firm evidence for the occurrence of constancies in infancy. For the perceptual resolution view (Day, 1969, 1972) and in large part for that of registration of invariants (Gibson, 1966) the question is empirical. The facts about the age at which the constancies occur simply have to be established.

One further point can be made in connection with theories about perception and its origins. Given that the involvement of perceptual and cognitive processes in the three classes of perceptual constancy varies, it should not be expected that all constancies are operational simultaneously. Furthermore, it is reasonable to suppose that although some constancies of relevance to adaptation and survival in infancy will occur during the first year, some possibly early in that period, others may not occur until considerably later. It is in the interest of this point that the classes and ranges of the perceptual constancies have been stressed and the possibility of varying involvements of sensory-perceptual and cognitive processes emphasized.

EGOCENTRIC CONSTANCIES IN INFANCY

It would be of interest to establish, as Hill (1972) has done for adult observers, whether the apparent direction of an object is relatively fixed when the eyes are deflected right or left so that the object image varies in retinal location. It would be equally interesting to know whether, as in adults, objects in the visual field are seen as stationary as the eyes sweep across them so that their images move across the retinas (Wallach & Kravitz, 1965). Were these constancies not to occur the visual world of the infant would be unstable indeed, for head and eye movements are well developed during the first few months. Neither of these egocentric constancies has so far been formally investigated in infants. Nor has distance constancy of the sort studied by Purdy and Gibson (1955) and Smith (1958) been considered. Although it seems unlikely that the visual environment is rendered unstable during head and eye movements and passive transport, this issue has yet to be studied.

When considering position constancy in infancy, two separate issues are involved. First, can infants locate objects in space, and second, are their perceived locations maintained with change in sensory stimulation? From the first week infants are able to distinguish the direction of movement of an object in relation to themselves. An avoidance response is elicited by a moving object on a hit path but not by a similar object whose trajectory would not result in direct collision (Ball & Tronick, 1971; Bower, Broughton, & Moore, 1971a). Thus radial location of moving objects is well developed. That infants also perceive the location of stationary objects in space is shown by early swiping at them (Bower, Broughton, & Moore, 1970), early prereaching (Trevarthen, 1974), and orienting of head and eyes toward a sound source within minutes of birth (Wertheimer, 1961).

Evidence for position constancy would be provided by accuracy of swiping following eye movements, and the consequent change in site of retinal stimulation, or orienting toward a sound source regardless of head position. Only incidental observations of the latter type are available.

Failure to investigate systematically the visual egocentric constancies in babies leaves a considerable gap in our knowledge about the stability of the visual environment as the head and eyes are moved. Given the frequency with which the infant's head is moved, it is also relevant to raise the question whether there is constancy of auditory direction as found by Day (1968) with adult subjects. In short, none of the egocentric constancies, visual or auditory, known to occur in adulthood, has been investigated in infancy.

VISUAL OBJECT CONSTANCIES IN INFANCY

Among the numerous object constancies briefly described only those of visual shape and size have so far been investigated in infants during the first year.

Visual Shape Constancy

Four experiments directly concerned with shape constancy in early infancy have been reported; three by Bower (1966) who used generalization of a conditioned head-turning response and one based on habituation of visual fixation (Day & McKenzie, 1973).

In the first of Bower's (1966) three experiments subjects 50 to 60 days old were reinforced by "peek-a-boos" to leftward head turns (CR) in the presence of a rectangular stimulus object (CS) 2 m away and rotated 45° in depth in relation to the infant's line of regard. In the test session generalization of the CR was observed under four conditions: (a) with the CS in its original orientation, (b) with the CS frontoparallel, (c) with a trapezoidal object frontoparallel so that the shape of the retinal image was the same as in (a) and (d) with the same trapezoid at 45°. The mean response frequencies for 30-second exposures to each condition indicated a strong tendency to respond to the same real shape (Condition b) and a much weaker tendency to respond to the same retinal shape and the same orientation (Conditions c and d), an outcome strongly in support of the occurrence of shape constancy.

It was intended in Bower's second experiment to establish whether shape constancy in infancy develops out of the correlation between projective shape and apparent slant in depth. Using essentially the same procedures as in the first experiment, infants aged 60 to 63 days were trained with a rectangle rotated 5° in depth as CS. These infants were then tested under each of the following conditions: (a) with the rectangle at 5° and in three other orientations, (b) with four frontoparallel trapezoids projectively equivalent to the stimulus objects in (a), and (c) with the rectangle at the four orientations of (a) viewed through a rectangular aperture so that the edges were not visible. Under each of these test conditions the responses to the rectangle at 5°, or its projective or slant equivalent, were reinforced. Information about projected shape differences and slant differences was available in Condition a but only one of these, either projected shape differences or slant differences, was available in Conditions b and c, respectively. Because generalization was greater in Condition a, despite increased information for difference, it was concluded that the hypothesis concerning correlation between projected shape and apparent slant as a basis for shape constancy had not been supported.

This last experiment has two limitations. First, as Bower (1966) recognized, all infants were originally trained with the rectangle at 5° as CS. It might be predicted, therefore, that greater generalization would occur under Condition a, the only test condition in which the rewarded stimulus object was identical to the original CS. Second, continuous reinforcement of one stimulus object and testing for generalization to other stimulus objects confounds reinforcement and generalization effects. These limitations do not apply to a third experiment in which generalization to the dimensions of real shape, projected shape, and orientation was assessed by using independent groups of subjects and without reinforcement during testing. When orientation and projective shape varied but real shape was constant, there was greater generalization than when only projective shape varied and when only orientation varied.. Although infants were able to perceive differences in slant and in projective shape, the response to constant real shape was clearly prepotent. Bower concluded that "young humans possess the capacity for shape constancy, the capacity to detect the invariants of shape under rotational transformation in the third dimension" (Bower, 1966, p. 833).

Essentially the same conclusion was reached by Day and McKenzie (1973) following an experiment with three groups of infants aged 6 to 16 weeks. Rate of habituation of visual fixation was observed over eight 20-second trials to a cube of constant orientation, the same cube in different orientations, and to cut-out photographs of the cube in different orientations. The latter were two-dimensional projections of the three-dimensional cube in the second condition. Control trials with an "interesting" object were included before and after the habituation trials as a check on alertness over the experimental session. Variations of the outline shapes and patterns of brightness of the cube in different orientations are shown in Figure 9-1. The logic of the experiment was that because habituation of fixation occurs with repeated presentation of the same stimulus then, if shape constancy obtains, that is, apparent shape is the same with variation of object orientation in depth, little difference in rate of habituation would be expected with the cube in constant and different orientations. Furthermore, if shape constancy is dependent on information for slant in depth, possibly deriving from motion parallax, habituation to flat frontoparallel representations of the cube in different orientations would be expected to be less than for the cubes themselves.

The results shown in Figure 9-2 clearly indicated a decline in fixation time over trials for infants viewing a solid cube in one orientation and for those viewing it in different orientations, with no significant difference between the rates of decline for the two groups. However, for infants viewing photographs of the cube in different orientations there was no significant decline in fixation

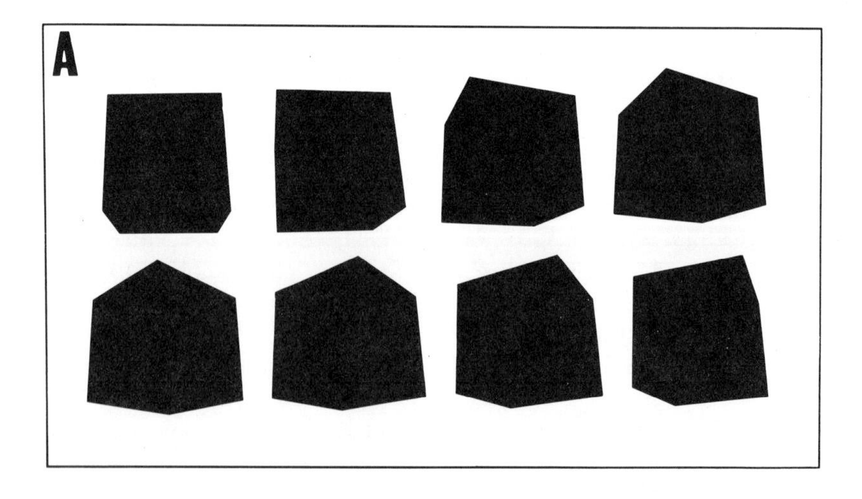

Figure 9-1 Outline shapes (A) and luminance patterns (B) of a cube when tilted at 0 (near face frontoparallel), 11, 23, 34, 45, 56, 68, and 81° in depth in relation to the infant. The two light sources illuminating the cube were on either side of the infant and slightly behind him (from Day & McKenzie, 1973).

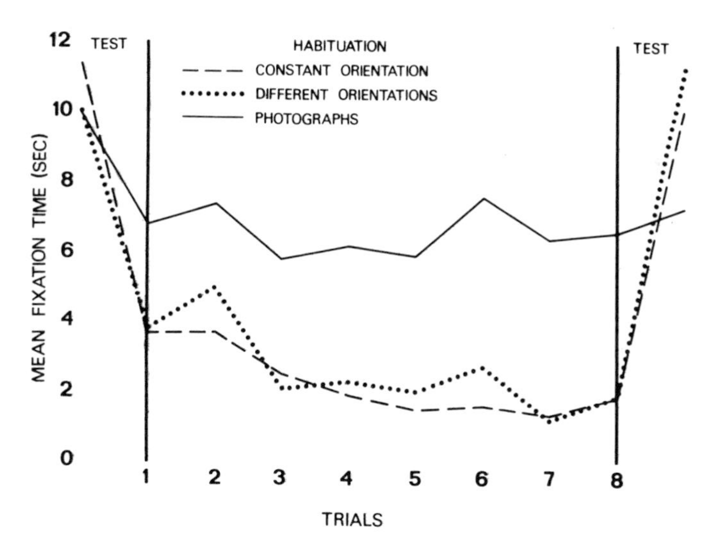

Figure 9-2 Habituation of visual fixation and recovery from habituation for three conditions: a cube (Figure 9-1) in randomly selected constant orientation, a cube in random variable orientations, and two-dimensional representation of a cube in random variable orientations. The orientations of the cube are shown in Figure 9-1. Initial and final trials were controls for general alertness during the session (from Day & McKenzie, 1973).

duration. It was concluded that shape constancy for solid objects rotated in depth obtains in early infancy and that stimulus information for slant, probably derived from motion parallax, is one of its determinants.

In general, conclusions derived from four experimental investigations with different research paradigms are essentially in agreement concerning the occurrence of shape constancy in the early weeks of life. Unfortunately such agreement does not obtain in respect to size constancy.

Visual Size Constancy

Two early investigations of size constancy (Cruikshank, 1941; Misumi, 1951) are discussed first. Before taking up recent studies, however, it is appropriate to consider whether infants are capable of discriminating object size and object distance, the two variables primarily involved in size constancy.

The study of size constancy in infancy, like that with adult observers, has a longer history than that of shape constancy. Let us review the earlier, frequently quoted studies and then turn to more recent enquiries involving somewhat different procedures and more sophisticated data analyses.

EARLY OBSERVATIONS.

Piaget and Inhelder (1969) assert that "constancy" of size makes its appearance at about six months. Once the child has been trained to choose the larger of two boxes, he continues to choose correctly if you move the larger box farther away so that its retinal image is smaller (Piaget & Inhelder, 1968, p. 31). This assertion was apparently based on the work of Cruikshank (1938, 1941) which led Piaget to believe that perceived constancy of size develops before object permanence but after the beginnings of coordination of vision and prehension. He argued that sensorimotor schemes force the variable information about size derived from vision to conform to the constant information derived from touch. Cruikshank's (1941) and Misumi's (1951) experiments represent the first attempts to investigate size constancy in infancy and are often quoted as evidence for it. It is worth noting straightaway, however, that because of design and procedural defects neither experiment is considered conclusive or even suggestive in revealing the presence or absence of size constancy in infancy.

The ages of Cruikshank's (1941) 73 *Ss* ranged from 10 to 50 weeks. The response used to indicate size constancy was reaching for and attempting to grasp an object the size and distance of which was varied. The results clearly showed that the frequency of reaching was markedly greater when a small rattle was near and declined when either a small or a large rattle was far. There was little difference between reaching frequency for different sizes at the greater distance, which suggested to Cruikshank that object distance was the principal determinant of the response. In fact, Cruikshank's experiment could not answer the problem of size constancy, because the response indicator was ambiguous. If distance perception is assumed, an infant may perceive that an object is the same size at two different distances, yet will reach more frequently at the nearer distance, just as an adult whose response is limited to reaching would not attempt to obtain an object of the same apparent size beyond his grasp. At the most, her experiment could show that reaching was not determined solely by retinal image size. It did not address the problem of perceived size. Furthermore, Cruikshank failed to establish initially whether there was differential reaching for one of the objects when they were equidistant. A consistent preference for, say, the larger object indexed by more frequent reaching would determine the interpretation of the same frequency for large and small objects at the greater distance. In the absence of this baseline comparison of responses at the greater distance was meaningless.

Misumi (1951) recognized the latter defect in Cruikshank's procedure. A total of 457 infants aged 2 to 59 weeks was involved in the experiments in which the two objects were presented together so that preferential reaching for one or the other could be observed. In general, a preference for the larger of two objects was exhibited by infants older than about 26 weeks and persisted when it was placed further away than the smaller, suggesting that relative size is maintained over changes in retinal size and distance. No such persistence of preference at the greater distance occurred with infants younger than 26 weeks of age. Two features of Misumi's work render the results questionable. First, the total number of reaching responses was small and insufficient for reliable conclusions. In Experiment 2, 13 babies made a total of only 52 responses and in Experiment 6 the same number made only four responses. Therefore the question is whether manual reaching to objects of different sizes at different distances is an appropriate technique. Second, for reasons that are unstated the stimulus objects were very small, a factor that may have contributed to the low frequency of reaching.

Both Cruikshank's and Misumi's experiments suffer because of the inadequacy of the reaching response. Differential reaching for objects of different sizes at a near distance may not be exhibited at a greater distance, not because the apparent size has changed but because the greater distance may inhibit reaching.

DISCRIMINATION OF OBJECT SIZE.

Regardless of whether apparent size remains constant with variation in object distance or changes in accordance with the retinal projection, capacity to discriminate object size is assumed. If this capacity is lacking the question of the occurrence of size constancy is hardly necessary. Although the assumption is justified for older children and adults, it requires independent indexing for young children and infants. Thus young children might be trained to choose the larger of two objects which are at the same distance. Success would indicate the capacity to discriminate object size. Maintenance of the successful choice when the retinal projection of the larger object is smaller or the same as that of the other object would indicate size constancy. The first is necessary for the second. The ability of young infants to adapt their behavior according to object size at a fixed viewing distance has frequently been claimed, but the validity of some of these claims is questionable. Bower (1972) suggested that the distance between thumb and fingers and between the two hands before contact with an object is related to object size as early as the second week. His method of single-camera recording of hand adjustments cannot yield an unambiguous measure of interhand or interdigit distance because only two-dimensional records of movement in three dimensions were obtained. A more convincing demonstration was provided by Bruner and Koslowski (1972) who used a gross measure of arm and hand movement in relation to objects of different sizes with infants 10 weeks old. They found that adduction of the hands to the midline of the body and greater activity of the hands in that position (both appropriate to grasping the object) were more likely to occur for a smaller, graspable object than for a larger, ungraspable one.

The use of duration of visual fixation as an index of size discrimination gives rise to equivocal findings. Early preference for visual inspection of large rather than small objects disappears after the first eight weeks (Fantz & Nevis, 1967a, b). Fantz and Fagan (1975) have recently observed that in patterned figures the salience of size of elements in relation to the number of elements in the pattern decreases from 5 to 25 weeks. These changes in response pattern may be related to changes in visual acuity, oculomotor coordination, and neural maturation.

Although weak, the evidence suggests that infants are capable of discriminating three-dimensional size changes of the magnitude involved in the size-constancy experiments to be described later. Discrimination of the size of two-dimensional objects is less certain. In an experiment involving 40 infants between 7 and 19 weeks old and using recovery from habituation as an index of discrimination, we adduced evidence that suggested that whereas the difference between 12-and 4-cm square three-dimensional objects produced response recovery that between two-dimensional objects of the same size did not. The significance of information derived from motion parallax is again suggested.

DISCRIMINATION OF OBJECT DISTANCE

Most recent demonstrations of distance discrimination by infants have involved response to depth-at-an-edge, that is, "downward" distance. Using the "visual cliff," Walk (1966) and Campos, Langer, and Krowitz (1970) have adduced evidence of depth discrimination. There are few studies of horizontal distance, however. Bower (1972) found more attempts to reach toward a small near object than toward a large far object among infants 7 to 15 days old. In a similar experiment Field (1975) observed that 5-month old infants adjusted their reaching efforts according to stimulus distance, but 2-month old infants did not. It seems that, at least by 5 months of age, infants make few attempts to reach beyond arm's length. Whether younger infants also reach selectively has yet to be established conclusively.

Recent findings by McKenzie and Day (1972), which are discussed in the next section in connection with size constancy, show that before 20 weeks relatively little attention is paid to stationary achromatic objects farther away than about 90 cm. Field (1975) also found in 5-month old infants a decrement in looking time as object distance increased. The extent of the visual field for moving objects, however, is greater than that for stationary objects (McKenzie & Day, 1976. Fixation duration to rotating patterned cylinders in the arrangement shown in Figure 9-3 did not decrease with increase in object distance from 30 to 90 cm but the previously observed decrease for stationary objects at greater distance was confirmed. The same pattern of responding was exhibited by 2-month and 4-month old infants, shown in Figure 9-4. Just as Tronick (1972) found that the angular size of the visual field is greater with moving than with stationary objects so the outward extent of the field from the baby is

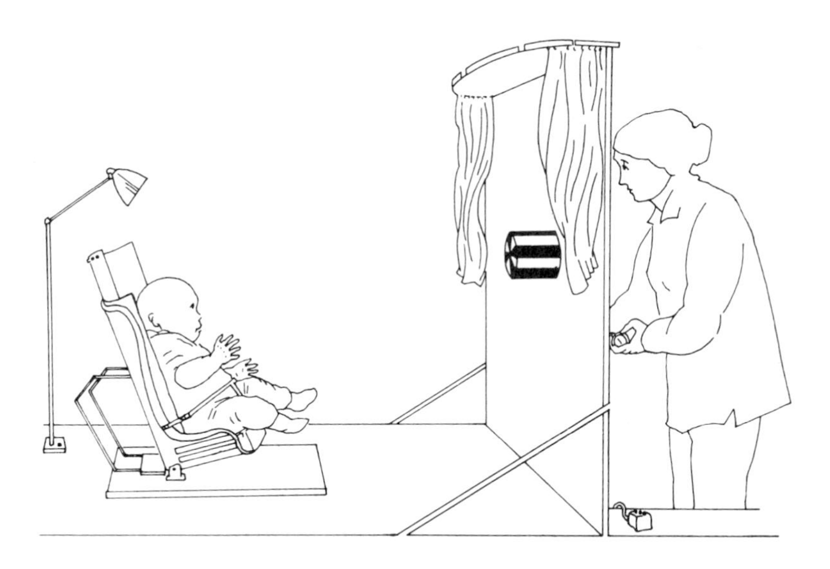

Figure 9-3 Arrangement for presenting a stationary or moving patterned cylinder to an infant at distances of about 90 cm. (from McKenzie & Day, 1976).

greater with moving objects, that is, with objects that remain fixed in position but are in motion.

A clear indication of distance perception is provided by responses to objects that move toward the baby to produce a looming stimulus. Avoidance responses (throwing head backward, raising hands, opening eyes) were exhibited to a small object at 8 cm from the infants but not to a larger object which projected a retinal image of equal size at 20 cm (Bower, Broughton, & Moore, 1971a). These responses occur to near objects that are on a hit-path, as indicated by a symmetrical optical expansion pattern (Ball & Tronick, 1971). It can be noted that essentially the same response was observed with infants aged 1 and 8 weeks. Because the same responses were observed when a pattern at a fixed distance expanded (but not when it contacted), it seems that the cue to distance is provided by symmetrical image expansion.

Is accommodation of the eye a factor in distance perception in infancy? Little is yet known about this factor. Fantz, Ordy, and Udelph (1962) found that changes in object distance from 5 to 20 in. did not change their measure of visual acuity in infants younger than 4 weeks old. Similarly, Salapatek, Bechfold, and Bushnell (1975) found little difference in acuity thresholds at each of four distances ranging from 30 to 150 cm. Others, however, have suggested rapid change in accommodative ability in the early weeks (Haynes, White, & Held,

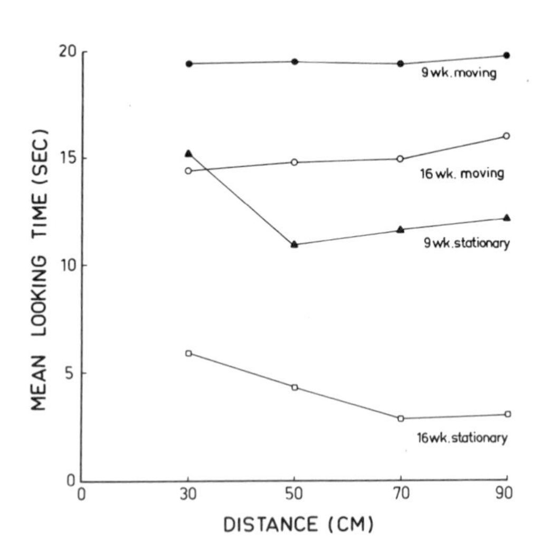

Figure 9-4 Mean visual fixation times with a stationary and moving patterned cylinder (Figure 9-3) exhibited by infants approximately 9 and 16 weeks old. Duration of fixation was greater with the moving cylinder than with the stationary cylinder for both age groups (from McKenzie & Day, 1976).

1965; White, 1971). Deficiencies in accommodation are evidently not of great enough magnitude to prohibit distance judgments.

In general, there appears to be good evidence for the detection of the distance of stationary unpatterned objects when they are relatively near. Moving and patterned objects capture attention at considerably greater distances.

VISUAL SIZE CONSTANCY

Visual size constancy in early infancy has been investigated with four techniques: generalization of an operant conditioned response, rate of acquisition of a conditioned response, recovery from habituation, and rate of habituation.

Bower (1964, 1965) has concluded that with babies about 6 to 12 weeks of age perceived size corresponds more closely to actual size than to retinal size. In general, his procedure involved training infants to respond to the presence or absence of an object at a standard distance and then testing generalization of this response to a larger object at a greater distance, a same-sized object at a greater distance, and a larger object at the original distance. After conditioning with a 12-in. cube at 3 ft, generalization tests were conducted with (a) a 12-in. cube at 3 ft, (b) a 12-in. cube at 9 ft, (c) a 36-in. cube at 3 ft, and (d) a 36-in. cube at 9 ft. Response rate was greatest for (a), the training situation, and least for (d), when the visual angle was the same as in (a). For both conditions (b) and (c) response rates were less than that of (a) but greater than in (d). It seems, therefore, that the infants were responding to at least some of the stimulus variables that specified the true size and distance of the object. Bower (1965) confirmed these results in a second experiment with infants aged about 6 to 9 weeks, notably younger than in his first experiment.

In assessing the significance of these results four points need to be considered. First, in comparison with objects used in experiments on size constancy with adult observers, the objects were very large; for example, the 36-in. cube at 3 ft subtended a visual angle of more than 53°. Second, the distances involved, 3 and 9 ft, were much greater than is normally used with infant subjects. For both reasons greater confidence in the size constancy hypothesis would be provided by a replication of results at shorter distances with smaller objects. Third, the difference between the number of responses to the CS and to the same-size cube at a different distance was large. This difference is to be contrasted with the very small difference obtained between the CS and the same-shape object in a different orientation. Shape constancy is sometimes regarded as a special instance of size constancy, and similar information and mechanisms are thought to be involved. The magnitude of the difference in response which indicates perceptual constancy in these two experiments is not consistent with this interpretation. Fourth, in the shape constancy experiment generalization was tested separately to the dimensions of real shape, orientation, and retinal shape. Comparable assessment of the degree of generalization to the variables of real size, distance, and retinal size would provide stronger evidence for size constancy. This investigation has yet to be carried out.

Data from experiments involving recovery from habituation are inconclusive with respect to size constancy but indicate the importance of distance as a factor controlling attention. In our experiment (McKenzie & Day, 1972) with infants aged 6 to 20 weeks the stimulus objects were 6- and 18-cm cubes, the stimulus distances were 30 and 90 cm, and the planned index of constancy was extent of recovery from habituation of visual fixation. Contrary to prediction, response habituation did not occur with repeated trials for the larger object at 90 cm. Fixation times increased only when either object was advanced to the shorter distance of 30 cm. These results, together with the previously observed finding of habituation to a cube at the shorter distance, suggested that neither the small nor the large object was an effective stimulus at the greater distance, that fixation was largely dependent on distance, and, therefore, that the degree of response recovery was not an appropriate index of apparent size for different distances of the object.

In an attempt to explore further the response to size invariance fixation time was recorded under two conditions: retinal size constant over four distances, 30, 50, 70, and 90 cm, and actual size constant over the same distances. Object distance varied randomly over trials in both conditions. It was argued that habituation would occur with retinal size constant if infants attended primarily to an object's projected size, that habituation would occur with acutal size constant if they attended primarily to real size, and that habituation would not occur if infants responded primarily to distance stimuli. The last of these outcomes obtained; there was no decline in fixation over trials under either condition. There was, however, a linear decline in fixation time with increasing distance under both conditions, as shown in Figure 9-5. These results are consistent with the hypothesis that attention is directed toward varying object distance rather than constancy of real or retinal size.

In a third experiment to investigate fixation responses to stationary and moving cylinders over similar distances of 30 to 90 cm (McKenzie & Day, 1976) response decrement over trials was exhibited only by 4-month old infants (Figure 9-6). This decrement was not associated differentially with either of the conditions in which real or retinal size was invariant over distance. Thus it was concluded that, although the visual field of infants is more extensive for moving than for stationary objects, attention to size invariance was demonstrated for neither.

The results of these three experiments and that of Field (1975) do not support the hypothesis that infants respond predominantly to real size rather than to distance or retinal size. Under the conditions investigated, attention appears to be dependent on variation in distance rather than on invariance of size, but because the method used a naturally occurring response with no deliberate

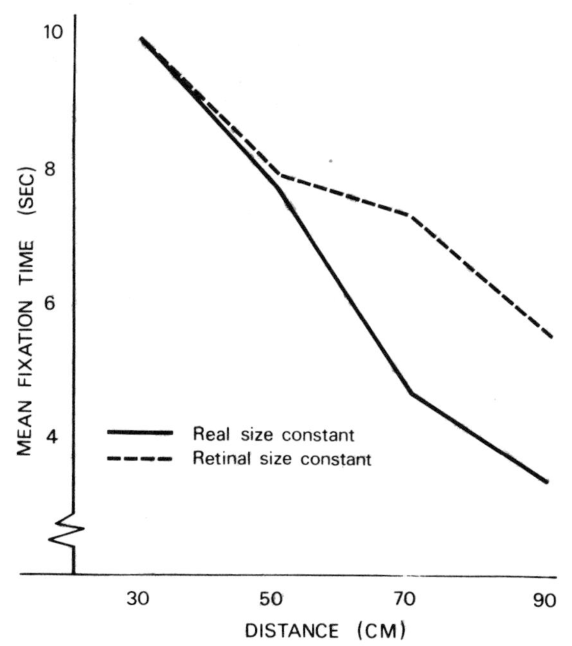

Figure 9-5 Mean visual fixation times for patterned cubes at four distances with real size and retinal size constant. A sharp decline can be noted in duration of fixation with distance. There was no difference in the rate of decline under the two conditions (from "Object Distance as a Determinant of Visual Fixation in Early Infancy," McKenzie, B.E., and Day, R.H., *Science*, Vol. 178, pp. 1108-1110, 8 December 1972).

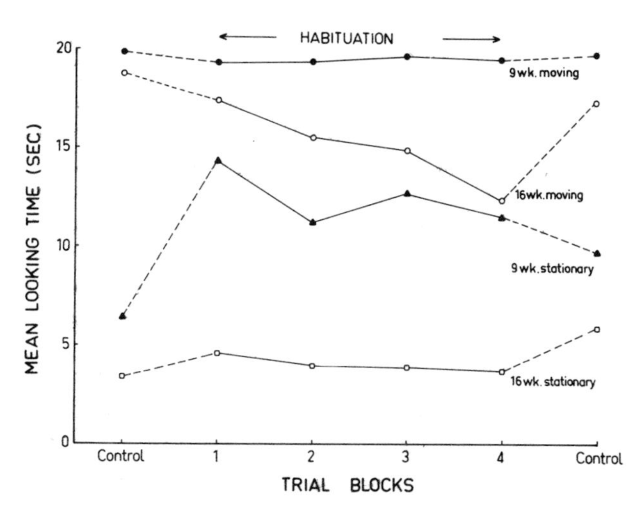

Figure 9-6 Mean visual fixation times over trials with a moving and stationary cylinder (figure 9-3) for infants approximately 9 and 16 weeks old (from McKenzie & Day, 1976).

response consequences it is possible that the obvious contradiction between these and Bower's (1964, 1965) findings might be attributed to differences in method. It is conceivable, of course, that operant training can direct attention to objects well beyond the preferred range of fixation found in previous experiments. For this to occur, however, the training procedure must also have altered the relative salience of stimulus properties that existed before training; attention must have somehow been made to focus on size and its perceptual invariance over distance as well as on distance itself.

An experiment described here for the first time was designed to study the rate of acquisition of a learned size discrimination as measured by direction of head turn under three different conditions; (a) real size constant with distance and retinal size different, (b) retinal size constant with real size and distance different, and (c) real size, retinal, size and distance different. Twenty-four infants aged 6 to 12 weeks were randomly allocated to one of the three experimental conditions and their head-turns recorded telemetrically by a method already described (McKenzie & Day, 1971; McKenzie & Sack, 1970). Reinforcement in the form of smiling, talking, and stimulation with a music-making toy was administered immediately a head-turn of appropriate direction and amplitude occurred. Equal numbers of males and females were included in each group and the direction of reinforced head-turns to each successively presented object was counterbalanced.

It was predicted that rate of response acquisition would reflect the difficulty of perceptual discrimination. If infants responded primarily to real size, the most difficult discrimination would be that involving a 12-cm cube at 20 cm and a 12-cm cube at 40 cm. If response was primarily controlled by retinal size, Condition b should respresent the most difficult discrimination: a 6-cm cube at 20 cm and a 12-cm cube at 40 cm. A 12-cm cube at 20 cm and a 6-cm cube at 40 cm constituted the third condition which, it was predicted, would be acquired most easily.

There was no significant difference between the three groups in the number of right and left head-turns emitted during a 2-minute baseline period before training. By the end of 40 trials administered over two consecutive days each group had emitted a greater number of correct responses (i.e., head-turns in the appropriate direction) than would have been expected by chance, $t(7) = 4.36$ 3.21 and 3.22, $p<.01$, for Conditions a, b, and c, respectively. The rate of acquisition of correct responses over trials was examined by a conditions (3) x blocks (4) analysis of variance with repeated measures on the last factor (where trials were grouped in blocks of 10). This analysis showed that for all groups the discriminative response was acquired within the first 10 trials, for neither of the factors nor their interaction yielded reliable effects.

Thus the data fail to support the size constancy hypothesis and do not suggest that contradictions in previous findings were brought about solely by differences between conditioning and habituation paradigms. On the other hand, the results support the hypothesis that object distance may be a salient variable in a conditioning situation. Infants attend primarily to distance and Conditions a, b, and c would be of similar difficulty because each condition involves discrimination at the same distances; that is, 20 and 40 cm.

SUMMARY OF THE FINDINGS ON SIZE CONSTANCY

In general, the results of recent experiments on size constancy do not support the earlier reports based on operant conditioning procedures. Object distance dominates object size with respect to measures of visual attention. The variables specifying the distance of the object for the baby clearly involve information other than that of retinal size. Rate of acquisition of a conditioned response is not clearly related to similarities in the physical size of the object. It is reasonable to conclude that size constancy in infancy has not yet been convincingly demonstrated. The weight of evidence suggests that unlike shape constancy it may not occur during the first half year. It can be noted that the shape and size constancy have sometimes been regarded as essentially similar phenomena. Change in retinal shape derives from tilt in depth so that there is a gradient of retinal size. The experiments reviewed here hardly support this view. Although shape constancy has been clearly demonstrated, the evidence for size constancy is less clear-cut.

IDENTITY-EXISTENCE CONSTANCY IN INFANCY

Objects can appear in different places in various orientations at different times. They can disappear briefly or for prolonged periods behind other objects. Other objects can pass between observer and object, thus obscuring the latter. For adult observers an object commonly retains its "sameness" or identity when it appears in various orientations and positions in the visual field. Likewise it is usually perceptually identical when it is observed on different occasions. When it is briefly obscured from view, it is regarded as being "still there." Such perceived identity of objects with variation in their time and place of appearance and assumption of continuing existence while occluded or obscured is commonplace among adult observers. Here we are concerned with whether such constancies of identity and existence occur during the first year.

Identity Constancy with Variation in Object Orientation

It is well known that when the orientation of frontoparallel figures is altered they frequently look different to adult observers and, in consequence are either not recognized or recognized as that particular object only with greater difficulty; that is, rotation of the figure about a median axis results in some loss of identity (Rock, 1973, 1975). It is surprising therefore that such constancy of identity with frontoparallel rotation occurs between 6 and 12 months.

McGurk (1972, 1974) observed responses to changes in orientation and form in a group of 144 subjects aged 10 to 54 weeks. There were four age groups: approximately 3 months, 6 months, 9 months, and 12 months old. One group of subjects in which all ages were represented was exposed for several trials to a two-dimensional figure in a constant orientation, another group to the same figure in different orientations, and a third to different forms in a constant upright orientation. The figures and conditions are shown in Figure 9-7. In the first and the second group subjects 6 months old showed response decrement over familiarization trials but although the first showed response recovery to identity change and orientation change the second showed recovery only to identity change. These data, which are shown in Figure 9-8, support the hypothesis that infants at 6 months or older recognize the identity of the same form in different orientations. McGurk, like some others, found difficulty in demonstrating habituation in his 3-month group, and hence was reluctant to conclude that his findings pertained to very young infants.

McGurk concluded that infants at 6 months and older, although capable of discriminating orientation differences, nonetheless recognized the identity of the object when it was presented in different orientations.

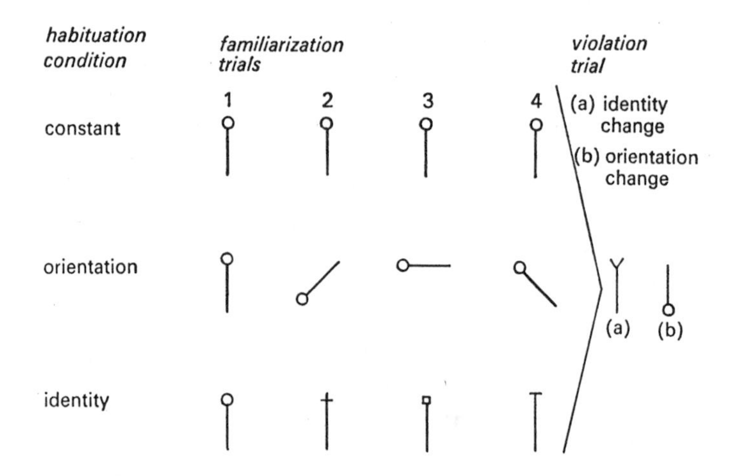

Figure 9-7 Examples of figures used during familiarization and violation trials by McGurk (1972). All infants were presented with the same stimulus in the first trial, but under form constant-orientation variable (FCOV) and form variable-orientation constant conditions the sequence of presentation of figures exposed during trials 2 to 4 were varied randomly. (from McGurk, 1974).

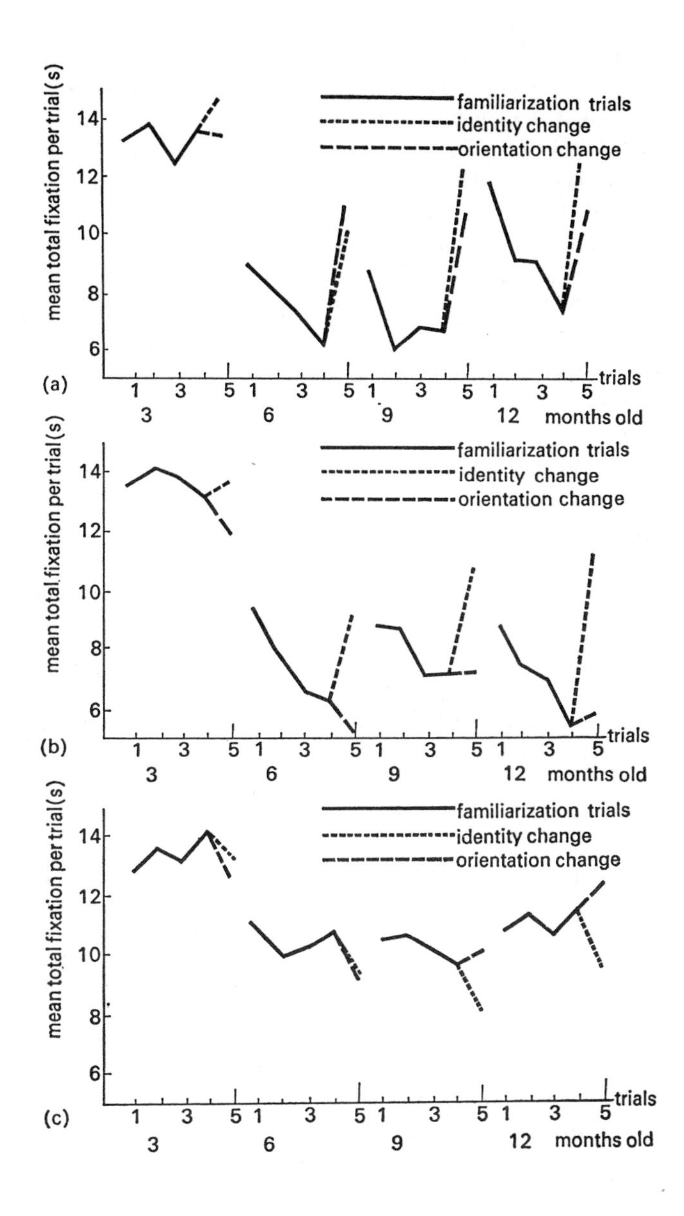

Figure 9-8 Duration of fixation per trial under (a) form constant-orientation constant (FCOC), (b) form constant-orientation variable (FCOV), and (c) from variable-orientation constant conditions in the experiment by McGurk (1972) (from McGurk, 1974).

Schwartz (1975), who used essentially similar methods, has recently shown that such constancy of object identity occurs also with a square as it is rotated. In addition, she has demonstrated that constancy occurs with infants younger than 6 months (see Day, 1974). These data show that McGurk's results are not confined to objects of the class in which the top is clearly indicated, as in Figure 9-7. The intriguing feature of Schwartz's experiments is that constancy of identity occurs in infancy with an object that does not exhibit it in adulthood; that is, an object in which the directions of top, bottom, right, and left cannot easily be assigned. Rock (1973, 1975) contends that it is the unconscious assignment of such directions that underlies this form of constancy. The issue clearly needs further experimental investigation.

Constancy of Identity with Variations in Time and Place

Object permanence, identity constancy, and existence constancy are terms used to describe the individual's belief in the object's continued and unchanged existence at a different time or at a different time and a different place. The sequence of stages in the development of this concept have been described by several research groups (Bell, 1970; Casati & Lezine, 1968; Corman & Escalona, 1969; Decarie, 1965; Gratch & Landers, 1971; Kramer, Hill, & Cohen, 1975; Miller, Cohen, & Hill, 1970; Piaget, 1952, 1954; Uzgiris & Hunt, 1975) and will receive no attention here. Discussion is focussed mainly on different methods for inferring identity-existence constancy and on the problem of what changes in location and movement an object can undergo and still be perceived by infants as the "same" object.

Clearly, the ability of an infant to identify an object perceptually as the same when it is presented at different times in different places must rely heavily on memory. Because memory rather than stimulus factors must constitute the main basis of this form of constancy its inclusion in a discussion of perceptual constancies can be questioned. The fact remains, however, that the young infant *perceives* an object as this or that even though the location and time of occurrence of the object varies. Perception in these situations, if it occurs, simply depends more on cognitive factors than on immediate stimulus information. This consideration justifies its inclusion in this discussion. The central question is, does identity-existence constancy occur in early infancy?

Indices of this class of constancy have included changes in heart rate (Bower, 1967, 1971), conditioned sucking (Bower, 1967), surprise reactions (Charlesworth, 1969), anticipatory visual tracking (Piaget, 1952, 1954; Bower, Broughton & Moore, 1971b; Bower & Paterson, 1972, 1973; Nelson, 1968, 1971, 1974; Mundy-Castle & Anglin, 1974), and, most frequently manipulative search (Bower

& Paterson, 1972; Bower & Wishart, 1972; Evans & Gratch, 1972; Gratch, Appel, Evans, Le Compte, & Wright, 1974; Gratch & Landers, 1971; Harris, 1971, 1973, 1974, 1975; Landers, 1971; Le Compte & Gratch, 1972; Miller, Cohen, & Hill, 1970; Webb, Massar, & Nadolny, 1972).

Bower (1971) reports that for infants as young as 6 weeks the nonappearance of a stationary object from behind a moving screen produces a greater change in heart-rate than its reappearance. This observation suggests that the infant is aware of the continuing existence of the object. However, the conclusion that some permanence is involved is not supported by studies using manual search responses until a much later age.

EVIDENCE FROM VISUAL TRACKING

Anticipatory visual tracking also appears to indicate a more precocious development than manual search. Piaget (1954) described the first steps in the development of identity-existence constancy in terms of eye movements directed toward the spot at which an object has disappeared but later toward the predicted object location. He argued that "so long as the search for the object consists merely in extending accommodation movements already made in its presence, the object cannot yet show either an independent trajectory in space or consequently intrinsic permanence" (Piaget, 1954, p. 89). More recently Bower, Broughton, and Moore (1971b), Bower and Paterson (1972, 1973), Gardner (1971), Nelson (1968, 1971, 1974), and Mundy-Castle and Anglin (1974) examined visual tracking of moving objects. Nelson (1971) showed clearly that for 3- to 9-month infants anticipation of the emergence of a train from a tunnel as indicated by anticipatory direction of gaze took several trials to develop and did not generalize when the train's direction of travel along the circular track was reversed. Learned anticipation of object movement was generalized, however, in other situations not involving reversal of direction (Nelson, 1974). Seven-month infants showed a small but rapid increase in anticipatory responses which generalized to a change in the speed and length of a trajectory and to a change in the object following the trajectory. It is noteworthy that the maximum probability of anticipation after 18 trials was only .3. Taken together, Nelson's findings suggest responses based on learned place associations rather than on the predicted trajectory of a single moving object.

Bower *et al.* (1971a) observed anticipatory tracking in which infants looked along the extension of a trajectory even when the object stopped in full view. They rejected the interpretation that infants were unable to inhibit ongoing head or eye movements because the error occurred as frequently for a rotary as for a linear trajectory and because analyses of eye-movements showed that the eyes halted for several hundred milliseconds before tracking the predicted path

(Bower & Paterson, 1973). They argued that this response error in anticipatory tracking indicated the infant's inability to identify a moving object with itself when stationary. A similar error was exhibited when infants searched for an object at a place in which it had been seen stationary after seeing it move in the opposite direction (Bower, 1971, 1974a). These errors cease to occur at about five months, Its similarity to the manual search error of Stage IV of the object concept (Piaget, 1952), when infants search in a location in which an object was previously found rather than where they have seen it hidden is obvious. Bower (1974b) has argued that visual tracking is a preliminary stage in a sequence of development of the object concept, for enriched experience with the visual task accelerates later developnent of the manual task.

Alternative interpretations of these visual errors are possible. A more detailed study of precise eye movements in the initial trial and throughout training and test trials would be helpful in selecting anong alternatives; for example, the continued tracking of a moving object which suddenly becomes stationary and the visual place error (searching in a previously successful location when the object is stationary for an unexpectedly long duration or when it moves in an unexpected direction) required the inhibition of recently established motor programs: eye movements in the direction of a moving object in the first example and a repeated sequence of left-right eye movements in the second. It might be expected that voluntary tracking, an acquired motor response, takes time to develop and to extinguish, and its rate of acquisition probably varies with the type of trajectory—its speed, extent, direction. and number of changes in direction. Inappropriate anticipatory search when an object becomes stationary or changes its established pattern of movement may simply reflect the difficulty of inhibiting this motor program. Thus these visual errors may be evident in younger infants and corrected on older infants when motor programs become more flexible and subject to greater control by stimulus events. The assumption that perceptual hypotheses about the nature of objects are also involved needs further justification. If the tracking of moving objects by infants is not smooth (as assumed by Bower et al., 1971a) but saccadic, involving a number of brief pauses (Trevarthen, 1974), the occurrence of a single pause assumes less significance. One pause among many is less suggestive than one during smooth following. This observation reduces support for the hypothesis that infants are unable to identify an object as the same when it is moving and stationary.

EVIDENCE FROM MANIPULATIVE SEARCH

Piaget's (1952, 1954) analysis of the development of identity-existence constancy was primarily based on manipulative search. He argued that the place error in Stage IV—searching in an earlier location rather than in the one in which the object was last hidden—was a consequence of assimilation of this observation to the scheme of "finding the-object-of-place." The contribution of this place error of memory factors (Gratch et al., 1974; Harris, 1973) of the number of preceding successful searches in a given location (Landers, 1971) and of the type of familiarity with the hidden object (Harris, 1971) have also been examined. It is clear that neither memory nor motor deficits alone can explain the occurrence of this error because infants search in a given location when an object is perfectly visible in another (Harris, 1974). A deficit in the cognitive understanding of objects and their location is space is indicated. It is not until the end of the second year that infants recognize that an object can occupy only one position in space at any one time.

DISCUSSION

This chapter began by asking whether infants, like adults, perceive the objective situation as its sensory representation varies; that is, whether perceptual constancies occur in human infancy. This review of experimental observations shows that the question cannot be fully answered and that a clear and definite answer is possible in the case of only a few of the visual constancies, notably those of visual shape and some visual identity constancies. The situation in regard to visual size constancy is insettled. Before discussing some general issues that arise from the review of experiments and their outcomes it will be helpful first to summarize the main findings brieflv.

Among the object constancies there is good evidence of the occurrence of shape constancy in infancy. The outcomes of three experiments by Bower (1966), mainly involving operant procedures, and one by us (Day & McKenzie, 1973), based on rates of habituation to solid objects, are in agreement in their indication that shape constancy occurs as early as 6 to 8 weeks. The situation in regard to size constancy in which the weight of evidence shows that it does not occur during the early months is confused. The studies by Cruikshank (1941) and Misumi (1951) which are frequently invoked in evidence for size constancy must be discounted on the grounds of inadequate method and unsuitable indexing response. Bower (1964, 1965) claimed the occurrence of size constancy before 12 weeks of age on the basis of evidence from operant conditioning experiments. McKenzie and Day (1972) were unable to confirm this result when they used rate of habituation as an index. In another experiment we again failed to adduce evidence for size constancy when rate of acquisition of head turning was used as an index. The subjects were aged between 6 and 12 weeks. Finally, we (McKenzie & Day, 1976) found that response decrement over trials to a patterned cylinder does not vary when real or retinal size is held constant. Thus the situation in regard to size constancy with infants is confused and unsettled.

We have consistently failed to find evidence for it. The only positive evidence so far is that reported by Bower (1964, 1965).

For most adult observers an object becomes increasingly difficult to recognize as it is tilted in a frontoparallel plane, especially when internal directions cannot be assigned. McGurk (1972) and Schwartz (1975) have shown convincingly, however, that for infants constancy of object identity occurs under this condition. Furthermore, it occurs even when directions in the figure (top, bottom, etc.) cannot easily be assigned. The most persistent difficulties for the identity-existence constancies appear to be associated with visible and invisible displacements of objects in space rather than with their continued existence in the temporary absence of direct receptor excitation. Infants from an early age bring an unseen but grasped object into the visual field (Piaget, 1952), exhibit greater heart-rate change to the nonappearance rather than the appearance of a previously seen object behind a moving screen (Bower, 1971), and reach for an object after the lights are turned out (Bower & Wishart, 1972). Yet at a later age the baby behaves as if a visible or invisible object may be in several locations simultaneously. Having seen and failed to obtain an object in one location, the baby will search in a prior location (Harris, 1974) just as, at an earlier age, having seen an object become stationary in one location, the infant searches visually in another location (Bower, et al., 1971). Harris (1975) argues that these difficulties are a consequence of the absence of recognition of a general rule that an object can occupy only one place at any one time. How the infant acquires this rule is yet to be determined.

Perhaps the most striking fact to emerge from this review of experimental literature is the complete absence so far of any work on the egocentric constancies in infancy, which, it will be recalled, are constancies of object position, movement, and stationariness in relation to the observer as sensory representations of these properties vary. Before about 6 months of age the young infant is incapable of independent locomotion. Nevertheless, head and eye movements are common; changes in the representation at eyes and ears of the location of things in the environment frequently occur. In the face of these changes do the perceived fixity and location of things remain constant? The question cannot yet be answered. This represents a considerable gap in our knowledge about perceptual stability in infancy. Simple naturalistic observations could go some distance toward telling whether such constancies occur. In addition, experiments along the lines of those reported with adult subjects on visual direction, stationariness, and distance constancy need to be carried out at different stages of development.

It can be noted also that until recently lightness and color constancy in infancy have been largely neglected. Attempts to demonstrate chromatic vision have necessarily involved control for variation in brightness sensitivity to different wavelengths. Following recognition of the problems involved in equating brightness, the procedure of systematically varying the luminance within hues has been adopted. Thus Schaller (1975) operantly conditioned infants to look longer at one hue than another. Because the luminance of the hue varied over trials, it could not have been the stimulus factor controlling the discriminative response. Color constancy was indicated by increased looking to the reinforced stimulus, despite changes in its luminance. It can be noted also that we (Day & McKenzie, 1973) incidentally found some evidence of lightness constancy with infants 6 to 16 weeks old. When the cubes used in our shape constancy experiment were rotated in depth not only did the outline projected shape change but also the pattern of light and dark on the surface, as shown in Figure 9-1. It will be recalled that habituation occurred for different projected outline shapes and therefore variable surface patterns, as it did for a single projected shape and pattern. It is reasonable to assume from this observation that lightness constancy prevailed. The projected pattern of light and dark seems not to have been perceived as such but as the front, top, and sides of a white unpatterned cube; variations in luminance merely signaled the parts of the object rather than different parts of a pattern. More detailed and comprehensive experiments on lightness constancy, with variation in ambient light intensity, need to be done before firm conclusions can be reached.

The grouping of perceptual constancies into three major classes—egocentric, object, and identity-existence constancies—outlined together with a description of effects falling into each class—serves to emphasize how limited the information is so far on these phenomena in infancy. Because the constancies can be viewed as adaptive patterns of behavior, information about their occurrence or nonoccurrence at various stages of development would serve as a measure of adaptation. In this connection it can be noted that there are few studies in which age has been systematically included as an independent variable.

The significance of the occurrence of perceptual constancies early in development for the major theoretical standpoints has been discussed briefly. How are these views affected by the findings reviewed here? The fact that at least one perceptual constancy occurs in infancy, of course, lends some support to those standpoints which stress the role of innate organization in the nervous system. This is meagre support, indeed. Size constancy has not been reliably shown to occur in early infancy. It seems, therefore, that before considering the support or nonsupport offered by findings on perceptual constancies in infancy to theories of perception a great many more data are required. In short, the issue is still an empirical one.

The problem taken up at the beginning of this chapter was defined by some questions about perceptual constancy in infancy. It was asked: do infants perceive an approaching or receding person as the same size, a feeding bottle as the same shape as it is tilted in depth and their hands as the same lightness when parts of them are in shadow? So far we cannot answer the first and third questions; the results bearing on the first are in conflict and the third has not even been tackled. The second can be answered in the affirmative with two studies confirming each other. In additional, we can say that as the feeding bottle is tilted laterally it retains its identity for the baby. Clearly there is still much more to find out.

REFERENCES

Ball, W. & Tronick, E. Infant responses to impending collision: optical and real. *Science*, 1971, 171, 818-820.

Bell, S. The development of the concept of object as related to infant-mother attachment. *Child Development*, 1970, 41, 291-311.

Berkeley, G. *A new theory of vision.* London: (Everyman's Ed., London, 1969), 1709.

Beryl, F. Ueber die Grossënauffassung bie Kindern. *Zeitschrift für Psychologie*, 1926, 100, 344-371.

Bower, T. G. R. *Development in infancy.* San Francisco: Freeman, 1974a.

Bower, T. G. R. Repitition in human development. *Merrill-Palmer Quarterly*, 1974b, 20, 303-318.

Bower, T. G. R. The object in the world of the infant. *Scientific American*, 1974, 225, 30-38.

Bower, T. G. R. Object perception in infants. *Perception*, 1972, 1, 15-30.

Bower, T. G. R. The development of object-permanence: Some studies of existence constancy. *Perception & Psychophysics*, 1967, 2, 411-418.

Bower, T. G. R. Slant perception and shape constancy in infants. *Science*, 1966, 151, 832-834.

Bower, T. G. R. Stimulus variables determining space perception in infants. *Science*, 1965, 149, 88-89.

Bower, T. G. R. Discrimination of depth in prenatal infants. *Psychonomic Science*, 1964, 1, 368.

Bower, T. G. R., Broughton, J,. M., & Moore, M. K. Infants' responses to approaching objects: An indicator of response to distal variables. *Perception & Psychophysics*, 1971a, 9, 193-196.

Bower, T. G. R., Broughton, J. M., & Moore, M. K. Development of the object concept as manifested in changes in the tracking behavior of infants between 7 and 20 weeks of age. *Journal of Experimental Child Psychology*, 1971b, 11, 182-193.

Bower, T. G. R., Broughton, J. M., & Moore, M. K. Demonstration of intention in the reaching behaviour of neonate humans. *Nature*, 1970, 228, 679-681.

Bower, T. G. R. & Paterson, J. G. The separation of place, movement, and object in the world of the infant. *Journal of Experimental Child Psychology*, 1973, 15, 161-168.

Bower, T. G. R. & Paterson, J. G. Stages in the development of the object concept *Cognition*, 1972, 1, 47-55.

Bower, T. G. R. & Wishart, J. G. The effects of motor skill on object permanence. *Cognition*, 1972, 1, 165-172.

Bruner, J. S. & Koslowski, B. Visually preadapted constituents of manipulatory action. *Perception*, 1972, 1, 3-14.

Brunswik, E. Organisimic achievement and environmental probability. *Psychological Review,* 1943, 50, 255-272.

Brunswik, E. Thing constancy as measured by correlation coefficients. *Psychological Review,* 1940, 47, 69-78.

Brunswik, E. *Wahrnehmung und Gegenstandswelt.* Vienna: Deuticke, 1934.

Brunswik, E. Die Zugänglichkeit von Gegenständer für die Wahrehmung and deren quantitative Bestimmung. *Archiv für die gesamte Psychologie,* 1933, 88, 378-418.

Brunswik, E. Ueber Farben-Grössen, and Gestalt-knostanz in der Jugend. In H. Volkelt (Ed.), *Bericht uber den XI Kongress fur experimentelle Psychologie.* Jena: Fischer, 1930, p. 52-56.

Campos, I. J., Langer, A. J., & Krowitz, A. Cardiac responses on the visual cliff in prelocomotor human infants. *Science,* 1970, 170, 196-197.

Casati, I., & Lezine, I. Les etapes de l'intellignece sensori-motrice. Paris: *Les Editions du Centri de Psychologie Applique,* 1968.

Charlesworth, N. The role of surprise in cognitive development. In D. Elkind & J. H. Flavell (Eds.), *Studies in cognitive development: Essays in honor of J. Piaget.* New York: Oxford University Press, 1969, p. 257-314.

Corman, H., & Escalona, S. Stages of sensori-motor development: A replication study. *Merrill-Palmer Quarterly,* 1969, 15, 351-361.

Cruikshank, R. M. The development of visual size constancy in early infancy. *Journal of Genetic Psychology,* 1941, 58, 327-351.

Cruikshank, R. M. The development of size-constancy in early infancy. *Onzième Congres International de Psychologie,* Paris: Imprimeria Moderns, 1938.

Day, R. H. Perceptual processes in early infancy. *Australian Psychologist,* 1974, 9, 15-34.

Day, R. H. Visual spatial illusions: A general explanation. *Science,* 1972, 175, 1335-1340.

Day, R. H. *Human perception.* Sydney: Wiley, 1969.

Day, R. H. Perceptual constancy of auditory direction with head rotation. *Nature,* 1968, 219, 501-502.

Day, R. H. & McKenzie, B. Perceptual shape constancy in early infancy. *Perception,* 1973, 2, 315-320.

Day, R. H. & Wade, N. J. Mechanisms involved in visual orientation constancy. *Psychological Bulletin,* 1969, 71, 33-42.

Decarie, T. G. *Intelligence and affectivity in early childhood.* New York: International Universities Press, 1965.

Engel, G. R. & Dougherty, W. G. Visual-auditory distance constancy. *Nature,* 1971, 234, 308.

Evans, W. F. & Gratch, G. The Stage IV error in Piaget's theory of object concept development: Difficulties in object conceptualization or spatial localization? *Child Development.* 1972, 43, 682-688.

Fantz, R. L. & Fagan, J. F. Visual attention to size and number of pattern details by term and preterm infants during the first six months. *Child Development.* 1975, 46, 3-18.

Fantz, R. L., Ordy, J. M. & Udelph, M. S. Maturation of pattern vision in infants during the first six months. *Journal of Comparative and Psysiological Psychology,* 1962, 55, 907-917.

Fantz, R. L. & Nevis, S. The predictive value of changes in visual preference in early infancy. In J. Hellmuth (Ed.), *Exceptional Infant,* Vol. 1. Seattle: Special Child Publications, 1967a, p. 349-414.

Fantz, R. L. & Nevis, S. Pattern preferences and perceptual cognitive development in early infancy. *Merrill-Palmer Quarterly*, 1976b, **13**, 77-108.

Field, J. The adjustment of reaching behavior to object distance in early infancy. Paper presented at the Second Experimental Psychology Conference, Sydney, 1975.

Gajzago, C. & Day, R. H. Visual uprightness constancy. *Journal of Experimental Child Psychology*, 1972, **14**, 43-52.

Gardner, J. K. The development of object identity in the first six months of human infancy. Paper presented at the Society for Research in Child Development, Minneapolis, Minnesota, 1971.

Gibson, J. J. *The senses considered as perceptual systems.* Boston: Houghton Mifflin, 1966.

Gratch, G., Appel, K. J., Evans, W. F., Le Compte, G. K. & Wright, N. A. Piaget's Stage IV object concept error: Evidence of forgetting or object conception? *Child Development*, 1974, **45**, 71-77.

Gratch, G. & Landers, W. F. Stage IV of Piaget's theory of infant's object concepts: A longitudinal study. *Child Development*, 1971, **42**, 359-372.

Harris, P. L. Development of search and object permanence during infancy. *Psychological Bulletin*, 1975, **82**, 332-344.

Harris, P. L. Perseverative search at a visibly empty place by young infants. *Journal of Experimental Child Psychology*, 1974, **18**, 535-542.

Harris, P. L. Perseverative errors in search by young children. *Child Development*, 1973, **44**, 28-33.

Harris, P. L. Examination and search in infants. *British Journal of Psychology*, 1971, **62**, 469-473.

Haynes, H., White, B. L. & Held, R. Visual accommodation in human infants. *Science*, 1965, **148**, 528-530.

Hill, A. L. Direction constancy. *Perception & Psychophysics*, 1972, **11**, 175-178.

Hillebrand, F. Theorie der scheibaren Grösse bei binocularem Schen. *Derkschriften Akademie Wissenschaftlichen Wien*, 1902, **72**, 255-307.

Hochberg, J. Perception. In J. W. Kling & L. A. Riggs (Eds.), *Woodworth's and Schlosberg's experimental psychology.* London: Methuen, 1971, pp. 395-550.

Howard, I. P. & Templeton, W. B. *Human spatial orientation.* New York:Wiley, 1966.

Kaufman, L; *Sight and mind: An introduction to visual perception.* New York: Oxford University Press, 1974.

Kramer, J. A., Hill, K. T. & Cohen, L. B. Infants' development of object permanence: A refined methodology and new evidence for Piaget's hypothesized ordinality. *Child Development*, 1975, **46**, 149-155.

Landers, W. F. Effects on differential experience on infants' performance in a Piagetian Stage IV object-concept task. *Developmental Psychology*, 1971, **5**, 48-54.

Le Compte, G. K. & Gratch, G. Violation of a role as a method of diagnosing infants' level of object concept. *Child Development*,1972, **43**, 385-386.

Martius, G. Ueber die scheinbare Grosse der Gegenstände und ihre Beziehung zur Grösse Der Netzhautbilder. *Philosophishe Studien*, 1889, **5**, 601-617.

McGurk, H. Visual perception in young infants. In B. Foss, *New perspectives in child development.* Harmondsworth, Middlesex: Penguin Education, 1974.

McGurk, H. Infant discrimination of orientation. *Journal of Experimental Child Psychology*, 1972, **14**, 151-164.

McKenzie, B. E. & Day, R. H. Distance as a determinant of visual fixation in early infancy. *Science*, 1972, **178**, 1108-1110.

McKenzie, B. E. & Day, R. H. Operant learning of visual pattern discrimination in young infants. *Journal of Experimental Child Psychology*, 1971, **11**, 45-53

McKenzie, B. E. & Day, R. H. Infants' attention to stationary and moving objects at different distances. *Australian Journal of Psychology*, 1976, **28**, 45-51.

McKenzie, B. E. & Sack, K. A. A telemetric method for measuring and recording head rotations in human infants. *Behavior Research Methods and Instrumentation*, 1970, **2**, 173-174.

Michotte, A. Perception et cognition. *Acta Psychologia*, 1955, **11**, 69-91.

Michotte, A. *La perception de la causalite.* Louvain: Institut Superieue de Philosophie, 1946.

Miller, D., Cohen, L. & Hill, K. A. A methodological investigation of Piaget's theory of object concept development in the sensory-motor period. *Journal of Experimental Child Psychology*, 1970, **9**, 59-85.

Misumi, J. Experimental studies on the development of visual size constancy in early infancy. *Bulletin of the Faculty of Literature.* Kyushu University, 1951, **1**, 91-116.

Mohrmann, K. Lautheitskonstanz im Entfernung wechsel. *Zeitschrift für Psychologie*, 1939, **145**, 146-199.

Mundy-Castle, A. C. & Anglin, J. M. Looking strategies in infants. In L. J. Stone, H. T. Smith & L. B. Murphy (Eds.), *The competent infant.* London: Tavistock, 1974, pp. 713-717.

Nelson, K. E. Infants' short-term progress toward one component of object permanence. *Merrill-Palmer Quarterly*, 1974, **20**, 3-8.

Nelson, K. E. Accommodation of visual tracking patterns in human infants to object movement patterns. *Journal of Experimental Child Psychology*, 1971, **12**, 157-169.

Nelson, K. E. Organization of visual-tracking responses in human infants. *Journal of Experimental Child Psychology*, 1968, **6**, 194-201.

Over, R. Distance-constancy. *American Journal of Psychology*, 1961, **74**, 308-310.

Piaget, J. *The origins of intelligence in children.* New York: International Universities Press, 1952.

Piaget, J. & Inhelder, B. *Psychology of the child.* New York: Basic Books, 1969.

Postman, L. & Tolman, E. C. Brunswik's probabilistic functionalism. In S. Koch, *Psychology: A study of a science* (Vol. 1). New York: McGraw-Hill, 1959, pp. 502-564.

Purdy, J. & Gibson, E. J. Distance judgment by the method of fractionation. *Journal of Experimental Psychology*, 1955, **50**, 374-380.

Rock, I. *An introduction to perception.* New York: Macmillan, 1975.

Rock, I. *Orientation and form.* New York: Academic, 1973.

Rock, I. Orientation of forms on the retina and in the environment. *American Journal of Psychology*, 1956, **69**, 513-528.

Rock I. & Heimer, W. The effect of retinal and phenomenal orientation on the perception of form. *American Journal of Psychology*, 1957, **70**, 493-511.

Rock, I., Hill, A. L. & Fineman, M. Speed constancy as a function of size constancy. *Perception & Psychophysics*, 1968, **4**, 37-40.

Salapatek, P., Bechfold, A. G. & Bushnell, E. W. Infant accommodation and acuity threshold as a function of viewing distance. Paper delivered at the Society for Research in Child Development, Denver, Colorado, 1975.

Schaller, M. J. Chromatic vision in human infants: conditioned operant fixation to hues of varying intensity. *Bulletin of the Psychonomic Society*, in press.

Schwartz, M. *Visual shape perception in early infancy*. Unpublished Ph.D. Thesis, Monash University, 1975.

Smith, O. W. Distance constancy. *Journal of Experimental Psychology*, 1958, 55, 388-389.-

Thouless, R. H. Individual differences in phenomenal regression. *British Journal of Psychology*, 1932, 22, 216-241.

Thouless, R. H. Phenomenal regression to the real object. *British Journal of Psychology*, 1931, 21, 339-359; 22, 1-30.

Trevarthen, C. The psychobiology of speech development. *Neurosciences Research Program Bulletin*, 1974, 12, 570-585.

Tronick, E. Stimulus control and the growth of the infant's effective visual field. *Perception & Psychophysics*, 1972, 11, 373-376.

Uzgiris, I. & Hunt, J. McV. Assessment in infancy: Ordinal scales of psychological development. Urbana: University of Illinois Press, 1975.

Vernon, M. D. *A further study of visual perception*. Cambridge: Cambridge University Press, 1962.

Wade, N. J. *The effect of body posture on visual orientation*. Unpublished Ph.D. Thesis, Monash University, 1968.

Walk, R. D. The development of depth perception in animals and human infants. *Monographs of the Society for Research on Child Development*, 1966, 31, 82-108.

Wallach, H. & Kravitz, J. H. The measurement of the constancy of visual direction and its adaptation. *Psychonomic Science*, 1965, 2, 217-218.

Webb, R. A., Massar, B. M. & Nadolny, T. Information and strategy in the young child's search for hidden objects. *Child Development*, 1972, 43, 91-104.

Wertheimer, M. Psychomotor co-ordination of auditory-visual space at birth. *Science*, 1961, 134, 1692.

White, B. L. *Human infants: Experience and psychological development*. Englewood Cliffs, New Jersey: Prentice-Hall, 1971.

Woodworth, R. *Experimental psychology*. London: Methuen, 1938.

Woodworth, R. S. & Schlosberg, H. *Experimental psychology*, New York: Holt, Rinehart & Winston, 1954.

CHAPTER

10

IN DEFENSE OF UNCONSCIOUS INFERENCE

IRVIN ROCK
Rutgers University

TWO THEORIES OF CONSTANCY

Two different kinds of explanation have been proposed for constancy in perceptual experience, the origins of which date back to Helmholtz and Hering. Both start with the fact that the proximal stimulus (or retinal image in visual perception) representing the object under consideration cannot by itself be the basis of what is perceived because it is subject to varation. Because an object's image will change as a function of the object's distance, slant, or illumination or of the observer's movement or orientation, any particular image considered alone is ambiguous as to what it might represent in the world.

According to one explanation (Helmholtz), the perceptual system takes account of factors such as distance or illumination in arriving at a perceptual

I wish to thank Carl Zuckerman, Alan Gilchrist, Arien Mack and William Epstein for their helpful suggestions in the preparation of the chapter.

judgment of what the retinal image represents in the world. There are two aspects of this theory that distinguish it from the other:

1. The information (or cue) taken into account derives from a source separate from the retinal image of the object; for example, cues to distance or illumination.
2. The view that the process is a cognitive operation or an unconscious inference, which in turn seems to imply a determining role of experience.

It will be helpful in the discussion to follow, at least for the time being, to separate these two aspects and to put the emphasis on the first as the major defining characteristic of this theory. Its essence, then, is the notion of coupling (Hochberg, 1974) or a combinatorial process (Epstein, 1973) in which the percept results from combining a property of the retinal image of the object with other information concerning the location and orientation of the object or the state of the observer. Stated in this way, the theory reduces to a testable hypothesis; whereas if the second aspect is also included in the statement, far more is implied concerning the origin and nature of the combinatorial process. Therefore it is best to postpone discussion of this second aspect. I will refer to this view as the taking-into-account theory.

According to the other kind of explanation the information that leads to constancy is given in the retinal image, but we must look beyond the local stimulus representing the object itself; for example in the perception of achromatic color it is not merely the intensity of the image of the surface under consideration that matters but the intensity of the images of neighboring surfaces as well (Hering). Another way of stating this is to say that it is not the absolute property of the image of the object that determines perception but the relationship of that image to other neighboring images.

There are several variations of this kind of theory that I propose to group together, at least for the moment. Thus, for Hering, the basis for the important role of neighboring images in the perception of "brightness" was what he referred to as opponent processes in the nervous system, anticipating what is now called lateral inhibition. The emphasis here is on neural events determined by extended patterns of stimulation. More recently J. J. Gibson (1950, 1966) argued that all the information necessary for constancy is given in the retinal image; therefore it suffices to specify the attributes of the proximal stimulus correlated with particular perceptions. Gibson's claim is that hitherto unrecognized features of the proximal stimulus entail a higher order level of analysis (e.g., visual angle relations deriving from perspective transformations in the formation of the image), which are directly correlated with constant features of perception.

Despite some differences of viewpoint, the essence of this second theory for present purposes is the proposition that constancy is directly based on information contained in the retinal image, be it in the form of certain relationships, invariants, or gradients. I refer to it as the higher order stimulus theory.

Which theory is right? It may turn out that neither can do justice to all the phenomena; one gives a better account of one constancy, the other of another constancy. It is obviously preferable, however, to seek a single unified theory and to retreat to separate explanations of each constancy only if forced to do so by the evidence. In what follows I discuss first the problem of achromatic (or neutral) color constancy in some detail, then consider briefly some of the other constancies, and finally treat the problem of size constancy. It is probably safe to say that the majority of investigators in perception believe that achromatic color is determined primarily by luminance relationships or ratios (the higher order stimulus theory) rather than by a process of taking account of the prevailing illumination. Conversely, the majority would probably agree that size perception depends on a process of taking account of distance.

ACHROMATIC COLOR CONSTANCY

It is not necessary to review all the evidence here, for fairly comprehensive surveys are available (Hochberg, 1972; Beck, 1972; Kaufman, 1974; Rock, 1975). It is enough to say that there is much evidence that whether a region appears white, some shade of gray, or black depends on the relation or ratio of the intensity of light in that region (its luminance) to the luminance in an adjacent region (for the clearest statement and evidence for this hypothesis see Wallach, 1948). This principle can explain a variety of facts. It explains constancy in daily life because typically a source of illumination will affect a particular region and its surroundings about equally. Therefore the ratio between any surface of a given reflectance and its surrounding surfaces will not be altered when illumination changes, for example, when a light is turned on or off. The ratio principle also explains anomalous illusions such as the Gelb effect in which an isolated black surface, suspended in midair by a string, looks white when only it is illuminated or a white surface looks dark gray when only it is shadowed. The principle explains the departure from constancy obtained in experiments when the surfaces to be compared, each under different illumination, are viewed through openings in a reduction screen. The reduction screen represents a new surround for each surface and because its color is *uniform* and is in *uniform* illumination throughout the ratio for the surfaces under comparison, each to its respective surround, is no longer preserved.

Not only does this principle have great explanatory power but the alternative theory of constancy based on a process of taking account of illumination is weak. The major difficulty of the latter concerns the question how information about illumination can be obtained. The light reflected by a surface is a function not only of the illumination on that surface but also of the reflecting property (or reflectance) of the surface. Therefore the luminance reaching the eye could result from any combination of surface reflectance and illumination. Unless the reflectance were already "known" there would be no way of "knowing" the incident light received in order to take it into account.

To be sure, there are certain instances in daily life in which the notion of taking account of illumination on a surface seems to make sense and when it is not obvious how the ratio principle could apply. Consider the best known example of this kind, namely, the situation in which surfaces are oriented differently with respect to the source of light. A surface orthogonal to a light source receives more intense light per unit of area than one slanted with respect to that source. This is conveniently illustrated by any dihedral angle, either opening toward or away from the observer (i.e. where the walls of a room meet at a corner) in which both surfaces are the same color. One surface will typically receive more light than the other. How do these surfaces appear? Were they to look alike, we might say that constancy obtains.

There is some disagreement about how to describe this kind of experience. Suppose both surfaces are white. Some observers will say that the two surfaces appear quite different—one light in color, the other dark in color. (I will use the less cumbersome term "color" to refer to achromatic color) Others will say they look different, but both appear to be white—one intense or "bright" and the other "dim." (What I mean by "bright" and "dim" here is best defined by appealing to the reader's own experience. A white surface in sunlight appears to be "brightly white"; indoors it appears to be "dimly white.") If, however, the corner is subjectively flattened out, either by closing one eye or by blocking from view the junction points of the corner with ceiling and floor, then all observers will agree that the two surfaces now look differently colored—one white, the other some shade of gray. In fact, after seeing the corner thus flattened the observers who had until then described the two surfaces as "light" and "dark" may now realize that they should have said "bright" and "dim."

What do the two theories predict? It is not obvious how any tendency toward constancy can be accounted for by the ratio principle, which refers to the ratio of luminances between the retinal images of the surfaces in question. The surfaces meeting at the corner and receiving different amounts of illumination give rise to images at the eye of different luminance. They should look as different from one another as two differently colored surfaces in the same illumination. Yet for some observers both look white. Although the perception of different colors that occurs with subjective "flattening" is predictable by the ratio principle, *the change* in perception that definitely does occur purely as a result of a change in depth perception is not at all predictable. Actual luminance relations are not affected by subjective reorganizations of depth perception.

How can the taking-into-account theory of constancy explain facts of this kind? The logical difficulty of independent access to information about illumination can be overcome, at least in part, in this example. Here depth and shadow cues that signify the direction of the source of light are available. Given tha. information and the "knowledge" of the relationship between the orientation of the surface to the light source and the proportion of light received, the perceptual system can then consider that orientation in arriving at the perception of the surface's achromatic color. By taking account of the different amount of light received by the surfaces meeting at the corner a tendency toward constancy is to be expected, but if the two surfaces are perceived to be in one plane the illumination must be assumed to be equal and the different luminances of the surfaces must signify different colors.

Although it would seem, therefore, that there is at least one situation in which the outcome (more or less) follows from the taking-into-account theory but not from the ratio principle, laboratory evidence has been weak. The now classic paradigm to study this problem experimentally was first used by Hochberg and Beck (1954)* The basic plan was to have observers view a surface that could be perceived in either of two orientations and to determine whether a change in its orientation would affect its apparent color. The surface is trapezoidal in shape and perpendicular to the top of a table. Viewed from the end of the table from one position with one eye, the trapezoidally shaped retinal image typically gives rise to an impression of a rectangle lying on the table. If other depth information is introduced (whether by binocular vision, head movement parallax, or kinetic depth via slight rotation of the trapezoid), the trapezoid will be seen veridically as perpendicular to the table top. Cues to the direction of illumination are provided by the location of shadows cast by cubes lying elsewhere on the table. Under each condition of viewing the observers match the apparent color of the trapizoidal surface to a comparison surface.

If the perceived orientation of the surface is relevant and light comes from above, the prediction would be that the trapezoid would look darker when it looks like a rectangle on the table than when it looks like a standing trapezoid. If it is standing upright it will receive relatively little light and the luminance reflected to the eye must derive from a relatively high reflctance property; if it is horizontal, it will receive much more light, and that same luminance reaching the eye must be based on a much lower reflectance value. The result in some experiments is in the direction of the prediction but it is quantitatively very small, of the order of one-half a step or less on the Munsell scale of calibrated reflectance values. (Hochberg & Beck, 1954; Beck, 1965; Flock & Freedberg, 1970). Others have found no difference in the perceived color of a surface as a

* Others such as Mach (1959), Katona (1935), and Evans (1948) had explored this problem but the Hochberg and Beck study is a clearer, better controlled, formal experiment.

function of its perceived orientation (Epstein, 1961; Flock, 1971). If surface slant were completely taken into account, the effect should be considerable.

From this evidence, therefore, we might draw precisely the opposite conclusion about the effect of the perceived slant of a surface than that suggested above. Rather than challenging the ratio principle and supporting a taking-into-account theory, the weakness or absence of an effect suggests that the ratio of luminances of adjacent retinal images is the major determinant of perceived color, regardless of the perceived spatial orientation of the corresponding surface. This would, however, leave unexplained the apparent effect of perceived orientation of surfaces in daily life as in the example of a corner of a room, the slight effect sometimes obtained in laboratory experiments, and another slight effect of spatial position to be described next.

Several investigators have raised the question whether the separation of a surface in depth from its background would have any effect on its perceived color. Because the retinal image of the target surface would remain surrounded by the image of the background surface, there should be no reason for any change in the perceived color of the target surface if the proximal stimulus relation determines the outcome exclusively. Gogel and Mershon (1969) investigated this question by placing a small white disk in front of a larger black disk. The black disk was illuminated in such a way that if it alone were visible it would have appeared white in the manner of the Gelb effect. The white disk had the effect of darkening the larger disk as is to be expected in terms of the ratio principle. The question at issue was whether the white disk would have a smaller darkening effect on the larger disk when it appeared well in front of it than when it appeared with it in the same plane. It was found that the effect was lessened by spearation of phenomenal planes by approximately one-half a Munsell step.* Others have obtained results of this kind (Mershon & Gogel, 1970; Mershon, 1972) but some have not (Gibbs & Lawson, 1974). Here again we have only a slight departure from what is to be expected on the basis of the ratio principle.

NEW EVIDENCE SUPPORTING A MODIFIED RATIO PRINCIPLE

Research just completed in our laboratory by Alan Gilchrist (1975) indicates that these slight experimental effects as well as the effects observable in daily life reflect the operation of an important principle; that is, perceived achromatic

* This is the effect that Gogel predicted and obtained in a variety of examples on the basis of what he calls the adjacency principle. This principle asserts that the effectiveness of cues between objects or parts of objects in determining perceived characteristics is inversely related to the relative separation of the objects or parts. This separation can be one of direction or depth (see Gogel, 1973).

color is indeed a function of the ratio of luminance values but the relationship of regions in a phenomenal plane governs the outcome. The reason that the effects of separating regions into different planes in the experiments described were so weak concerns the way in which these experiments were performed. In the Hochberg and Beck experiment the trapezoid viewed monocularly appears in the plane of the table, and according to the new or coplaner ratio principle its phenomenal color ought to be determined entirely by its luminance in relation to that of the surrounding table. Let us assume that this was the case. When, however, the trapezoid is viewed with depth cues, it appears to be standing up and *it is now in a plane all by itself*. Therefore, according to the principle, no other region is in that plane to create the necessary ratio. This, then, could lead to a state of indeterminacy concerning color; Gilchrist suggests, however, that under such special conditions the isolated region will tend to form a ratio with the retinally adjacent region, regardless of the perceived plane of the surface that produces the retinal image. If this is so, the outcome will not be very different for the two perceived orientations of the trapezoid. A slight difference, however, would be understandable, for when the trapezoid appears in the plane of the table the coplaner ratio principle applies directly whereas when it appears to be vertical the same ratio tends to apply but only, so to speak by default.

A similar argument applies to the experiments separating parallel surfaces from one another in depth. When the surfaces appear to be in the same plane, the coplanar ratio principle applies fully. When they are separated, it should not apply at all, but because each surface is now without another surface in its plane with which to establish a ratio the perceptual system falls back on the only ratio available, that deriving from the retinally adjacent region. It does not, however, do so fully or effectively as when the surfaces are coplanar. Finally the same argument applies to the example of viewing a corner of a room. It will be recalled that when one succeeds in "flattening out" the corner the adjacent regions look different in color. This is to be expected: two regions are now phenomenally localized in one plane and the ratio of their luminances governs the outcome. On the other hand, when the corner is seen veridically in depth, the ratio of luminances between the two surfaces does not govern the outcome because they are not perceived as coplanar. Still, constancy sometimes obtains because there may be other regions present within each plane that can form ratios with each of the surfaces under consideration. If constancy does obtain, the luminance difference between the surfaces at the corner gives rise to an impression of a difference in phenomenal intensity or brightness rather than color.

I now present the evidence on which the coplanar ratio principle is based and which supports the interpretations of the experiments given above. As the reader will appreciate, what is required to test the hypothesis is a situation in which a target surface can form a ratio with either of two other surfaces—one that appears in its plane and one that does not. Two kinds of experiments were done,

one in which the planes were parallel but were separated in depth, the other in which they were contiguous along one edge but differed in slant.

Figure 10-1 illustrates the experiment on parallel planes. The point of the experiment was to determine the effect on perceived color of localizing a target surface in a far or near plane. In the far plane it appears to be adjacent to a square surface of much higher luminance and in the near plane, to a square surface of much lower luminance. The change in localization is achieved by making use of interposition as a depth cue and so arranging things that the target square appears to overlap a square in the near plane (near condition) so that it appears in the near plane, or it appears to be overlapped by that square as well as by the distant square so that it appears in the far plane (far condition). This target is, in fact, in the near plane in both conditions, partly covering the adjacent square in the near condition and partly covered by it in the far condition. By cutting out a notch in the upper left corner of the target square and correctly positioning it, it was made to appear to be in the far plane partly covered by the adjacent square in the far condition. [Figure 10-1A illustrates the general arrangment of surfaces but specifically illustrates the far condition. The notch in the target and overlapping by the black cardboard occurs only in the far condition. B (left) shows what the subject perceived in the near condition.] The target was white cardboard. The adjacent square in the far plane was also white cardboard, that in the near plane was black. The far plane was much more highly illuminated than the near plane (by hidden light sources). As a result the relative luminance values were as follows (see B): the near adjacent black square was assigned a value of 1; the target square being white and in the same illumination, 30, the far adjacent white square, being in a much stronger illumination, approximately 2167. Separate groups of subjects were used for the near and far conditions. The subject matched the target square to one of a series of 16 values on a Munsell achromatic color chart.

Before examining the results, it would be helpful to consider what might be predicted in such an experiment. From the standpoint of the retinal image there is only a negligible difference between the near and far conditions, as shown by a comparison of the two lower configurations in B. Therefore there is no basis for predicting a difference in the perceived color of the target in the two conditions. Beyond that it is difficult to predict precisely what color that should be. The image of the target is near several images of different luminances, some of which are greater and some, lesser. In terms of the coplanar ratio principle, however it should be predicted that in the near condition the target will look white because it stands in the ratio of 30:1 with respect to the phenomenally adjacent black square in the near plane. Because the target *is* white, that outcome perhaps will not occasion much surprise. In the far condition it should be predicted, that the target will look black, because it stands in the ratio of 30:2167 (or 1:72) with respect to the white phenomenally adjacent square in that plane. The results were as follows: the median observer match for eight subjects in the near condi-

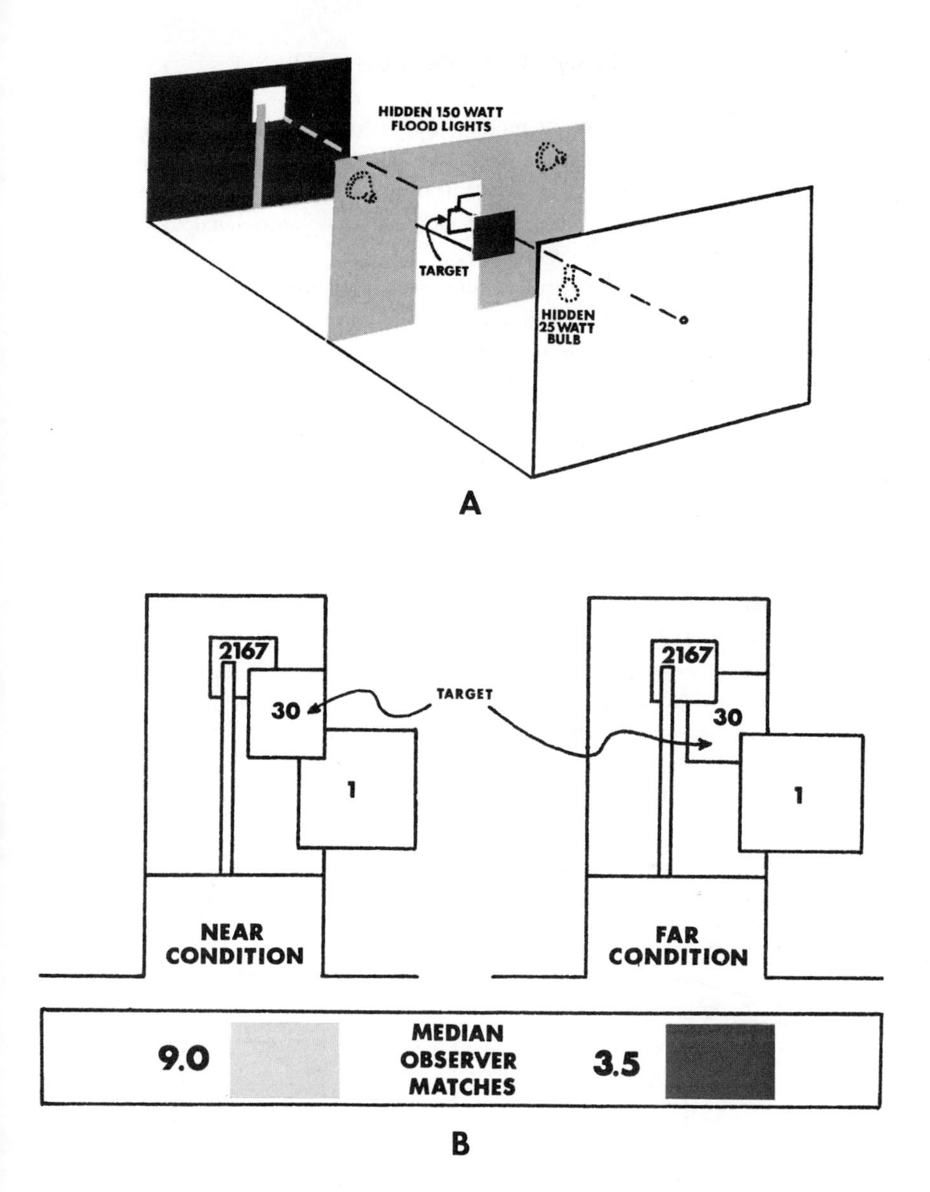

Figure 10-1 Arrangement used in parallel plane experiment. A: Three-dimensional illustration of the physical arrangement of surfaces. Here the upper corner of the target surface is shown cut out and the near black cardboard is shown overlapping the target, leading to the impression shown in B, (right) (far condition). For the near condition this notch is not cut out and the target overlaps the near black cardboard, thus leading to impression shown in B, (left). B: The view seen by the observer in the nar and far conditions, the relative luminance values of the three surfaces, and the median observer matches in the two conditions.

tion was 9.0, which is white; the median match for eight different subjects in the far condition was 3.5, which is a dark gray, almost black. Thus, as a result of a change in phenomenal depth localization only (interestingly enough created by a pictorial cue to depth), a target surface changes its phenomenal color from one end of the scale almost to the other.

The other kind of experiment is concerned with planes meeting at one edge and forming a 90° dihedral angle. The major and essential difference between this and previous experiments on the effect of the perceived slant of surfaces on phenomenal color is that in each of the two planes two regions of different luminance values are introduced. The basic idea is illustrated in Figure 10-2. One plane is horizontal and receives stronger illumination than the other. The inner squares are the target surfaces whose colors are to be judged. By eliminating or manipulating certain cues to depth the entire array can then be made to appear flat as one plane (B) or the two inner target squares can be made to appear coplanar with one another while the background surfaces continue to appear to be perpendicular, and so on. Gilchrist performed a variety of experiments, but for present purposes I shall describe only one, one in which prediction in terms of relationships between luminance values of regions considered in terms of the retina *is opposed* to prediction in terms of relationships between regions perceived to be in the same plane.

The plan of the experiment can be readily grasped by examining Figure 10-3. The observer views the array from above. In the binocular condition the figure

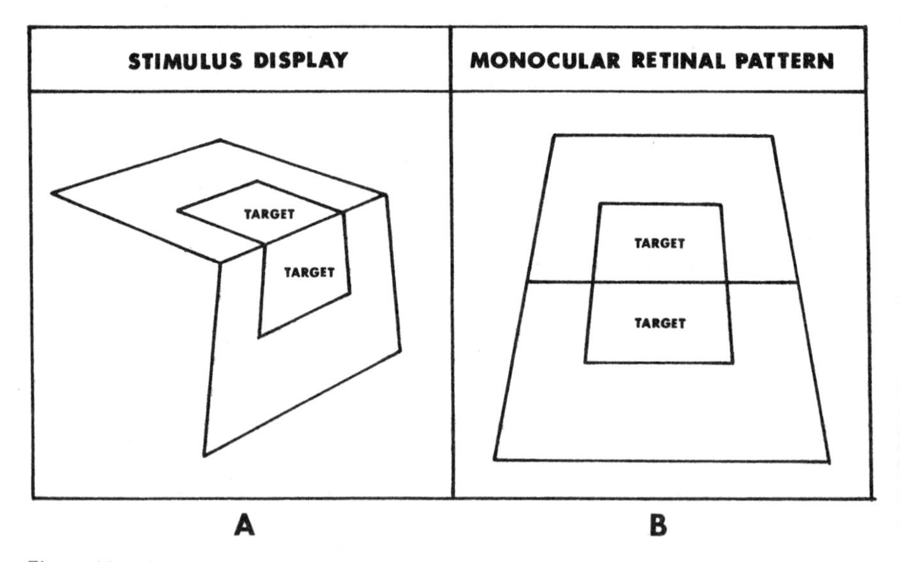

Figure 10-2 Display used in other experiments by Gilchrist (1975). A: Three-dimensional illustration of the physical arrangement of surfaces. B: The monocular retinal pattern.

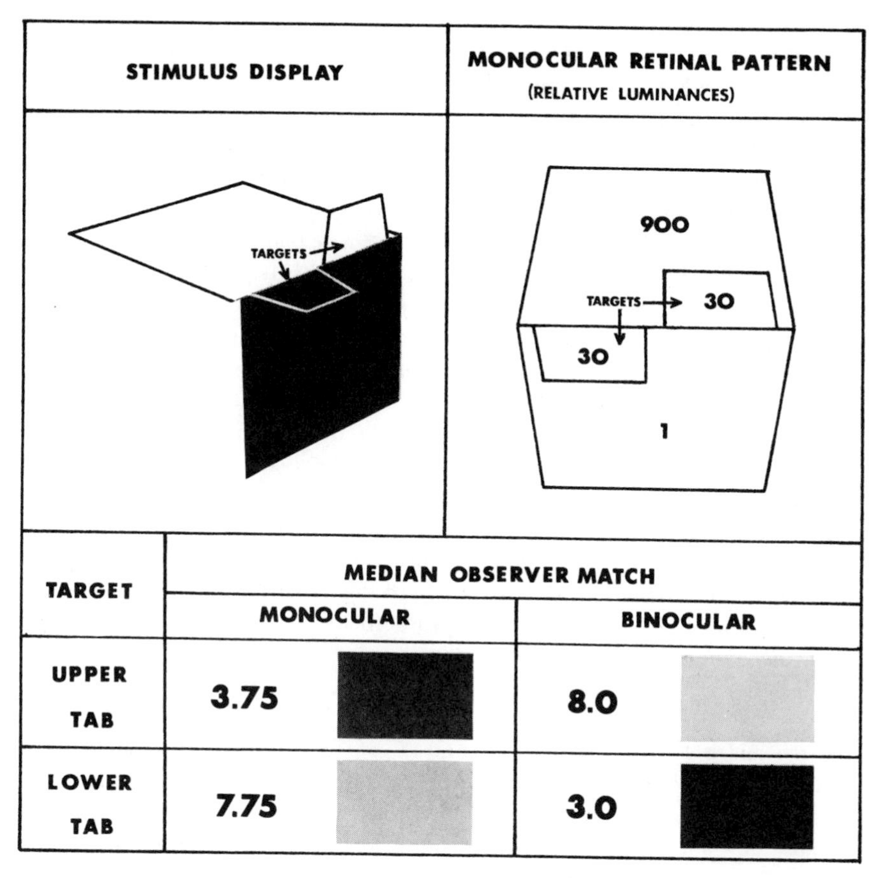

TARGET	MEDIAN OBSERVER MATCH	
	MONOCULAR	BINOCULAR
UPPER TAB	3.75	8.0
LOWER TAB	7.75	3.0

Figure 10-3 Display used in the experiment by Gilchrist (1975) described in the text. A: Three-dimensional illustration of the physical arrangement of surfaces. B: The monocular retinal pattern, the relative luminance values of the four surfaces, and the median observer matches in the two conditions.

on the left illustrates what the subject sees, namely, tabs (the targets) attached to the background surfaces in which their edges meet. If what matters for perceived color is the ratio formed between a target surface and another surface in that plane, one prediction follows, but if the ratio that governs perception of color is between adjacent retinal images an entirely different prediction follows. This is so because the retinal images of the tabs are surrounded on three sides by stimulation deriving from the surface *not* in the plane of the tabs [Figure 10-3, (right)].

The horizontal plane receives 30 times as much illumination as the vertical plane. Given the actual reflectances of tabs and background, as shown in Figure

10-3 (left) the relative luminance values are as shown in Figure 10-3 (right). Note that the difference in illumination compensates exactly for the difference in color of the two tabs (one white; the other black) such that the two are of equal luminance. These tabs are trapezoidial in shape. When the array is viewed monocularly, they look like squares in the planes of the background surfaces, but the background surfaces are perceived veridically as meeting at a right angle. That being the case, the coplanar ratio principle here predicts that the ratio formed by each tab to the surface perceived to be in that same plane will determine its outcome. Because the background luminances are so different from one another, the horizontal one being 900 times as intense as the vertical, the phenomenal color of each tab should undergo rather drastic change if the effective ratio for it entails the horizontal or vertical background surface.

The median matches for eight subjects in each condition in Munsell units were as follows [see Figure 10-3 (bottom)]: binocular condition, upper tab, 8.0 (or almost white), lower tab, 3.0 or black; monocular condition, upper tab, 3.75 (almost black), lower tab, 7.75 (almost white). Thus the outcome for both conditions is exactly what the coplanar ratio principle predicted. We might try to explain the results of the monocular condition in terms of retinal ratio but to do so requires the assumption that the image of the background surface adjacent to a tab on only one side has essentially no effect. This is a tenuous assumption from the standpoint of the traditional approach. Therefore even these results are much better accounted for in terms of the coplanar ratio principle because the background surface on the fourth side of the tab is not in the perceived plane of the tab and accordingly should have no effect. In any event, the results of the binocular condition are striking, for in this case it is the surface on that fourth side, that is, on one side of the tab only, that completely governs the outcome because only on that one side is the background in the phenomenal plane of the tab. It does so despite the fact that the prediction based on luminance ratios for the other three sides of the tab go in the opposite direction. The fact is that the phenomenal color of each tab changes from almost black to almost white simply as a function of change in depth perception. It is noteworthy that this strong effect occurs in contrast to the weak one of previous experiments on perceived slant of surfaces, despite the fact that here no information concerning the direction of illumination is provided by cues such as the location of shadows cast by objects.

Let us then accept the coplanar ratio principle as a descriptive law that best explains many of the known facts concerning the perception of achromatic color. Which of the two kinds of theory of constancy does it support? In some sense it is compatible with both theories and in another sense it is incompatible with them. As a principle concerning ratios rather than absolute stimulus values it clearly fits the higher order stimulus theory, but it is not simply the retinally determined ratios that govern the outcome. Yet that theory refers to relations within the retinal image. On the other hand, to the extent that the perceived spatial structure of the array is relevant, the principle fits the idea of considering information other than the luminance value(s) of the image(s) of the surfaces. However, taking account of depth relations is not the same as taking account of illumination.

Why, we might well ask, should luminance ratios within planes and only within planes determine phenomenal color and thereby lead to constancy? The question becomes more meaningful if we ask why *retinal* luminance ratios would determine phenomenal color. The answer to this question favored by most investigators is the mechanism of lateral inhibition, to be described later on in greater detail. A relevant fact is the discovery that neural signals based on the luminance differences across an edge is the crucial information for detecting the luminance differences in the entire areas separated by the edge (Craik, 1966; O'Brien, 1958; Cornsweet, 1970; Ratliff, 1972). This fact may be interpreted to mean that lateral inhibition modifies the edge signal and thereby affects the entire areas on the two sides. It may also be interpreted to mean that, apart from any contribution of lateral inhibition, the signal at the edge is based on the relative luminance difference and not on absolute levels of luminances. If the coplanar ratio principle is correct, these explanations cannot be sufficient because they are based on relative luminance values at the level of the retina. I therefore return to this question toward the end of the chapter.

OTHER CONSTANCIES

If the coplanar ratio principle is correct, the higher order stimulus theory of constancy has been seriously challenged in the domain in which it has always been considered basically correct and therefore in which the strongest case can be made for it, namely the perception of achromatic color. What about the other constancies? With respect to constancy of direction, position, and orientation, a rather good prima-facie case can be made in favor of the taking-into-account theory.

The phenomenal direction of a point in the field (its egocentric location with respect to the observer's mideye position as origin) clearly depends on combining information about the retinal locus of the image of the point with information about the position of the eyes in the head (see Hill, 1972). It is difficult to see how a theory based entirely on retinal stimulation is possible in this case. A similar argument holds for the related fact of position constancy, the apparent stability of the visual world despite eye or head movement. Here, too, such movement must be taken into account by the perceptual system in order to assess the implications of displacement of the retinal image correctly. Gibson (1966) suggested that the shifting of the entire retinal image is the stimulus for phenomenal stability (and the simultaneous experience of one's own movement). In this view eye or head motion need not be considered. This explanation, however, is incorrect. If the entire visible array is suddenly moved, it will appear to move but will not yield an induced impression that the eye or body is in motion in the opposite direction.* The visual field appears to move during head movement in patients with vestibular dysfunction. The entire array appears to move when one pushes one's eye with a finger. Conversely, it appears to move but when one tries to move one's eyes the eyes are prevented from moving, (Brindley & Merton, 1960). Moreover, position constancy obtains even for an isolated point in an otherwise homogeneous field when the eyes are moved normally. Therefore the evidence is strong for a process of taking head movement or eye movement (or, more carefully stated, intended eye movement) into account.

By orientation constancy I mean the constancy of the perceived orientation in the environment of an extended object, such as a line, regardless of head orientation. Such constancy occurs, within certain limits of accuracy, for a single line in a dark field so that it clearly depends on a process of taking account of head orientation in assessing the perceptual significance of an image in a given orientation. With the entire scene visible it is true that constancy is now complete, that is, the orientation in the environment is perceived verdically, regardless of the observer's posture. Hence we might argue that relationally defined stimulus information is relevant as well. The line is now seen to be parallel to those contours of the room taken to be vertical, and this parallelism is no doubt a source of information that is utilized by the perceptual system. Information of this kind, however, is not sufficient, for when a conflict is created by tilting a room or the frame of reference a compromise occurs. The line does not look upright for most observers when it is parallel to the edges of the room. Gravity information is taken into account.

SIZE CONSTANCY

I believe it fair to say that for these constancies there can be little doubt that the mechanism is one of considering information from other sources in interpreting the perceptual significance of the retinal image. What can we say about size constancy that has been the subject of so much research in the last 50 years? It is well known that size perception is intimately related to distance perception, as tance are the independent variables and perceived size, the dependent variable. Emmert's law also illustrates the same functional relationship and the moon

* It is true that under the appropriate conditions movement of the surrounding visual field gives rise to the impression that the field is stationary and we are moving (induced movement of the self). This fact is the source of Gibson's hypothesis.

illusion can be explained as a special case of Emmert's law (Kaufman & Rock, 1962; Rock & Kaufman, 1962).

It is possible, however, to argue that size perception is not a direct function of a process of taking account of distance but a function of certain information contained within the retinal image that happens to covary with distance. One hypothesis is that phenomenal size is determined by the size of the retinal image of an object in relation to that of neighboring object images. that is, a size ratio or proportionality principle. Because the size of object A in relation to that of neighboring object B (both at the samé distance from the observer) remains constant for all distances of the pair of objects, such a principle could also explain size constancy and it would not depend at all on taking account of distance.

Sheldon Ebenholtz and I directly tested this possible explanation some years ago (Rock & Ebenholtz, 1959). The subject sat in the center of a dark room. In one direction was a line within a rectangle of fixed length. In the opposite direction was a variable-length line within a much larger rectangle. The subject had to swivel around 180° to see the other rectangle so that the two rectangles could not be viewed simultaneously. Only lines and rectangle perimeters were visible in the otherwise dark room. The two line-in-rectangle pairs were equidistant from the subject, who was asked to adjust the variable line until it appeared to be the same length as the standard line. Needless to say, without the rectangles present, the subject had little difficulty in equating the lines, but with rectangles visible an appreciable tendency toward a proportionality match was evidenced; for example in one experiment the variable line in the larger rectangle was set on the average at more than twice the length of the standard. The result fell short of a complete proportionality effect, which in the experiment referred to would have required a match of 3:1 because one rectangle was three times larger than the other.

Before commenting further about the implications of this kind of hypothesis and the evidence cited in support of it, it is necessary to consider a similar hypothesis advanced by Gibson (1950). Consistent with his overall theory that the information for all aspects of perception is contained within the proximal stimulus, Gibson proposes that information exists for phenomenal size, regardless of an object's distance. The very texture-density gradient that is said to be the stimulus for the perceived slant of a plane in the third dimension is also held to be the basis of size perception. All objects of equal size at any distance on this plane will cover the same number of texture elements. Therefore this equivalence of number of subtended textural elements is said to be the information that directly leads to veridical size perception or constancy.

As I understand this hypothesis, the argument is that the texture-density gradient directly leads to an impression of a plane at a given slant or slope in *which all texture elements are perceived on the average as equisized and equidistant from one another.* In Gibson's terms, the *scale* within the plane is constant, for if the underlined phrase is not part of the claim size constancy does not follow at all. If, instead, the argument is only that the texture-density gradient is the stimulus for a plane receding into depth, why should equality of the number of texture elements subtended determine equality of size of the object? Information concerning distance or depth relations is not ipso facto information about size; it is only information that can potentially be used in assessing size.

Therefore Gibson's hypothesis differs from the proportionality hypothesis described above in several respects. First, Gibson's explanation links size with distance perception, whereas the proportionality hypothesis makes no reference whatsoever to the perception of the third dimension. If, for whatever reason, a testure-density gradient failed to lead to an impression of a plane at a slant (but instead to the perception of a fronto-parallel plane), would Gibson predict that equality of number of texture elements subtended by objects would determine equality of size of the objects? The scale within the plane would no longer be constant. Second, as made clear by the previous statement, Gibson's proposed explanation is not a relational one in which ratio or proportion of one thing to another is the determining factor.*

There are strong arguments and evidence against both versions of the higher order stimulus theory of size constancy.

1. It is not a necessary explanation because constancy is present under dark field conditions in which only a single isolated object is visible. Distance cues such as accommodation, convergence, and head-movement parallax are enough to achieve constancy, at least at distances within which these cues are effective. Here only the taking-into-account kind of theory can be applicable.

2. The higher order stimulus theory is not a sufficient explanation because, with distance held otherwise constant, as in the experiments by Ebenholtz and myself, the effect achieved falls far short of what would be necessary to explain the constancy that obtains in daily life or in experiments permitting distance cues. (For details, see Rock & Ebenholtz, 1959; Rock, 1970). I am not aware of any experiments that directly tested Gibson's hypothesis; the large number of experiments performed on the texture-gradient explanation of depth perception focused on the perceived inclination of the plane rather than on the perceived size of objects in the plane. We do know that in these experiments the perceived

* It seems to me that Gibson's hypothesis begs the question. What we want to know is why the smaller visual angle of and between distant elements of the textured surface leads to the perception of the same phenomenal extent as the larger visual angle of and between nearby elements of that surface. To say that the scale is constant, that is, all elements in the texture gradient appear equisized and equidistant from one another, is simply to assert what is true if constancy obtains rather than to explain why it is true.

inclination of the plane falls considerably short of the predicted inclination and this means that constancy would have to be far from complete.*

3. Logically the higher order stimulus theory cannot explain the perception of absolute (or specific) size. By these terms I mean that we have an impression of the linear size of an object at a distance specificable in units such as inches or feet. For this to be possible we must "know," that is, have the information, about the object's absolute distance from us. What the higher order stimulus can explain, however, is the equivalence of an object at one distance with that at another distance.† This is very clear in the case of the proportionality principle but I believe it is also true for Gibson's hypothesis. Texture-density gradients do not convey information about absolute distance. Therefore on purely a priori grounds a higher-order stimulus theory cannot provide a sufficient explanation of size perception.‡

4. Then there is the question of ecological validity; for example, for the size proportionality principle to lead to constancy the size of an object relative to neighboring objects must be invariant for all distances. Thus, if an automobile is in front of a small house at a distance but in front of a larger house nearby, the

* The account of Gibson's approach given in this chapter does not do justice to his current thinking (Gibson, 1966) because he now emphasizes the information that can be "picked up" from the flux of light rays in the environment by a moving eye rather than what impinges momentarily on the retina. Thus, for example, in depth perception he now places greater weight on motion perspective than on static texture-density gradients. It is not obvious how motion perspective as information concerning the inclination of planes can ipso facto be information for verdical size perception, perticularly if texture density is eliminated as a factor. Moreover, movement of the observer is certainly not necessary for size constancy.

† I am indebted to William Epstein for having called this limitation of higher order or relational hypotheses to my attention in connection with the problem of speed constancy (see Epstein, 1973). Perhaps this analysis explains the failure of Gogel and Sturm (1972) to obtain the size transposition effect in an experiment essentially repeating the one done by Ebenholtz and me. They instructed their subjects to judge the absolute size of each line in linear units rather than to match one line with the other. Not only does this method work against the proportionality effect but it requires the subject to judge a property (absolute size) that there is no reason to believe is a function of relational effects. We have had no difficulty in our laboratory in obtaining the kind of results originally reported.

‡ However if one analyzes what absolute size means at a deeper level, it would appear that it reduces to something relative. The absolute size of an object is a function of comparing it to a whole set of familiar objects, in particular to the self. Thus, for example, an apple is larger than a grape, smaller than a grapefruit, a certain size in relation to one's hand, etc. (The appreciation of sizes of units of measurement also depends upon such comparisons, as for example the "foot".) But this analysis does not contradict what was said above, that the absolute size of an object as thus defined would remain indeterminate without information concerning its absolute distance.

prediction must be that it will *not* appear to be the same size. In this case the nearby car would have to appear *smaller.* To what extent does the environment contain such necessary invariants? It seems doubtful that it does to an extent sufficient to explain the prevalence of constancy. Moreover, the relationship is not reciprocal in that the size of the larger object (or frame of reference), for example, the rectangle in our experiments, is not at all a function of its size in relation to the smaller one (the line in our experiments). This leaves size constancy unexplained for the largest objects in any given grouping of objects in the environment, as the house in the foregoing example.

For Gibson's hypothesis uniformity of texture in the plane is an ecological prerequisite for veridical depth and size perception. Although such uniformity is certainly frequent, the question is whether texture is always visible. Thus beyond a short distance the texture of snow, ice, sand, or the water in a lake is undoubtedly below the acuity threshold. Is size and depth perception less accurate under such conditions than when texture is visible? I would think not.

5. Recent research on perceptual effects known to be a function of stimulus relationships has rather consistently shown that if the objects are to have a strong effect on one another they must be localized in the same plane. This is, of course, parallel to what we now know to be true about achromatic color perception. Thus Gogel and Koslow (1972) have shown that induced movement of a point is governed primarily by a moving frame of reference localized in the same plane as the point. Luminous rectangular frames were located at two different distances from the observer, one in front of the other, and moved in opposite directions. The point was perceived to be in the plane of the near or far frame. The direction of induced motion of the point was determined primarily by the direction of motion of the frame in the phenomenal plane of the point. Yet the retinal image consists of point and two frames moving in opposite directions. Therefore it would seem that induced motion is not simply the result of relative image displacement. Displacement in relation to a frame of reference within the plane seems to be the essential factor. Although to my knowledge the experiment has not yet been done, it would seem a likely prediction, given these findings, that the transposition effect on perceived velocity (Brown, 1931) will also turn out to depend on localization of moving objects and frame of reference in the same plane in depth. My reason for this prediction is that this transposition effect has generally been interpreted as a function of perceived rate or relative displacement. Because the higher order stimulus explanation of speed constancy is predicated on the transposition effect (Wallach, 1939), this outcome would imply that the correlate of perceived speed cannot be stated in terms of proximal-stimulus relationships only.*

* I have not covered the topic of speed constancy, although it lends itself to an explanation in terms of the higher order stimulus theory and the evidence in support of this view is impressive. In addition to the point made here concerning the issue of phenomenal planes, there is the further fact that speed constancy has been shown to obtain under dark field conditions based on a process of taking distance into account (see Rock, Hill, & Fineman, 1968).

More to the point of the topic of size perception under discussion is another recent experiment by Gogel (1975) which followed up earlier findings of other investigators on the Ponzo illusion. Observers were presented with two sets of converging lines, one in a near plane converging in one direction, the other in a far plane converging in the opposite direction, but so placed that retinal images of both sets of lines crossed over one another. The horizontal test lines were then localized in either plane. It was shown that the illusion was a function primarily of the inducing lines in the phenomenal plane of the test lines. From a purely retinal point of view there would be no basis for an illusion because the two sets of inducing image lines would lead to opposite effects and therefore cancel one another.

Following up these implications of this experiment, we might predict that the size proportionality effect would also prove to be a function of adjacency of object and frame in a phenomenal plane. The observer would be presented with two sets of luminous frames, one behind the other in each set (Figure 10-4). In one set the luminous test line is localized in the plane of the smaller rectangle, whereas in the other it is localized in the plane of the larger rectangle. The two test lines are equidistant from the observer. From the standpoint of the retinal image there is no basis to predict any difference in the perceived size of the lines because each line is surrounded by identical images of two rectangles. In terms of perceived belongingness there is one line *in* a small frame and another *in* a large frame. We have tried out this procedure informally in our laboratory with preliminary results in line with the prediction. Therefore even a weak version of the higher order stimulus theory—that it is a sufficient (or alternative) explanation of size constancy (and admittedly not a necessary one)—is questionable. The size relationships that do affect phenomenal size under certain conditions are apparently perceived extents of objects in relation to frames of reference to which they belong, not simply relative retinal sizes.

In summary, formidable arguments and evidence support a theory of size constancy based on a process of taking account of distance and against an alternative explanation based on higher order attributes of the proximal stimulus. This conclusion leaves us with an unsolved problem of explaining the precise meaning of the size proportionality effect if it is not to be understood as the basis of size constancy.

DUAL ASPECTS OF PERCEPTION

One major fact concerning all constancies which has been more or less neglected is relevant to the discussion of the two kinds of theory of constancy. Although it is true that perception is generally in agreement with the objective properties of things and events rather than with absolute features of the proximal stimulus, one aspect is correlated with these features. Following the recent suggestion of Arien Mack (1974) on terminology, I refer to these two aspects of perceptual experience as the constancy mode and proximal mode, respectively. Some investigators, such as Brunswik (1956) and Gibson, have recognized this dual aspect of perception. Gibson (1950) distinguished the two by the terms visual field and visual world, but from the beginning he thought of visual field experiences as the result of a rather special introspective "pictorial" attitude. Therefore, although present at times, such experiences were not considered relevant to an understanding of our perception of the "visual world." In his most recent book (1966) we find only one reference to the "visual field." Here he is even more emphatic that although such field "sensations" exist they result from a rather sophisticated attitude of seeing the world in perspective, as a picture; this attitude is a result of exposure to pictures and of learning to draw. I believe that the presence of proximal mode experience is not merely of interest as a phenomenological nicety but rather has important ramifications for a thoroughgoing theory of perceptual constancy. It will be helpful first to examine some different examples.

Consider size perception. Although an object at varying distances does appear to be the same objective size, its changing visual angle is by no means a fact without representation in consciousness. We are aware of, even if not attending to, the fact that at a greater distance the object does not fill as much of the field of view as it does when it is nearby. What I call the perceived *extensity* of an object is a function of its visual angle. This fact has been demonstrated in experiments on

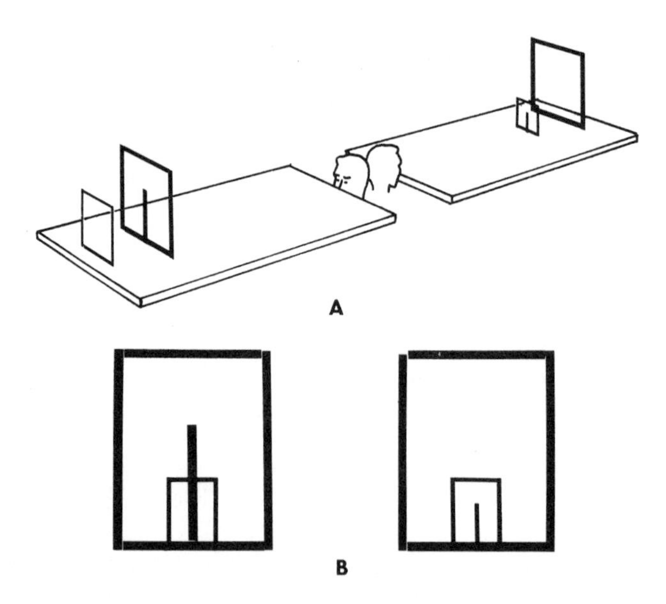

Figure 10-4 Arrangement for proposed experiment on size proportionality. A: Three-dimensional illustration of the rectangular frames at two different distances in each direction. Only the luminance frames and line are visible in an otherwise dark room. B: The retinal patterns of the frames shown in A.

size perception when the instructions are to match on the basis of visual angle (in whatever way this is explained) rather than in terms of objective size or simply the size that the objects spontaneously appear to be (e.g. Gilinsky, 1955; Epstein, 1963). Typically, matches obtained under such instructions are more nearly in correspondence with visual angle than those obtained under the customary instructions. That size matches nevertheless still depart from visual angle equivalency in the direction of constancy suggests that it is often difficult to isolate extensity from objective size experiences under daylight conditions when information concerning distance is adequate.

Another approach has been to require observers to match the sizes of objects at different distances under reduction conditions, that is, in a dark field in which no information whatsoever about distance is available (Holway & Boring, 1941; Hastorf & Way, 1952; Rock & McDermott, 1964; Epstein & Landauer, 1969). There is no question that this task can be achieved with fair accuracy. Because a definite objective size can only arise in experience when distance is determinate, such findings would seem to prove directly that the perception of extensity is possible and a function of visual angle. However, a controversy has developed concerning the meaning of these findings. Gogel (1969, 1971, 1973) has asserted that accurate matches under reduction conditions are based not on the perception of pure extensity, which he apparently considers to be impossible, but rather on the assumption by the observer that both objects are equidistant. If equidistant, distance can be equally taken into account and can provide the basis for size judgments along more traditional lines.

It is difficult to resolve this issue experimentally. There is evidence for and against Gogel's position. In my research with McDermott we tested the equidistance hypothesis directly by including measurements of perceived relative distance. The subjects viewed two reduction triangles successively, separated from one another by 90°. One was the standard and the other the triangle that appeared equal in size to the standard for that subject. The task was to indicate the relative distances of the two triangles by positioning two luminous beads, each of which slid along strings over luminous lines at right angles to one another. Among all the judgments of the relative distances of the triangles thus obtained, few indicated equality or near equality.

We also demonstrated that a comparison object can be equated with the standard reduction object when cues to distance are provided for the comparison object but not for the standard. The reduction triangle was 14 in. high at 32 ft. and viewed with one eye through an artificial pupil. When the comparison triangle was viewed at a distance of 2 ft with binocular vision, it was set on the average to 1.16 in. in height, which is quite close to a visual angle match, namely, .87 in. When, however, the comparison triangle was at 8 ft, a group of different subjects set it on the average to 3.08 in. in height; this is quite close to a visual angle match for these distances, namely, 3.5 in. These findings suggest that we can compare objects on the basis of extensity even though we are simultaneously aware of objective size. After all, the subjects appreciate the objective size of the comparison object, and in a control experiment it was shown that they correctly matched a binoculary seen triangle at 8 ft with one seen at 2 ft. A further control experiment indicated that the visual angle matches obtained were not based on an impression of equidistance between standard and comparison objects.

Logically, a tendency to perceive reduction objects as equidistant can only explain the *equivalence* of size perception; that is, the objects will appear equal in size. What is the phenomenal size of such equated objects? My impression and that of our subjects is that it is indeterminate as far as linear size is concerned.* Regardless of how this controversy concerning reduction objects is resolved, the simple fact is, as already noted, that in daily life we can clearly distinguish between perceived objective size and perceived extensity. A tree a hundred yards away may appear as large as another of the same size nearby, but we certainly are aware it fills far less of our visual field. The facts concerning the other constancies to be reviewed also add great weight to the claim that in the case of size we perceive in the proximal as well as in the constancy mode.

Consider next the perception of shape. A circle seen from the side, let us say at a 45° angle, may in one respect be said to continue to look circular, but its elliptical retinal image is not without some perceptual representation. Again instructions to match in accord with the "projected shape" rather than the objective shape will result in matches that are closer to the shape of the retinal image than to the object shape. Although it is difficult to describe the nature of this aspect of shape perception, perhaps the term *extensity relations* will suffice. We are aware that one diameter of the circle has a greater extensity in our field of view than the other while nevertheless simultaneously experiencing the objective sizes of these diameters as equal.

Turning now to the perception of movement, it is a fact that when we move, or when only our eyes move, stationary objects do not appear to move. Yet this image movement is in some sense perceived because we are aware that objects are changing their location in the field of view. Now the chair is to my left, now straight ahead, now to my right, as I swing my head around sharply to the left. It is true that the chair does not look like *it* is moving; therefore the designation position constancy is certainly a correct one. Nevertheless, there is some experience correlated with a changing retinal location and that can be termed "field movement" or "pseudo-movement" (Carr, 1935). In some situations this type of experience is more directly noticed, as when we move rapidly in a vehicle. We say then that the trees and road markings are "sweeping by," although

* Gogel (1969, 1973) would invoke his notion of a specific distance tendency to respond to this question. He believes that in the absence of absolute distance information objects tend to appear at the specific distance of 4 to 8 ft. However, this answer clashes with the central fact that under reduction conditions objects appear indeterminate in linear size, yet produce an impression of a certain intensity.

surely they are not seen as moving in the objective sense; that is, they do not appear to be in actual movement.

An interesting example relevant to the distinction made here is an illusion first noted by Filehne (1922) [see also Stoper (1973)]. If we track a target moving across the field, the stationary background is then said to appear to move in the opposite direction. If true, this would be a violation of position constancy. In a careful study of this effect Mack and Herman (1973) have demonstrated that a very slight illusory movement of the background does occur, thus implying a slight departure from constancy. This slight movement may be explicable in terms of an underregistration of rate of eye movements; in other words the rate of eye movement taken into account is not so great as their actual rate.

Still, if we try out this procedure informally—for example, by tracking a slowly moving finger—it is clear that we are aware of the *total displacement* of the background in a direction opposite to that of moving target, not merely a slight displacement. In other words, in some sense we experience as much motion of the background as its retinal image is displacing. I would therefore say that, for the most part, we are observing field or pseudo-movement, not objective movement. In support of this interpretation is the fact that the tracked target is perceived to move. If it moves, let us say 10°, and is perceived to do so, the image of a stationary object that displace 10° over the retina away from the tracked target must be perceived as stationary. The perceptual system typically follows "rules" in an internally consistent manner.

It is also relevant in this connection to consider the question of perceived velocity. The apparent speed of an object moving across the field is more or less constant with variations in the distance of the object from us. Are we unaware of the difference in rate of displacement of the retinal image? I think not. Although in one respect we perceived the distant automobile to be moving rapidly, in another respect we are simultaneously aware that it displaces across our field at a slower rate than when it is seen nearby.

Closely related to these facts about movement are others about perceived direction. Does the perceived constancy of radial direction, despite changes in the location of the retinal image with changes in eye position, mean that retinal locus has no unvarying significance for perceived direction? No, because we can speak of perceived field location. By field location I mean the apparent location of a point within the *momentary visual field*, that is, whether central, left, right, up, or down; for example, a foveal stimulus will always appear in the "center" of a momentary field of vision, regardless of how the eyes are positioned and despite the changing radial direction of the object. An object straight in front of the observer's head and torso will continue to appear straight in front radially even when viewed with eyes turned sharply to the side, *but* the observer is then aware that the object is in the extreme periphery of the momentary field of view.

Less well known are the experiences that occur in the perception of orientation. To be sure, we continue to perceive a vertical line as more or less vertical, whether we view it from an upright posture or with the head tilted.* Is there any perceptual awareness correlated with such changing retinal orientations? Yes. The line is perceived as having varying orientations in relation to ourselves. Thus, for example, when the head is tilted 90° to the side, an objectively vertical line is perceived to be egocentrically "horizontal," by which I mean that the observer is aware or can easily become aware that the line is parallel to the horizontal axis of his head. As in the other examples given of proximal mode perception, the observer may not realize he is aware of this egocentric aspect of orientation, but a question would immediately elicit evidence of its phenomenal reality.

The perception of egocentric orientation can be conveniently isolated from the perception of environmental orientation by requiring an observer to view a line that rotates in a horizontal plane. Thus, if a supine observer views a liminous horizontal rod in the dark, which rotates about a vertical axis above his head, he can make fairly accurate judgments about its egocentric orientation (Rock, 1954). Under these conditions the line never changes its orientation with respect to the direction of gravity; it remains horizontal. Therefore we see that perceived egocentric orientation is directly correlated with retinal-image orientation, whereas perceived environmental orientation, that is, whether an object appears vertical, horizontal or tilted in space, is a joint function of retinal-image orientation *and* other information about body position.†

As a final example consider the perception of achromatic color. To be sure, a surface will continue to appear white, gray, or balck, regardless of the illumination it receives. Therefore it would seem that the absolute intensity of reflected light or liminance is not correlated with perception. The fact is that a white object in bright sunlight does look different from one in dim room light in one respect; namely, it appears *more intense* or *brighter*. Therefore changes in the absolute intensity of reflected light are in some sense directly correlated with changes in this aspect of perception. In a reduction situation, such as a single region in an otherwise dark field or in a Ganzfeld situation, there is no experience of surface color at all, and only an impression of light intensity that is correlated with luminance occurs.

* I say "more or less" because certain errors do occur, namely the Aubert (1861) effect and the E effect discovered by Müller (1916), but this does not alter the main thrust of the argument.

†The dual aspects of perceived orientation are also relevant to the problem of the perception of disoriented figures where the phenomenal and functional reality of egocentric orientation has been demonstrated (see Rock, 1973).

So much for the facts concerning these dual aspects of perception.* The phenomenal reality of the proximal mode of perception to which I have called attention here has been denied or considered to be an insignificant epiphenomenon which emerges only under special or artificial conditions. It remains to be seen what role if any these experiences play in perception in daily life. However, the emphasis I have given to these attributes in the above survey is not intended to argue against the fact that the constancy mode of perception is not only salient in experience but most relevant to guiding behavior. It has an objective character in that it characterizes the physical properties of things. Conversely, the proximal mode is not salient in experience, it is not in any obvious way relevant to behavior, it is often hard to describe, and phenomenally has a subjective or egocentric character; that is, it is experienced as stemming from the relation of the object to the self or to the momentary conditions of observation. Compare phenomenal size and perceived extensity, phenomenal movement and pseudo-movement, or environmental orientation and egocentric orientation.

Factors Facilitating or Impeding Perception in the Proximal Mode

There would seem to be certain conditions in which the proximal mode of perception becomes more accessible. Whatever facilitates comparison between one image and another facilitates such perception; for example, in the case of extensity, when the objects to be compared are adjacent, the differences become obvious. Who is not aware of the smaller size of field subtended by a window of a building seen across the street through one's own window, although in terms of objective size, constancy may prevail? Similarly, the awareness of the ever-diminishing extensity of the ties of a railroad track with distance is perfectly clear. The fact that the ties are parallel to one another probably also facilitates awareness of extensity differences. Yet if we compared one such object in one direction with another in a different direction and at a greater distance we would

*Although not germane to the topic of this chapter, there are other examples than those pertaining to constancy in which dual aspects of perception can also be distinguished; for example, in the case of interposition as a depth cue we say that we perceive one rectangle behind another and the rectangle behind is seen as completed. This is the dominant, spontaneous impression. We certainly are also aware that we do *not* perceive parts of the completed rectangle or, to put it differently, we are in some sense aware of the L shape corresponding to that rectangle's actual image. In the case of linear perspecive we experience the converging lines as parallel and receding into depth but we are also aware of the convergence. In the case of motion parallax the relative shifts in the retinal images of points at different distances leads directly to an impression of depth but we are also often aware of these retinal changes.

be far less aware of the extensity difference and would be primarily aware of the objective size quality. The obviousness of extensity changes in the case of expanding and contracting images leading to what is now called the "looming effect" may also derive from the adjacency of the successive images. The dominant impression is of an object approaching and receding, but we are also simultaneously aware of expansion and contraction.

When illumination in one region is different from that in one that is adjacent, as in a cast shadow or on two walls meeting at a corner, differences in perceived intensity (or brightness) become perfectly obvious. Although the region of a white wall in shadow may continue to look white, it also appears to be in shadow and, in the latter sense, "darker." Conversely, adjacency can facilitate the perception of equality of proximal stimulation when unequal colors are in unequal illumination such that they produce images of equal luminance. In Gilchrist's experiment (Figure 10-3) we can be aware that the two tabs look equal in a certain sense, in phenomenal intensity, although one appears to be white, the other, black.

Great differences in the proximal stimulus also facilitate awareness of the proximal mode. Thus it is difficult to note differences in extensity between identical objects at different distances within a room (unless their images are adjacent). Yet objects out of doors at very great distances are readily seen in terms of their reduced extensity. People look diminutive and houses look like toys at a distance of a mile or so. I believe these descriptions refer to phenomenal extensity and not to perceived objective size. Rapid-movement in a vehicle makes field displacement of objects (or pseudo-movement) evident in comparison to slow movement, as in walking.

Finally, under reduction conditions the proximal mode of perception moves into the center of the stage. There is no longer a dominant objective percept that supersedes it.

On the other hand, proximal mode attributes are not readily available when contextual and relational factors affect the way these components are perceived; for example, in movement perception, whenever the outcome depends on relational information, it would seem that we do not experience or are not capable of experiencing the behavior of an object as a direct function of the behavior of its retinal image. In induced movement the object whose image is stationary appears to move (Duncker, 1929). Similarly the perceived path of moving objects depends primarily upon their displacements relative to one another and not upon their retinal-image displacement (Johansson, 1950).* Perceived velocity is strongly affected by the frame of reference within which a moving object is

* In induced movement, however, we may be aware that the stationary object that appears to move remains in the center of our visual *field.* The seeming contradiction is explained by an induced impression that we are tracking a moving object. In Johansson's effects it is interesting that the behavior of the retinal image of each point is fully accounted for if we consider *all* the vectorial components of the motion perceived.

displacing, and there does not seem to be an awareness of velocity as a function of absolute rate of retinal displacement when conditions that lead to the transposition effect obtain (Brown, 1931).

In the perception of the orientation of an object in the environment the relation of the object to a visual frame of reference has a profound effect (Asch & Witkin, 1948). It is doubtful if, in viewing a rod within a tilted frame, that veridical perception of the rod's egocentric orientation (as a function of its retinal orientation) is possible. In fact, the perception of egocentric orientation is itself affected by a frame of reference. In a supine position an observer's impression of the orientation in a horizontal plane of a luminous rod in relation to himself will be affected by the presence of a luminous rectangle surrounding the rod. (An experiment conducted by Sheila Hafter, described in Rock, 1966). In the strong proportionality effect that has been shown to obtain in size perception it would not seem to be correct to say that we can abstract from the effect of the frame of reference and perceive veridically on the basis of visual angle.

The geometrical illusions represent another set of phenomena in which there does not seem to be a proximal mode of perception distinct from the dominant perception. Perceived extent, direction, or curvature is not perceived correctly in such illusion patterns. It does not seem appropriate to speak of dual aspects of perception in the case of these illusions.*

Despite these examples in which it is difficult or impossible to perceive in the proximal mode,† such perception does occur in connection with all the constancies and in certain other situations as well. It remains now to consider the implications of the existence of the dual aspects of perception for a theory of constancy.

Implications of the Existence of the Proximal Mode in Perception

CHARACTERIZATION OF PHENOMENAL EXPERIENCE

First and foremost, we cannot properly describe sensory experience without including reference to the proximal mode of perception. Consider again the example of the railroad track. How shall we describe its appearance? Are the

* If the geometric illusions result from a process of misapplied constancy, as some have suggested, it ought to follow that the proximal mode of perception would be present. If so, with the appropriate attitude, it should be possible *not* to perceive the illusions by emphasizing the perception of extensity and the like. Yet this does not seem to be the case.

† There are other situations not relevant to constancy in which the proximal stimulus attributes are not accessible to conscious awareness. The difference between the two retinal images in binocular disparity, although producing impressions of depth, is not itself experienced. The binaural difference between time of arrival of an auditory stimulus is not experienced, although it leads to accurate sound localization. The movement of the retinal image when the eyes move saccadically from one position to another is not directly experienced.

tracks parallel or do they seem to converge toward the horizon? The fact is that both descriptions are true, a fact that has been referred to as the paradox of converging parallels. If we stress constancy of size, as has been true in the literature since the Gestalt revolution, we cannot explain the vivid impression of convergence that every observer will tell you he has. By the same token, the texture-density gradient in a plane can be said to be perceived at the same time that the plane appears to recede into the distance and the spacing between the texture element is perceived to be the same everywhere.

Consider again the circle seen at a slant. Many students in introductory classes in psychology are likely to shake their heads when the instructor points out that the circle looks circular and not elliptical. They often say "it looks elliptical but I know it is circular." In saying this, they are advocating the classical thesis the Gestaltists opposed so vigorously and successfully that perception is in accord with the proximal stimulus and that constancy is a fact of interpretation, not of perception. In this I believe they are mistaken. We do *perceive* the circle at a slant as circular but we *also* are aware that its projected extensity relations are "elliptical." The point is that we would be seriously distorting the phenomenal facts if we chose to speak *only* of the constancy aspects of perception. In this respect the students have been right and we have been wrong.

It is difficult to imagine what the phenomenal world would be like if these proximal stimulus attributes did not enter our awareness. Objects at varying distances, slants, and the like would look as much alike as the same objects seen at identical distances and slants. The fact is, however, that even when constancy is present a distant or slanted object does not look exactly like a near or unslanted object of the same size and shape.

As far as I can see there is no basis whatsoever for the proximal mode of perception according to the higher order stimulus theory of constancy. Therefore the phenomenal experiences that I have referred to as extensity or intensity of light are generally not referred to at all by those advancing explanations of size or achromatic color constancy. The fact that we are aware of a difference in light intensity between regions of the same phenomenal color is particularly revealing, for if color is determined by the relation of adjacent luminance values to one another (and if, further, this is determined by lateral inhibition) there is no basis for the perception of *any* differences between surfaces that appear to have the same color. To put it differently, the theory is designed to explain one mode of perception of light, not two.

As far as the taking-into-account theory is concerned, again viewed solely as a combinational process, it would seem that there is no place in the theory for proximal-mode perception. The presumption would seem to be that the information concerning the proximal stimulus is fused with the information that is taken into account and that the conscious experience of the attributes of the proximal stimulus is simply not available. A modern version of the combinatorial hypothesis is that neural units that "detect" features such as size or orientation are modified by feedback from other neural units that "detect" distance or body orientation (see Spinelli, 1970; Richards, 1967; Horn & Hill, 1969). Therefore the size or orientation detectors are in effect tuned to the distal stimulus rather than to the proximal stimulus; that is, they are constancy detectors. If so, there is then no basis for detecting the proximal stimulus feature per se (at least not in that particular detector mechanism). In any event, even if some basis is found for explaining proximal-mode perceptions, such perceptions would remain irrelevant as far as constancy is concerned, or they are viewed as epiphenomena, requiring special conditions, attitudes, or instructions to become manifest. In any case, they do not constitute a necessary stage in the achievement of constancy.

By contrast the earlier, classical view of constancy was that such proximal-mode experiences are the fundamental sensations. Constancy is the result of interpreting these sensations on the basis of further information. This view was, of course, strongly and effectively criticized, chiefly by the Gestalt psychologists. I believe the classical view is closer to the truth, that proximal-mode experiences are fundamental, but the sensation-interpretation dichotomy need not be reintroduced. Proximal mode experiences are best thought of as perceptions rather than as sensations. Constancy is clearly perceptual rather than merely a fact we have learned about an object. The notion may well be correct that constancy perception is based on a stage of prior detection of a feature directly correlated with the proximal stimulus and a central process of interpretation of that perceptual experience by the perceptual system on the basis of other relevant information. Otherwise the existence of the proximal mode would seem to be a curious, redundant, unexplained phenomenon.* I return to this point shortly.

EXPERIMENTAL IMPLICATIONS

A second implication has to do with the results of all empirical investigations of constancy in the laboratory. As is well known, the result is often one of compromise, of "regression to the real object" as Thouless (1931) put it, of less than complete constancy. The usual interpretation of such findings is to say that the information necessary for constancy is inadequately registered. Thus in the case of perceived position some slight motion of stationary objects is experienced when we track a moving point, an effect that is probably the result of inadequate registration of the actual rate of eye movement (see p. 343 and Mack & Herman,

* If, as I have argued, we do perceive in the proximal mode as well as in the constancy mode, there must also be memories faithful to such proximal mode experience. I believe such memories exist and have certain functional consequences, as, for example, in providing a basis for adaptation to distorting optical devices (see Rock, 1966).

1973). In size perception we say that we fail to register the true distance and that, predictably, phenomenal size cannot be veridical.

This explanation is no doubt correct under certain conditions; for example, in viewing an airplane in the sky, its diminutive phenomenal size is to some extent the result of poor cues to distance. After all, if the sky is cloudless and the head stationary only accommodation and convergence would seem to be operating in such a situation and these cues are ineffective at great distances. The small perceived size of the elevated moon can be explained in the same way. More typically, such quasi-reduction conditions do not obtain. In viewing an object across the terrain, in which pictorial cues such as texture gradients and perspective are presumed to be present, is there any reason why information about distance should not be accurate?

I should like to suggest a different explanation of the results of constancy experiments. Given the dual aspects of perception, a potential conflict is always present. The observer is aware of the discrepancy between the proximal mode appearance of the standard and comparison object. He is then no longer sure how he should match the objects. Should he concentrate on objective size or shade of gray and ignore extensity or brightness? What he actually does is to make a decision, and this decision may be to compromise between the two aspects. Instructions are therefore extremely important, and it follows that if the instructions could make sufficiently clear that the observer is *not* to be concerned with extensity or brightness that constancy would then indeed be complete. Some early investigations point in this direction. Another approach to eliminating the ambiguity would be to change the task for the observer so that, for example, in matching a distant and nearby object visual angle would be held constant as the objective size of the comparison object varies. This was accomplished in our laboratory by Begelman (1966) by moving the nearby comparison object along a track, As it receded its size increased in such a way that it held its visual angle constant. Both normal and schizophrenic subjects were easily able to match sizes in this way and, incidentally, did not differ from one another in degree of constancy achieved. (The converse procedure for varying visual angle without varying objective size was also employed. Here the comparison object moved back and forth along a track but remained constant in size.)

Research on size perception in recent years has reflected an increased sophistication concerning these dual aspects of phenomenal experience. In fact, even in the earliest experiments the objects to be compared were separated from one another, thus reflecting the realization that extensity relations are more readily perceived when objects are adjacent and that such proximal mode perception would interfere with constancy perception. Instructions to match on the basis of spontaneous impression of size (rather than on the basis of visual angle or inferred objective size) do result in more or less perfect constancy. Research on the perception of achromatic color has often been naïve with respect to this issue.

Either the instructions fail to make clear whether one is to match on the basis of perceived lightness of color or of perceived brightness or they incorrectly call for brightness matching when achromatic surface color is the phenomenon the investigator wishes to study.

In one well known test of the hypothesis that color is determined by the ratio of regions of differing luminances observers were instructed to match "brightness" (Jameson & Hurvich, 1961). It would seem, however, that little effort was made in this experiment to clarify what was meant by "brightness." Without changing the relative luminances of a set of adjacent regions the absolute luminance of the entire array was varied. The results were interpreted as a refutation of the ratio hypothesis because matches to each standard region varied with extreme change of its absolute luminance. Because such changes do produce varying impressions of phenomenal intensity or brightness and the instructions called for brightness matching, what is surprising about the result is that over a considerable range of luminance there was *little* change in the subject's judgments. This suggests that to some extent the subjects matched achromatic color or lightness rather than brightness either because they interpreted "brightness" to mean "lightness" or because lightness is the more salient aspect of the phenomenal experience.

In one experiment of another well-known study the subject was required to match a standard disk within a ring of varying luminance values with a comparison disk presented in a dark surround (Heinemann, 1955). It is known that a single region in an otherwise dark field always looks luminous and does not have the appearance of any surface color in the white to black continuum. Although the subject could not possibly match on the basis of perceived color, the results were considered relevant to the ratio hypothesis of achromatic color perception. When one is clear about the distinction between phenomenal lightness and phenomenal intensity, observations give different results. A very sophisticated observer and I viewed a disk that appeared to be gray (since it was surrounded by a ring of somewhat greater luminance) directly and through a pair of neutral density filters. Each filter reduced the intensity of light transmitted to the eyes by a factor of 100 so that together there was a reduction of luminance of 10,000. There was no apparent change in the gray color of the disk, although, of course, it appeared much dimmer through the filters.

It is as legitimate and desirable to study the perception of phenomenal intensity of light as it is to study the perception of achromatic color, but only confusion can result if we fail to distinguish which of these is under study or, worse, if we intend to study one while creating conditions in which the other is necessarily what is being measured, In a recent book on vision the distinction between achromatic color and brightness is never made (Cornsweet, 1970).

A related problem in experiments on the constancies concerns the effect of increasing the difference between the proximal stimuli of the objects under comparison; for example, in experiments on size perception we can increase the distance of one of the objects; in experiments on shape perception we can increase the slant of one of the objects; in experiments on color perception we can increase the difference in luminance of the regions under comparison by increasing the difference in illumination. In many earlier experiments on size and shape perception it was found that constancy "falls off" as distance or slant increases. In most experiments on color perception it is found that constancy holds for middle ranges of absolute luminance differences but "falls off" at extreme differences. In later, more sophisticated experiments on size perception it was found that constancy holds up under instructions to match in terms of spontaneous impression of size but that overestimation occurs (overconstancy) under instructions to match in terms of objective equality, with the distant object being judged increasingly large as a function of its distance.

The traditional interpretation of the "falling off" of size constancy is that the information which necessarily must be taken into account is the more inadequately registered, the greater the distance of the object. Thus it is believed that cues to distance become increasingly ineffective, the greater the distance of the object judged. As noted above, there are conditions of observation in which this is no doubt true, but under typical conditions it is not obvious why this should be the case. At great distances one would think that the pictorial cues are of primary importance. If so, why should one think that perceived distance would fall off as a function of objective distance? In the case of shape perception is it really true that we underestimate the slant or slope of surfaces, the greater their angle with respect to the frontal plane?

Another way of looking at data of this kind is that the greater the difference between the proximal stimuli of the objects under comparison, the more the differences between the correlated proximal mode of perception obtrude on our consciousness (see the discussion of this point on, p. 346). Thus, for example, we can hardly help noticing that people seen at a great distance subtend miniscule extensities in our visual field; we can hardly help noticing that a circle at an angle of 80° to the frontal plane subtends a much smaller extent along one axis than the other. That being the case, it is not surprising that effects of such extreme differences in proximal stimulation would show up in experiments. The potential dilemma for the subject between matches on the basis of objective properties and proximal stimulus properties is exacerbated. If, however, pains are taken to eliminate the dilemma by careful instructions or an unambiguous task, then perhaps little if any "falling off" of constancy will occur.

As to "overconstancy," it seems unlikely that such experimental findings reflect facts concerning perception. It simply is not the case that an object at a great distance looks larger than when it is nearby. Nor is it the case that distance is increasingly overestimated, which is what would be theoretically required to explain such findings were they genuinely perceptual in nature. Therefore overconstancy is clearly an artifact of the experimental situation and, as noted, seems to occur under the "objective match" instructions to which the subject seeks to get the "right answer." He seems to be applying this rule: "an object far away looks smaller than it is in fact; therefore to be correct I must compensate by judging it larger." In applying this rule he errs by judging the distant object too large.

How does the subject come to know that "objects far away look smaller than they are" if size constancy does not fall off with distance? One investigator suggested that we know this from experience with pictures and photographs (Carlson, 1960). Another answer is that "looks smaller" refers to the extensity aspect of size perception. Therefore the more this aspect becomes salient within the experimental situation, the more such errors of judgment will tend to occur, and it becomes salient with great differences between the proximal stimuli.

As a test of this reasoning, Robyn Posin (1962) conducted an experiment in our laboratory in which she duplicated the proximal stimulus differences between standard and comparison objects of a well-known study by Gilinsky (1955). In the Gilinsky study an object at 100 ft was compared with another at 100, 200, 400, 800, 1600, and 4000 ft. Gilinsky found that under the objective match instructions, the greater the distance of the distant object, the more its size was overestimated. At these distances the ratios of image sizes of equal-sized objects were 1:1, 2:1; 4:1; 8:1; 16:1, and 40:1, respectively. In Posin's study these same ratios were created, not by placing the distant object at increasingly greater distances, but by placing the nearby object at increasingly closer distances. Although one object was always at 20 ft, the other was at 20, 10, 5, 2.5, or 1.25 ft. Under these conditions the subject should be more inclined to be aware of the differences in extensity, the greater the difference in visual angle, although both objects are always well within the range of distances in which complete constancy almost always obtains. The more aware the subject is of the extensity differences, the more this will lead to a tendency to make overestimation judgments of the distant object. The results more or less paralleled those of Gilinsky indicating increasing overestimation of the object at 20 ft, the closer to the observer was the other object.

One other result of experiments on the constancies may be explicable in terms of the presence of dual aspects in perception. Typically, individuals differ from one another in their judgments and consistently so. There is evidence also that age is a factor, young children presumably perceiving objects as less constant than older children or adults. Why, according to prevailing theory, should individuals differ in such perception? Do some individuals perceive distance or slant more accurately than others? Is it not this difference that is relevant but the extent to which they apply such information in assessing objective size and shape?

I should like to suggest that individuals do not differ in such perceptions. There is no evidence that they do so in daily life, although admittedly it can be argued that they do, but it can only be revealed by careful measurement. At any rate, it is proposed that the differences obtained in experiments are artifacts resulting from the presence of the dual aspects under discussion. Some observers choose to give more weight than others to the proximal mode. This might conceivably result from the greater attention paid to this attribute by some individuals for reasons not yet understood. If true, to that extent, we might wish to characterize the differences as perceptual, although that is not what is usually implied by the individual differences in constancy experiments. However, the difference might also be one of attitude or interpretation of instructions, particularly when they are not made explicitly clear.

As to the findings with children, it is plausible to suppose that uncertainty, confusion, or some other difference concerning the dual aspects is responsible. The young child may be in greater conflict between the two aspects of his perception than the older child or adult and thus his judgments may reflect more of a compromise. The young child would also have more difficulty understanding instructions than the adult and this too would lead to differences in judgments. Of course, it is also possible that young children do genuinely differ in their perception of object properties, that they display less of a tendency toward constancy than adults, which in turn may reflect the role of experience as a basis of constancy. There are some reasons for skepticism about this conclusion. Evidence now exists that size and shape constancy occur in very young infants (Bower, 1966). One problem with the interpretation I have offered, it must be admitted, is that one might think that children should pay *less* attention than adults, not *more*, to proximal mode attributes, for they would be expected to be less introspective. Perhaps because children are more egocentric and less certain that they are supposed to match on the basis of objective properties my interpretation is correct.

PHYSIOLOGICAL EXPLANATIONS OF CONSTANCY

Ultimately the explanation of perceptual constancy, as for all perception, will be in terms of neural mechanisms. If the analysis to be offered here is correct, these will be mechanisms mediating inference processes. Of course, we have no knowledge of what such neural mechanisms would be like. Alternative physiological explanations have been suggested that have already been mentioned, namely, feature-detector mechanisms in the case of some of the spatial constancies and lateral inhibition in the case of achromatic color constancy. Some further comment on these suggested explanations is in order in the light of the arguments and evidence presented in this chapter.

Neural units responsive to the proximal state of affairs, such as those triggered by contours in a particular orientation (Hubel & Wiesel, 1962), cannot of course, do justice to constancy phenomena. The recent discoveries of units tuned to the distal state of affairs (discussed on p. 349) can, in principle at least, be invoked to explain constancy. One difficulty is that constancy is a function of taking account of various kinds of information, not only of the kind rooted in known physiological mechanisms. Although it is at least possible that feedback from accommodation of the lens or from convergence of the eyes could modify the nature of the receptive field of a neural unit, it is not at all obvious and seems unlikely that such a mechanism could explain size constancy based on pictorial information about depth relations.

Moreover, size perception can change without the introduction of a specific "cue" merely by a change in perceived depth. The perspective reversal of a drawing of a wire rectangle will often result in certain concomitant changes in phenomenal size of the different faces of the rectangle. Even more striking than this example are the dramatic size transformations that occur when one succeeds in reversing a three-dimensional wire cube. It is simply inappropriate to speak of "cues" here. Rather it is quite clear that a central change in one perception (depth) without any change in stimulation or other sensory information brings about a change in another perception (size). * One implication of these effects is that the perceptual system is indifferent to the source of information that is taken into account (see Epstein, 1973). What seems to matter for size perception is the perception of a particular distance, not the particular source of information (or, I would add, whatever central reorganization) that leads to that perception of distance. In the light of this implication it seems unlikely that at this stage of our knowledge a physiological theory based on feature-detectors is defensible.

Another example that illustrates how one percept depends on a prior percept concerns constancy achieved in viewing pictures. Consider the case in which an ellipse in a picture represents a circular object in a horizontal plane. The long axis of the ellipse is horizontal in the picture, but if we view it from the side the long axis of the retinal image of the ellipse may be vertical. Nonetheless we correctly perceive a circular object in a horizontal plane. This outcome suggests the following stages of processing:

1. The horizontally compressed image is "corrected" on the basis of information that we are viewing the picture from the side and results in the perception of an ellipse in the picture whose long axis is horizontal.

* The theoretical importance of the consequences of perspective reversal has been noted by Hochberg (1974) and I am indebted to Carl Zuckerman for pointing out the significance of this kind of effect on three-dimensional objects.

2. This new perception is now "corrected" or interpreted on the basis of the pictorial depth cues indicating the slope of the surface of the circle and results in the perception of the object as circular. What makes this example particularly interesting is that it is the *phenomenally experienced* (pictorial) *shape* that is assessed and corrected on the basis of information concerning slant, not the retinal-image shape per se.

Earlier I alluded to another difficulty with the attempt to explain constancy in terms of constancy detector units, namely, the problem of explaining proximal mode perceptions. Of course, one might argue that two kinds of unit exist—one responsive to the proximal state of affairs, for example, orientation of retinal contours (mediating perceived egocentric orientation), size of retinal image (mediating perceived extensity), and the like, and one responsive to the distal state of affairs; for example, object orientation and size. Together these units could be said to account for constancy and proximal mode perceptions.

No one to my knowledge has suggested this possibility, but if it were to be seriously entertained I am not sure that it could do justice to the facts. What has to be explained is not merely the dual aspects of perception but the *relationship* between them. The proximal mode of perception remains as an essential ingredient of the overall experience. Perceived extensity is an important feature of the overall impression of an object's size. A distant object is perceived as one of a certain physical size, *at* a distance, and of *small* extensity. A shade of gray is simultaneously a gray of a certain brightness. Therefore any attempt to explain each aspect of perception independently of the other would be wide of the mark.

As for the mechanism of lateral inhibition as an explanation of achromatic color constancy, it is understandable why this theory is favored by many investigators. There is first of all direct physiological evidence from research on animals for lateral inhibition and considerable evidence that the phenomenal color of a region is determined by the relation between the luminance of that region and the luminance of adjacent regions. Other things being equal, the greater the luminance of the adjacent region, the darker the perceived color of the critical region. This, then, can explain simultaneous contrast. Lateral inhibition can explain constancy by assuming that there is a trade-off between the increase in luminance of the critical region and the increased inhibiting effect of the adjacent region as illumination over the entire area increases.

Implicit in this kind of explanation is the belief that the ultimate determinant of perceived color is the absolute rate of discharge of neurons. If that rate is high, the corresponding region will look bright; if low, it will look dark. Because of lateral inhibition, this rate may remain constant despite changes in illumination. There are various difficulties with this assumption (see Rock, 1975, pp. 535-541), but one of particular importance is that there is no basis for explaining the differing phenomenal intensities of two surfaces in differing illuminations that appear to have the same color. If lateral inhibition leads to the same rate of discharging neurons from the critical region in the two cases, that region should look exactly the same in *every respect,* but it does not.

Therefore it is possible that lateral inhibition is not the mechanism that explains why constancy is a function of ratios of differing luminance values. Until now, however, it has been impossible to separate lateral inhibition from determination by ratio, for whenever the latter would apply, so would the former. Gilchrist's research described earlier does seem to provide a basis for separating the two kinds of explanation.

Lateral inhibition is a retinal effect; that is, it is the activity in the cells in the retina stimulated by light that inhibits the discharging of ganglion cell stimulated by adjacent retinal cells. With this in mind, consider the experiment illustrated in Figure 10-3. The critical tabs shown on the left yield images surrounded on three sides by images luminance values are shown on the right. Yet it is not these luminance ratios that govern the outcome, at least when veridical depth is achieved. In fact, this experiment can be viewed as a contest between what is to be expected on the basis of lateral inhibition and what is to be expected on the basis of ratios within phenomenal planes. At least it is if we make the plausible assumption that the effects of lateral inhibition on three sides of the target would dominate the effect on the fourth side. Therefore the two principles here lead not only to different predictions but to opposite predictions. According to lateral inhibition, the upper tab should look darker than the lower; in fact, the upper tab is seen as almost white and the lower one as black.

These findings in support of the coplanar ratio principle suggest that it is the luminance ratio itself, not the absolute luminance of one region as modified by neighboring luminance values, that governs achromatic color perception. The effect of lateral inhibition in the preception of achromatic color may be to modify the luminance differences between adjacent regions at a level in the visual nervous system before the level at which these differences are "detected." These "detected" luminance relations, rather than the ratios between the physical luminance values that reach the retina become functionally effective; they are then further assessed according to their perceptual significance, depending on the perceived three-dimensional structure of the field.

IN DEFENSE OF UNCONSCIOUS INFERENCE

So far I have presented evidence and arguments against the higher order stimulus theory and in favor of the taking-into-account theory of constancy. In discussing the latter earlier in this chapter, I deliberately restricted its definition to the notion of combination or coupling of retinal-image information with information from other sources. As noted at the outset there is the further implication that the process of taking-into-account is a cognitive operation, an unconscious inference. The question I should now like to address is whether there is reason for favoring a deeper interpretation of this theory beyond the acceptance of the combinatorial hypothesis.* The fact is that to speak only of combining information is not to offer a theory at all. By unconscious inference I mean that the process of arriving at the percept is one much like reasoning in which conclusions are drawn from premises, except that in perception the process is not conscious and the outcome is a percept rather than a conclusion. I do *not* argue as Helmholtz did that such a process is necessarily a direct result of experience. That is a separate question.

Consider the case of orientation constancy. We view a luminous line in the dark with head tilted and perceive its orientation in the environment more or less veridically. We also perceive its orientation relative to ourselves. It makes a good deal of sense therefore to view the achievement of constancy as resulting from a process analogous to reasoning:

1. The perception of the line's egocentric orientation on the basis of its image's orientation (proximal mode).**
2. Information available concerning the orientation of the head (or body) in the environment.
3. The interpretation of the line's orientation in the environment by the perceptual system of the basis of 1 and 2.

A concrete example may be helpful. If the observer is laterally tilted 50° clockwise and he achieves perfect constancy, the image of a line judged to be horizontal in the environment is retinally oblique. Therefore in terms of our analysis we have the following:

1. The line is perceived as egocentrically oblique at a 40° clockwise angle with respect to the head (proximal mode).
2. Information is available that the head is tilted 50° clockwise with respect to gravity.
3. Therefore the line in the environment producing the image must be 90° from the direction of gravity or horizontal.

* A few arguments that suggest the necessity for such a deeper interpretation have already been made. See, for example pp. 349-350 and 357-358 in this volume.

** As noted earlier, the observer may not realize that he is perceiving or detecting the line's egocentric orientation, although he would be able to report about it if questioned. In any event, this proximal-mode aspect of perception is immediately superseded by the perception of the line's orientation in the environment (constancy mode).

The process is then much like syllogistic reasoning. Perhaps a better term than syllogism would be general predicate logic, for the premises and conclusions entail relations. In this example we might say that the process has the form of a transitive deductive inference. (A is tilted 40° clockwise from B; B is tilted 50° clockwise from C; therefore A is tilted 90° from C.) This analysis assumes that the perceptual system "knows" how body orientation affects image orientation, a point to which I return shortly.

A similar process may underlie size constancy. The perceived extensity correlated with visual angle is interpreted by the perceptual system as signifying a particular objective size on the basis of information available concerning the object's distance. The components here would be as follows:

1. A particular extent is perceived based on visual angle (proximal mode).
2. Information is available that the object producing that visual angle is at a certain distance.
3. "Therefore" the object must be a particular size.

Again, however, this analysis presupposes that the perceptual system "knows" how visual angle changes as a function of distance. Therefore so far in the analysis certain additional principles that must be "known" in order for such inference processes to occur have not been made explicit.

Other spatial constancies can be understood in much the same way. To apply our analysis to achromatic color constancy in an exactly parallel form it would have to be argued that the process begins with the proximal mode perception of the intensity of a particular region, which presumably would be correlated with its luminance. This percept would then be centrally interpreted on the basis of information concerning illumination. We have already rejected this explanation because there is no way of obtaining independent, unambiguous information about illumination, for the luminance of a surface is the combined product of its reflectance property *and* the illumination falling on it. Moreover, we have seen that the coplanar ratio principle can explain many facts about color perception without recourse to any such notion of taking into account the precise illumination falling on a surface. The question now is why the perceptual system "interprets" ratios only within planes and not between them as signifying color differences.

A possible explanation of the basis of the coplanar ratio principle concerns an "assumption" about illumination. When regions of differing luminance are phenomenally localized in one plane, the perceptual system operates on the assumption that they are receiving equal illumination. This assumption would have high ecological validity. Nearby regions in a plane would receive unequal illumination only when one resulted from a cast shadow and that state of affairs is generally indicated by a penumbra. If such an assumption were made, it would "follow" that luminance differences must represent surface color differences. The logic here is as follows:

1. Regions that differ in luminance can derive from color differences or illumination differences
2. Regions in a plane receive the same illumination
3. Therefore regions of differing luminance within a plane represent regions of differing colors.

On the other hand, when regions are not localized in the same plane, they may or may not be receiving equal illumination. In fact, for surfaces slanted differently from one another, such as those forming dihedral angles, it is improbable that they are. Also, at least indoors, it is unlikely that parallel surfaces at different distances from the observer are receiving equal illumination because they cannot be equidistant from any light source. In any event, there is no reason for assuming that surfaces in differing planes are receiving equal illumination and no basis for knowing what the illumination difference is, if any. Therefore, in general, for regions not in the same plane there is indeterminacy as to whether the illumination is the same or different or an assumption that the illumination differs but indeterminacy as to the precise quantitative difference. Consequently the luminance difference between such regions can represent the same color with differing brightnesses or different colors, the exact difference between which can vary. There is therefore no logical basis for an inference concerning color differences across planes.

That information concerning reflectance value depends on the relative luminance of two or more regions should not be surprising, given the fact that the reflectance characteristic of a surface is itself a relative fact. A surface reflects a certain proportion of the light it receives and what matters is whether it reflects more or less light than other surfaces. What phenomenal whiteness "means" is that this proportion is high in relation to another surface. It is this property that must be detected if we are to achieve a perception correlated with it. How could that relational property be picked up by the perceptual system? Relative luminance between surfaces seems to be the direct source of that information. This information is valid if and only if it can be assumed that the surfaces receive the same amount of illumination. According to the explanation offered here, achromatic color and its constancy is not determined by a neural correlate based on the *interaction* of the regions of differing luminance in which, regardless of absolute luminance levels, that correlate remains invariant. Rather the luminance differences (modified somewhat by lateral inhibitory effects) are simply detected (i.e., perceived as light intensity differences) and are *interpreted* as signifying color differences if localized in one plane.

To summarize the argument concerning achromatic color perception, the first stage is the stimulation of the retina by regions of differing relative luminance. These differences are probably modified somewhat by a lateral inhibitory process and modified luminance differences or ratios are then detected at a higher level of the visual nervous system. We can consider perception at this stage to be in the proximal mode. If the regions are phenomenally coplanar, without a penumbra along the boundary separating the regions, they are assumed to be receiving equal illumination. Consequently they are interpreted as differing in color. Precisely what those colors are is a further problem, but clearly the magnitude of the relative luminance difference will be the major determinant here. If the regions are not phenomenally coplanar, the situation is ambiguous. In any event, the phenomenal color of a region is achieved as a result of taking account of equality or inequality of illumination in assessing the perceptual meaning of luminance differences.

Assuming that constancy prevails in each plane and that illumination is different on the different planes, the luminance differences or similarities between regions across planes remain more or less available to experience. If the regions in different planes are perceived to be the same in color—by virtue of the coplanar ratio principle governing color perception in each plane—the luminance difference between them is perceived as one of phenomenal intensity of light or brightness. This experience is closely linked with an impression of prevailing illumination; that is, the brighter region is receiving stronger illumination. If, however, the regions in the different planes happen to reflect equal luminances to the eye because the one receiving more light has a much lower reflectance value, they will be perceived as differing in color but can also be perceived as identical (or at least similar) in phenomenal intensity.* (Such perception of intensity differences or similarities is faciliated by adjacency as noted earlier.) Therefore, as in spatial constancies, the process begins with the detection of the proximal state of affairs (based on luminance differences among all regions of the field), and perceptual experience in the proximal mode (of varying phenomenal intensities) remains available even after constancy is achieved.

* This analysis suggests two possible meanings of "brightness," neither of which refers to achromatic color. The one I have used in this chapter is the experience of phenomenal intensity directly correlated with luminance. It therefore refers to the proximal mode of perception. It is evidenced either in situations in which color is not experienced (as in an isolated region in an otherwise dark field) or in which color is experienced but variation of illumination alters the luminance (as on a white surface in dim or strong illumination). "Brightness" can also refer to the impression of illumination on a surface. A black surface in strong illumination may look black but "bright" in this second sense, whereas a white surface in weak illumination may look white but "dim"; yet in the first sense of the term the two surfaces may appear equally intense.

"Knowledge" of the Rules of Proximal Stumulus Change and the Role of Experience

"Unconscious inference" for Helmholtz and others since has meant that what we infer to be present in the environment is based on experience. Through such experience we presumably learn that objects at a distance are much larger than they at first appeared to be, that objects receiving weak illumination are lighter in color than they at first appeared to be, and so on. A generalization arrived at inductively in this way then would constitute the major premise in the subsequent inference process. (The specific proximal stimulus and relevant cues would constitute the minor premise.) Of course, it makes sense to believe that such premises, referring as they do certain characteristics of the environment, are acquired by commerce with the environment, but it is not logically necessary to assume that they are acquired.

There are difficulties with the view that constancy is based on learning. The argument seems to assume some kind of direct access to the veridical state of affairs under some privileged conditions of observation; for example, when we approach the distant object, we find that it is large, not small. This presupposes that no learning is required for the perception of the size of a near object, for if there were how could we perceive its size when it is near? A possible resolution of this problem is that we can compare a nearby object with other familiar objects at the same distance, particularly with the body itself. Thus we can learn that a visual angle created by an object at a given distance is the same as that of one part of the body (e.g., the size of a foot) and different from other parts (e.g., larger than a hand); it is the size of a magazine, but smaller than a newspaper, and so on. This would seem to be all that absolute size can possibly mean (see footnote, p. 000).

One way of formulating the learning hypothesis is to say that what we learn *is how the proximal stimulus changes* as a result of our behavior. Thus we could learn that visual angle is a function of distance, that image orientation is a function of head orientation, that image motion is a function of eye, head, or body motion, and so on. In order for such learning to have any useful meaning it must be presupposed that the proximal stimulus transformations do not result from simultaneous object transformations. Hence the learning hypothesis presupposes an a priori cognitive assumption about the nature of the environment, namely, that stimulus changes resulting from our behavior are not caused by object changes; that is, under such circumstances things remain constant. Given that assumption, however, we could in principle learn precisely *how* the proximal stimulus varies as a function of change in our vantage point. In the special dynamic case in which the observer is *in* motion such an assumption may be based on the rejection of coincidence on the part of the perceptual system. The stimulus change begins and ends with the observer's movement and the rate of change is perfectly correlated with the rate of movement. Therefore it is plausible for the homunculus to conclude, "I am causing this stimulus to change rather than any change in the environment." Otherwise it must be acknowledged that the perfect correlation is pure coincidence.*

In order for the perception that is achieved to be veridical the perceptual system must take account of the relevant information in a manner that is *quantitatively correct;* for example, it is not enough to say that the perceptual system operates on the basis of the rule that visual angle is a function of distance. Assuming for the moment that the distance of the object is correctly registered, the mental computation requires that the correction be appropriate to the law of the visual angle. In other words, because visual angle varies inversely with distance, the computation requires a process equivalent to multiplication of visual angle by distance. This suggests that the perceptual system in some sense "knows" the law of the visual angle.

In stereocopic-depth constancy (Wallach & Zuckerman, 1963) the perceived depth relation between a pair of contours remains more or less constant despite the absolute distance of the pair and thus despite variation in the magnitude of retinal disparity. Distance is taken into account, but retinal disparity varies inversely with the *square* of the distance. To correct for the lessened disparity as a function of distance the computation must entail a process equivalent to multiplying the disparity by the square of the distance. Therefore to achieve verdical perception of the depth relation by stereopsis the perceptual system must in some sense "know" this inverse square law.*

It would seem necessary to assume that the organism "knows" various rules concerning proximal stimulus change and that these rules constitute the major premises from which constancy is inferred. It does *not* seem to me to be nec-

* If this analysis to correct, then in at least one constancy, position constancy, no learning at all is necessary because all that is implied by position constancy is that stationary objects do not appear to move during movements of the observer. If follows that a *different* behavior of the images of stationary objects concomitant with observers movement would also be discounted; for example, a slower or faster rate of displacement. We now know that in time constancy is achieved under such conditions (Posin, 1966; Wallach & Kravitz 1965a,b.) The reason why constancy is not immediate, that is, the reason for the necessity of adapting to the changed state of affairs, is undoubtedly that the specific relationship has been learned and preserved by memory.

* There has been some dispute about this claim. That it is indeed correct is easily demonstrated by viewing an anaglyph at varying distances. Here, because the disparity is *in the display*, retinal disparity varies inversely with distance but not with the square of the distance. When the distance to the anaglyph display is increased, the depth perceived among the parts of the display should remain constant if distance is taken into account as it is for size constancy. In fact, the perceived depth of these parts *increases* with the distance of the display.

essary to assume that these rules are learned. The evidence is not yet decisive on the origin of constancy and in fact is somewhat contradictory (Bower, 1966; Heller, 1968; for a review, see Rock, 1975). Whether or not the rules derive from learning does not preclude a theory that asserts that constancy results from a process of inference. In other words, the process of inference itself need not be thought of as deriving from experience, regardless of the origin of "rules" and "assumptions" utilized by the perceptual system.

To illustrate with a concrete example of size perception the inference would be of the following form:

Major premise: Visual angle is inversely proportional to distance.

Minor premise: Visual angle is 1° (producing a particular perceived extensity); distance is 50 ft (producing a particular perceived distance).

Conclusion: Object is equivalent to one that would yield a visual angle of 25° at 2 ft (or 5° at 10 ft, etc).

What about the case of achromatic color perception? The rules in this case are not so much concerned with the observer's locations or orientations as with conditions of illumination. For a cognitive theory of the kind I have proposed to be viable the organism would have to "know" the following: (a) that luminance differences are caused by reflectance-property differences or by illumination differences, (b) that illumination tends to be equal for nearby regions in a plane unless a penumbra is present along the border of a region, and (c) that illumination may or may not be equal for planes that are not parallel. Given the psychological reality of such premises, the major facts concerning the perception of achromatic color can be explained. Unlike the situation that obtains for the other constancies, however, it is difficult to imagine how such rules could be learned.

In suggesting that constancy can be explained by logical inference, as in the above examples, I obviously do not mean to imply that the process is one in which each premise is explicitly present in awareness or even that the brain event underlying it always occurs as an explicit state or that the order of events follows the formal logical order. After all we do not even say that about thought itself. We say only that thinking can be translated into a form of logical inference, as *if* it occurs in precisely that way, not that it does follow that form. A related matter is the fact that in thinking we do not believe that we must go through the same process each time the same problem comes up. Analogously in perception, it is possible that the process runs off more or less directly by virtue of repeated instances of the same kind of situation, based on memory of the earlier solution.

Other Examples of Inference Processes in Perception

Throughout this chapter I have restricted the discussion to the phenomena of constancy. If we are to consider other examples of perception, it seems fair to say that wherever we look we find instances in which the perceptual outcome smacks of intelligence. In some cases the intelligence is of the nature of applying a general rule to a novel situation. Consider the example of viewing an anaglyph (or vectograph). If we move our heads while doing so, the phenomenal object that appears in front of the background as a result of stereopsis also appears to move (although, of course, we *know* it is not moving). The explanation of this effect would seem to be as follows: ordinarily when one object is in front of another and we move, certain parallax changes occur within the retinal image. Therefore, if they do not occur as in the case of an anaglyph, it can be only because the object in front is moving with the head at the same rate. Another example of this kind is the apparent movement of an afterimage during eye movements. If the rule is that the image of stationary objects displaces over the retina when the eyes move, then a stationary image must signify a moving object.

In some cases the intelligence is of the nature of creatively solving a problem posed by the proximal stimulus or proximal stimulus transformation rather than simply the application of a rule; for example, stroboscopic movement perception can be understood as a solution to the problem of what in the environment is causing the sudden appearance and disappearance of two stimuli. The appearance and disappearance is inexplicable because no other stimulus information can explain it. Therefore movement is a plausible solution. Eric Sigman and I have shown that if other information is provided, for example, a phenomenally opaque figure that moves back and forth coincident with the appearance and disappearance of two dots, the dots do not appear to be moving (Sigman & Rock, 1974). Instead we see them as permanently present objects undergoing covering and uncovering. The covering object must, however, be in the appropriate position in its terminal location in order to "account for" the disappearance of the dots. Moreover, the covering object can itself be a phenomenal construct resulting from subjective contours, so that no physical contours need occlude the dots as long as they appear and disappear as the phenomenal object moves over and beyond each dot.

Subjective contour is another example of intelligent problem solving in perception. This phenomenon has been receiving a good deal of attention in the last few years as a result of Kanizsa's work (1955, 1974) (see Figure 10-5). We perceive a contour that has no physical basis in the retinal image, and the phenomenal figure that emerges from such subjective contours acquires an achro-

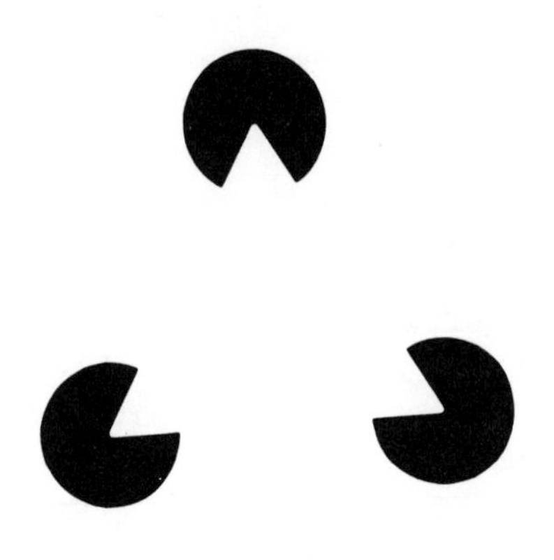

Figure 10-5 A figure similar to one devised by Kanizsa (1955).

Figure 10-6 A figure similar to one devised by Gregory (1972).

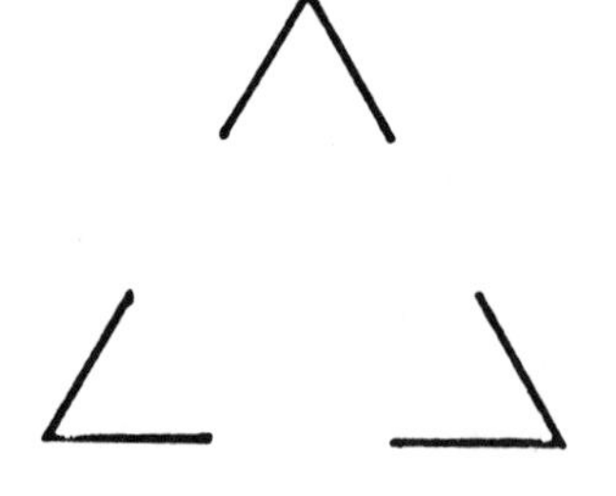

Figure 10-7 A figure that does not produce an impression of subjective contours.

matic color appearance lighter or darker than the background of the same physical luminance. Thus there is no question that the effect is perceptual, yet it would seem to be the end product of an intelligent construction on the part of the perceptual system. There are many reasons for ruling out explanations such as the triggering of feature-detector mechanisms representing the subjective segment of the contour by the physically present partial contours; for example, Gregory (1972) has shown that a curved contour is perceived even if the partial contours that are present are straight, provided the latter are so oriented that only a curved contour can connect them (Figure 10-6). Moreover, the mere presence of some partial physical contours in the appropriate locations does not guarantee the effect (Figure 10-7).

There are really two separate questions here, and Richard Anson and I are investigating them in our laboratory. One is the stimulus conditions that will lead to the hypothesis of a subjectively contoured figure. Factors such as alignment of the partial contours or recognizable incompletion of regions in the physically present configuration are relevant here. The fact is that the effect does not inevitally occur. A figure-ground reversal must occur because at the very outset the region that is to become the subjective figure is organized as ground. The components that are present must "suggest" the solution by their obvious incompletion or coincidental alignment and background "noise" will interfere with achieving the effect. Conversely, the appropriate set will facilitate it.

Perhaps of even greater interest is the question whether features of the stimulus configuration "fit" and "support" the hypothesis. The kind of pattern Kanizsa has devised (Figure 10-5) is ideal because once the incompleted corner figures—three-quarter solid circles or the like—suggest the possibility of some opaque figure covering them everything present fits that perceptual solution. If the incomplete regions are black, the solution is that a white opaque figure is covering these incomplete figures. In our research we have found that if stripes are visible at some depth behind the region in which the figure ordinarily would appear it will not emerge, for that would constitute a logical contradiction that the perceptual system abhors. A region cannot be figural and opaque and at the same time transparent. On the other hand, stripes seen in front of the same array will not interfere with achievement of the effect. (For a related finding see Gregory & Harris, 1974.)

Because in configurations of the kind Kanizsa created the background is the same physical color as the region of the subjective figure, the absence of physical contours is rationalized; for example, a white figure on a white background would not produce a retinal-image contour based on reflectance differences. The subjective color effect referred to above may have its explanation along such logical lines. If the entire figure, including the absent contours, is perceptable, it must have at least a slightly different shade than its background. Therefore a slight difference is "constructed." The reason why straight-line segments alone at the corners do not produce the effect (Figure 10-7) is that no rationale is provided for their discontinuation. There is no stimulus support for the perception of an opaque figure or even a complete outline figure, whereas the incomplete circles (Figure 10-5) do provide that rationale.

These examples of inferencelike processes in perception will suffice to buttress the argument that similar processes underlie the phenomena of constancy.

SUMMARY

Explanations of perceptual constancy are of one of two kinds. Either it is claimed that all the information for constancy is present in the retinal image in the form of higher order stimulus relationships or it is claimed that information from other sources must be taken into account to assess the meaning of the image of the object.

The first theory has always been considered to be on strong ground in explaining the constancy and other phenomena pertaining to achromatic color. Not only can many facts be explained by assuming that achromatic color is determined by luminance ratios (the underlying mechanism of which is held by many to be lateral inhibition) but the alternate theory has always been on logically weak ground. However, some examples of constancy in daily life and some laboratory data concerning the perceived slant of surfaces with respect to light sources or the separation in depth between surfaces do not follow from the ratio principle. New evidence was presented in support of a coplanar ratio principle, namely, that it is the luminance ratio between regions within phenomenal planes but not between regions in different phenomenal planes that determines achromatic color. This means that it is not the ratio between the retinal images of differing luminance values that governs the outcome. Therefore this modification in the ratio principle has profound theoretical ramifications.

The second theory of taking information into account has always been considered to be on strong ground in explaining the spatial constancies, and logical considerations and evidence examined here supports this conclusion. A higher order stimulus theory of size constancy has been shown to be neither necessary nor sufficient, logically inadequate, of doubtful ecological validity, and challenged by recent findings that certain higher order stimulus effects are governed by relationships within phenomenal planes rather than simply by retinal image relationships.

A review of most of the visual constancies suggests that in addition to the constancy mode of perception, which is admittedly dominant and salient, there is also a proximal mode of perception. The latter, while sometimes acknowledged, has been neglected by contemporary investigators. Certain conditions facilitate and certain others impede perception in the proximal mode. Descriptions of perceptual experience are incomplete, if not distorted, without inclusion of proximal-mode characteristics, yet neither major kind of theory of constancy does justice to them. It was suggested that their presence in experience can explain certain facts and experimental findings, such as the effect of instructions, underconstancy, overconstancy, individual and (possibly) developmental differences, and the increased departure from constancy when the objects compared are viewed under maximally different conditions. Physiological theories such as those based on feature-detector mechanisims and lateral inhibition were examined and found wanting. Lateral inhibition cannot do justice to the absence of an effect on color of ratios across planes and to the fact of dual aspects of experience of light, namely, surface color or lightness (the constancy mode) and phenomenal intensity or brightness (the proximal mode). It is possible, however, that lateral inhibition somewhat modifies the functionally effective luminance differences between adjacent regions which are then interpreted on the basis of the coplanar ratio principle.

Concerning all of these facts, it was suggested that the taking-into-account hypothesis is essentially correct for all constancies and moreover is based on a process of unconscious inference. In spatial constancies perception results from a process analogous to syllogistic reasoning. The major premise consists of certain "known" rules concerning transformation of the proximal stimulus (e.g., the law of the visual angle). It is not necessary to assume that these rules are acquired by experience. The minor premise consists of the present input, one componant of which leads to a percept in the proximal mode (e.g., perceived extensity), the other of which leads to other correlated perceptions (e.g., perceived distance). In achromatic color constancy perception is based on "knowledge" of the rule that nearby regions in a plane receive equal illumination.

Evidence was presented from other areas of perception supporting a cognitive theory.

REFERENCES

Asch, S. E. & Witkin, H. A. Studies in space orientation I and II. *Journal of Experimental Psychology*, 1948, 38, 325-337, 455-477.

Aubert, H. Eine scheinbare bedeutende Drehung von Objekten bei Neigung des Koppes nach rechts odor links. *Virshows Archives*, 1861, 20, 381-393.

Beck, J. *Surface Color Perception,* Ithaca, New York: Cornell University Press, 1972.

Beck, J. Apparent spatial position and the perception of lightness. *Journal of Experimental Psychology,* 1965, **69**, 170-179.

Begelman, D. Size perception and schizophrenia. Ph.D. Thesis. Yeshiva University, 1966.

Bower, T. G. R. The visual world of infants. *Scientific American,* 1966, **215**, 80-92.

Brindley, G. S. & P. A. Merton. The absence of position sense in the human eye. *Journal of Physiology,* 1960, **153**, 127-130.

Brown, J. F. The visual perception of velocity. *Psychologische Forschung,* 1931, 14, 199-232.

Brunswik, E. *Perception and representative design of psychological experiments.* Berkeley: University of California Press, 1956.

Carlson, V. R. Overestimation in size-constancy judgments. *American Journal of Psychology,* 1960, **73**, 199-213.

Carr, H. *An introduction to space perception.* New York: Longmans, Green, 1935.

Cornsweet, T. N. *Visual perception.* New York: Academic, 1970.

Craik, K, J. W. *The nature of psychology.* Cambridge: Cambridge University Press, 1966, pp. 94-97.

Duncker, K. Uber induzierte Bewegung. *Psychologische Forschung,* 1929, 12, 180-259.

Epstein, W. The process of taking-into-account in visual perception. *Perception,* 1973, 2, 267-285.

Epstein, W. Attitudes of judgment and the size-distance invariance hypothesis. *Journal of Experimental Psychology,* 1963, **66**, 78-83.

Epstein, W. Phenomenal orientation and perceived achromatic color. *The Journal of Psychology,* 1961, **52**, 51-53.

Epstein, W. & Landauer, A. Size and distance judgments under reduced conditions of viewing. *Perception & Psychophysics,* 1969, **6**, 269-272.

Evans, R. M. *An Introduction to Color.* New York: Wiley, 1948.

Filehne, W. Uber das optische wahrnehumung von Bewegungen. *Zeitschrift für Sinnesphysiologie,* 1922, **53**, 134-145.

Flock, H. R. Achromatic surface colorand the direction of illumination. *Perception & Psychophysics,* 1971, **9**, 187-192.

Flock, H. R. & Freedberg, E. Perceived angle of incidence and achromatic surface color. *Perception & Psychophysics,* 1970, **8**, 251-256.

Gibbs, T. & Lawson, R. B. Simultaneous brightness contrast in stroboscopic space. *Vision Research,* 1974, **14**, 978-983.

Gibson, J. J. *The senses considered as perceptual systems.* Boston: Houghton Mifflin, 1966.

Gibson, J. J. *The perception of the visual world.* Boston: Houghton Mifflin, 1950.

Gilchrist, A. *Perceived achromatic color as a function of ratios within phenomenal planes.* Ph.D. Thesis, Rutgers University, 1975.

Gilchrist, A. & Rock, I. Lightness perception as a function of ratios within perceived planes. Paper read at Meeting of *Psychonomic Society,* November, 1975.

Gilinsky, A. The effect of attitude upon the perception of size. *American Journal of Psychology,* 1955, **68**, 173-192.

Gogel, W. Depth adjacency and the Ponzo illusion. *Perception & Psychophysics,* 1975, **17**, 125-132.

Gogel, W. The organization of perceived space: I. Perceptual interactions. *Psychologische Forschung*, 1973, **36**, 195-221.

Gogel, W. The validity of the size-distance invariance hypothesis with cue reduction. *Perception & Psychophysics*, 1971, **9**, 92-94.

Gogel, W. The sensing of retinal size. *Vision Research*, 1969, **9**, 1079-1094.

Gogel, W. C. & Koslow, M. A. The adjacency principle and induced motion. *Perception & Psychophysics*, 1972, **11**, 309-314.

Gogel, W. C. & Mershon, D. H. Depth adjacency in simultaneous constast. *Perception & Psychophysics*, 1969, **5**, 13-17.

Gogel W. C. & Sturm, R. D. A test of the relational hypothesis of perceived size. *American Journal of Psychology*, 1972, **85**, 201-216.

Gregory, R. Cognitive contours. *Nature*, 1972, **238**, 51-52.

Gregory, R. & Harris, J. P. Illusory contours and stereo depth. *Perception & Psychophysics*, 1974, **15**, 411-416.

Hastorf, A. H. & Way, K. S. Apparent size with and without distance cues. *Journal of General Psychology*, 1952, **47**, 181-188.

Heinemann, E. Simultaneous brightness induction as a function of inducing-and test-field luminances. *Journal of Experimental Psychology*, 1955, **50**, 89-96.

Heller, D. Absence of size constancy in visually deprived rats. *Journal of Comparative and Physiological Psychology*, 1968, **65**, 336-339.

Hochberg, J. Higher-order stimuli and inter-response coupling in the perception of the visual world. In R. MacLeod & H. L. Pick Jr. (Eds.), *Studies in Perception: Essays in Honor of J. J. Gibson*, Ithaca, New York: Cornell University Press, 1974.

Hochberg, J. Color and shape. In J. W. Kling & L. A. Riggs (Eds.), *Woodworth & Schlosberg's Experimental Psychology*, (Vol. I), New York: Holt, Rinehart & Winston, 1972.

Hochberg, J. & Beck, J. Apparent spatial arrangement and perceived brightness. *Journal of Experimental Psychology*, 1954, **47**, 263-266.

Holway, A. H. & Boring, E. G. Determinants of apparent visual size with distance variant. *American Journal of Psychology*, 1941, **54**, 21-37.

Horn, G. & Hill R. M. Modifications of receptive fields in the visual cortex occurring spontaneously and associated with bodily tilt. *Nature*, 1969, **221**, 186-188.

Hubel, D.H. & Weisel, T.N. Receptive fields, binocular interaction, and functional architecture in the cat's visual cortex. *Journal of Physiology*, 1962, **160**, 106-154.

Jameson, D. & Hurvich, L. Complexities of perceived brightness. *Science*, 1961, **133**, 174-179.

Johansson, G. *Configuration in event perception.* Upsala: Almkwist and Wiksell, 1950.

Kanizsa, G. Contours without gradients or cognitive contours. *Italian Journal of Psychology*, 1974, **1**, 93-112.

Kanizsa, G. Margini quasi-percettivi in campi con stimolazione omogenea. *Rivista di Psicologia*, 1955, **49**, 7-30.

Katona, G. Color contrast and color constancy. *Journal of Experimental Psychology*, 1935, **18**, 49-63.

Kaufman, L. *Sight and Mind: An Introduction to Visual Perception* (Chapter 8). London: Oxford University Press, 1974.

Kaufman, L. & Rock, I. The moon illusion, I. *Science*, 1962, **136**, 953-961.

Mach, E. *The Analysis of Sensations.* New York: Dover, 1959.

Mack, A. Perceptual modes, I. Paper delivered at Conference on Perceptual Modes, University of Minnesota, 1974.

Mack, A. & Herman, E. Position constancy during pursuit eye movement: an investigation of the Filehne illusion. *Quarterly Journal of Experimental Psychology*, 1973, **25**, 71-84.

Mershon, D. H. Relative contribution of depth and directional adjacency to simultaneous whiteness contrast. *Vision Research*, 1972, **12**, 969-979.

Mershon, D. H. & Gogel, W. C. The effect of stereoscopic cues on perceived whiteness. *American Journal of Psychology*, 1970, **83**, 55-67.

Müller, G. E. Über das Aubertsche Phanomenon. *Zeitschrift für Psychologie und Physiologie der Sinnesorgane*, 1916, **49**, 109-244.

O'Brien, V. Contour perception, illusion and reality. *Journal of the Optical Society of America*, 1958, **48**, 112-119.

Posin, R. Perceptual adaptation to contingent visual-field movement; an experimental investigation of position constancy. Ph.D. dissertation, Yeshiva University, 1966.

Posin, R. Size overestimation and the disparity of retinal image size. Unpublished paper, Yeshiva University, 1962.

Ratliff, F. Contour and contrast. *Scientific American*, 1972, **226**, 90-101.

Richards, W. Apparent modifiability of receptive fields during accommodation and convergence and a model for size constancy. *Neurophychologia*, 1967, **5**, 63-72.

Rock, I. *An introduction of perception.* New York: Macmillan, 1975.

Rock, I. *Orientation and form.* New York: Academic, 1973.

Rock, I. Toward a cognitive theory of perceptual constancy in A. Gilgen (Ed.), *Contemporary scientific psychology.* New York: Academic, 1970.

Rock, I. The nature of perceptual adaptation. New York: Basic Books, 1966, pp. 71-72.

Rock, I. The perception of the egocentric orientation of a line. *Journal of Experimental Psychology*, 1954, **48**, 367-374.

Rock, I. & Ebenholtz, S. The relational determination of perceived size. *Psychological Review*, 1959, **66**, 387-401.

Rock, I., Hill, A. L., & Fineman, M. Speed constancy as a function of size constancy. *Perception & Psychophysics*, 1968, **4**, 37-40.

Rock, I. & Kaufman, L. The moon illustration, II. *Science*, 1962, **136**, 1023-1031.

Rock, I. & McDermott, W. The perception of visual angle. *Acta Psychologica*, 1964, **22**, 119-134.

Sigman, E. & Rock, I. Stroboscopic movement based on perceptual intelligence. *Perception*, 1974, **3**, 9-28.

Spinelli, D. N. Recognition of visual patterns (Chapter VIII). In D. A. Hamburg, K. H. Pribram, and A. J. Stunkard (Eds.), *Perception and its disorders.* Williams & Wilkins, 1970.

Stoper, A. E. Apparent motion of stimuli presented stroboscopically during pursuit movement of the eye. *Perception & Psychophysics*, 1973, **13**, 210-211.

Thouless, R. Phenomenal regression to the real object. *British Journal of Psychology*, 1931, **21**, 339-357; **22**, 1-30.

Wallach, H. Brightness constancy and the nature of achromatic color. *Journal of Experimental Psychology*, 1948, 38, 310-324.

Wallach, H. On constancy of visual speed. *Psychological Review*, 1939, 46, 541-552.

Wallach, H. & Kravitz, J. H. The measurement of the constancy of visual direction and of its adaptation. *Psychonomic Science*, 1965, 2, 217-218; Rapid adaptation in the constancy of visual direction with active and passive rotation. *Psychonomic Science*, 1965, 3, 165-166.

Wallach, H. & Zuckerman, C. The constancy of stereoscopic depth. *American Journal of Psychology*, 1963, 76, 403-412.

CHAPTER

11

SPATIAL CONSTANCY AND MOTION IN VISUAL PERCEPTION

GUNNAR JOHANSSON
University of Uppsala

In the study of visual space perception we traditionally distinguish between perception of static displays and perception of moving or kinetic displays. The first of these terms refers to displays or targets that do not change over time in either position or state and the second to displays or targets that undergo spatio-temporal change while being observed. With few exception, however, the century-long experimental effort in this field has been concerned with static perception. The dominance of static displays has even resulted in a use of the term space perception as equivalent to static space perception. It has been only during the last two or three decades that total reliance on the analysis of static displays has been abandoned.

Emphasis on static displays has been characteristic of the research on constancies in space perception. Nearly all experimental studies on size and shape constancy have treated static conditions. The traditional paradigm compares percepts from stimulus patterns representing static object of different size, at different distances from the observer, or at different slants. Consult the classical sources like Thouless (1931), Holaday (1933), Koffka (1935), Boring (1942), Stavrianos (1945), Woodworth (1938) or the newer ones like Woodworth and Schlosberg (1954), Epstein and Park (1963), Epstein, Park, and Casey (1961), Graham (1965), Day (1969), or Murch (1973). Rarely do we find treatment of kinetic constancies; that is, spatial constancy during perceived motion. This neglect must be regarded as rather astonishing. In everyday life perceptual object constancy during motion plays a most essential role for purposive behavior. The stronghold of the static model is nowhere more evident than in Brunswik's writing. Brunswik stressed ecological validity, yet he neglected to consider kinetic situations.

Fortunately the fact that research on visual constancies has largely ignored kinetic space constancy does not mean that relevant experimental data are lacking. A number of studies centered on the problem of visual perception of motion in the third dimension contain an impressive amount of data of direct relevance to an understanding of kinetic space constancy. Our main concern in this chapter is to collect and integrate a representative part of these findings from the viewpoint of perceptual constancy. In some of these studies the central problem for the investigator does not coincide with our problem. In these cases we abstract from the experiments those findings that are especially informative for our discussion; for example, we treat only cursorily the problems raised by veridical perception and the ambiguity of perceived shape and motion when the alternative percepts represent constant objects in motion.

Hopefully our study will contribute to a better understanding of the great advantage that can be gained from introducing kinetic components in the paradigms for research on space constancy problems. Our first step compares in a descriptive way some basic facts from the study of kinetic size and shape constancy with experimental findings on their static counterparts. This descriptive material affords a basis for an analysis of perceptual mechanisms underlying the kinetic constancies. Finally an outline of a general theory for spatial constancies is sketched.

DESCRIPTIVE PART

As an introduction to our descriptive studies it may be advantageous to examine the distinction between the perception of static and kinetic displays.

Motion Perception and Static Perception

As a first formal definition let us agree that in the following discussion the term *motion perception* denotes every visually recorded spatial change over time of a structure or of the perceiver, whereas *static perception* always stands for a stationary perceiver's perception of objects at rest relative to the environment. As illustration I describe my own perception during a short moment of relaxation when writing this page. I just stop writing and lean back in my chair. In this position I look steadily for a few seconds at my desk with its writing utensils and a bookshelf behind it. Nothing in my field of vision changes its position and I am sitting still. In this situation I perceive: (a) a fully static environment and (b) my own head as stationary in space, or at least approximately stationary. Perceptions with this double characteristic: stationary perceiver and a fully static environment—and only those—are called static perceptions.

Under critical experimental conditions, however, we must be even stricter. If I had tried to attend carefully to the position of my head, I probably would have noted some small irregular movements. In experiments that study static perception freed from kinetic cues such head movements must be prevented.

At right angles on my left hand there is a window. I now swing around about 90° in my chair and look out. In two respects this motion introduces perception of spatial changes. First I perceive my own partial rotation in relation to the room. This is a visual perception of *locomotion.* Next, and already during the last part of the swing, I see the view through the window. I see a part of a street with a car passing, but also, close to the window, a naked tree, the branches of which are swinging to and fro in the hard wind. A couple of great tits are eagerly pecking on a piece of tallow attached to the window ledge. All these motions seen throught the window represent various types of *object motion perception,* the second subcategory under motion perception. We have in this example the simple type of mechanical motion represented by the moving car, the complex motion of the tree with components of elasticity in it, and finally the intricate biological motion in the pecking great tits.

In this way we observed examples of two major types of motion perception, locomotion and object motion, and both represent in themselves broad spectra of variations. Furthermore, the examples make clear that locomotion and a number of object motions can be perceived simultaneously without interfering with each other.

We can also distinguish a third category of perceived motion beside locomotion and object motion proper. This is a degenerate or improper kind of perceived object motion namely *perception of form change.* In the following we pay some attention also to this type of space perception.

Kinetic Size Constancy

ABOUT RELATION TO STATIC SIZE CONSTANCY

The best way of characterizing kinetic size constancy perhaps is to compare it with the static constancy under similar conditions. For several reasons I have chosen Holway and Boring's (1941) classical investigation of the effects of cue reduction on static size constancy as a starting point for our comparison. In this experiment *S* was sitting in the junction of two long dark corridors. The task was to adjust the size of a circular disk in one corridor to be perceptually equal to a standard disk presented in the other. The standards were set at a number of different distances from *S*, but it always subtended one degree of visual angle. Results were obtained from a number of viewing conditions. Three of these were carried out under good but not perfect darkness. These conditions were (a) binocular vision, (b) monocular vision, and (c) monocular vision with artificial pupil. In completing the experiment, a tunnel producing total darkness was introduced, but in other respects the condition was an iteration of (c).

The result was that binocular regard gave a slight overconstancy and monocular regard with natural pupil resulted in an almost perfect size constancy, whereas introducing an artificial pupil had as a consequence a curve midway between perfect constancy and the opposite extreme: the law of visual angle. Adding also a reduction tunnel brought about a further drop in the curve and gave a function rather close to the one representing the law of visual angle.

Boring interpreted these results in the framework of Titchener's corecontext theory, a classical variant of the cue theory. Boring himself has summarized this theory in the following way:

> For descriptive purpose it is convenient to say that the sensory data that contribute to a perception can be divided into a core and its context. The *core* is the basic sensory excitation that identifies the perception, that connects it most directly with the object of which it is a perception. The *context* consists of all the other sensory data that modify of correct the data of the core as it forms the perception. The context also includes certain acquired properties of the brain, properties that are specific to the particular perception and contribute to the modification of its core. In other words, the context includes knowledge about the perceived object as determined by past experience, that is, by all the brain habits which affect perceiving.*

With this background Boring stresses that a reduction of the stimulus to contain only "core" information results in extremely low or no degree or perceptual

* Boring, The Perception of Objects, *American Journal of Physics*, 1946, 14. 2, 99-107. Quoted from Leibowitz, 1965, 69-70.

size constancy. Under these conditions the perceptual size varies with visual angle. Adding enough stimulus information also of "context" type, however, brings about nearly full size constancy. The consequent conclusion therefore is, according to Boring (and a large number of later handbooks and reviews), that the primary size response to visual stimulation is in terms of visual angle subtended by the stimulus pattern but that in real life this is perfectly compensated for by concomitant "context" (secondary) information.

Holway and Boring's experimental findings have been confirmed in a number of investigations: Hastorf and Way (1952), Renshaw (1953), Zeigler and Leibowitz (1957).

Thus the position that the perceptual size constancy is dependent on availability of secondary cues is founded on a lot of experimental data and is commonly regarded as an established fact by most theorists (e.g., cf. Kilpatrick & Ittelson, 1953; Gogel, 1969). For the sake of discussion let us accept Boring's interpretation as a starting point for our analysis. Thus we will transpose the problem of static size constancy and put it in terms of kinetic perception. Trying to repeat the Holway-Boring experiment under kinetic conditions means that the condition with unchanging visual angle of the disk under change of the distance disk to observer must be given up. Motion in depth relative to the observer of necessity means that the visual angle of an object of constant size must vary. This is just a characteristic difference between static and kinetic conditions. From the point of view of stimulus generation, however, it is not necessary to work with really moving objects. In an experimental situation we can produce the same proximal change by keeping the stimulus surface stationary and varying its size instead. Thus our reformulation of the Holway-Boring problem in kinetic terms can be given in the following wording: under "core" conditions will a continuous change of visual angle bring about perception in accordance with the law of visual angle, that is, perception of a stationary object changing its size, or will it result in perception of an object of constant size moving sagittally in space?

An assumed functional equivalence between static and kinetic conditions when it is a question of size constancy under lack of secondary cues implies a positive answer to the first alternative in this question. However, the results from a number of experiments on perception of sagittal motion under reduced cue conditions favor the second alternative. In these experiments the only stimulus information available has been a stationary bright surface continuously changing its size in an otherwise totally dark environment. Therefore the immediate conclusion must be that there is no parallel between the static and the kinetic conditions when it is a question of size constancy under highly reduced stimulus conditions. In order to get a more differentiated view, however, let us get in touch with some of the relevant experiments.

HILLEBRAND'S AND KOFFKA'S STUDIES

Hillebrand, the famous perceptionist, in a paper published in 1894 described the apparent motion in depth caused by continuously changing the size of the opening in an iris diaphragm (Hillebrand, 1894). He found that the diaphragm opening was seen as a circular disk of constant size which perceptually receded from the observer when contracting and approached when expanding.

Koffka in his "Principles" (1935) also briefly describes an experiment (probably an informal preexperiment) on the "illusory" depth motion produced by the expanding-contracting bright circle seen through an iris diaphragm. (Koffka, 1935, pp. 82, 277). Significantly, Koffka describes this experiment not—as we would expect—in his famous chapter on the constancies but in an earlier section titled "Why do things look as they do?" Koffka describes the perceptual effect thus:

> . . .the observers see either a forward or backward movement of the light circle, or its expansion or contraction, or finally a joint effect in which expansion and approach, contraction and recession are combined. (Koffka, 1935, p. 82).

Nothing is said about the viewing conditions.

ITTELSON'S EXPERIMENT

Ittelson (1951) carried this research on motion perception from a continuously changing diaphragm opening an important step further by a clever quantification of this effect under different combinations of monocular and binocular conditions. This was done in connection with his formulation of the size-distance invariance hypothesis. Ittelson compared this type of "apparent" motion with a corresponding real motion back and forth of an identical object (a diamond). He found that under strict monocular conditions there is hardly any difference at all between the two types of stimulation in terms of perceived length of the motion track and also that from a phenomenological point of view the two types of motion stimulus were almost identical. Introducing a situation of cue conflict by presenting the diaphragm device in a "well defined binocular field" also resulted in motion perception but of an anomalous type and comments like "it seems to be moving without getting anywhere."

In accordance with the transactional theory Ittelson interpreted the results in terms of past action and experience while simultaneously stressing the compelling character of the percepts.

JOHANSSON'S EXPERIMENT

Johansson's (1964) study of the percepts associated with size and form changes in rectangles included an experiment with continous change of size in a square. In these experiments the control of the stimulus conditions was highly elaborated. The stimulus square was generated on a CRT screen. The square was viewed monocularly and continuously changed its size in the proportions 4° : 2° visual angle with a frequency of about .5 p sec $^{-1}$. Information about the localization of a screen or of the object was excluded by presenting the stimulus as a virtual image on infinite distance through a special optical device of telescope type, neutralizing accommodation and convergence cues. The stimulus appeared to be suspended freely in space. Therefore in this experiment no secondary distance cues were available for the perceiver: the total field of view was fully homogeneous except the stimulus square, and there was no effective distance information due to convergence or accommodation. The instruction to *Ss* just said "Look into this tube and describe what you see." Under these conditions 66 out of 67 subjects (43 university students and 24 children aged 10-11 years) spontaneously reported seeing a frontal square moving in depth. All children and most of the naïve students took for granted that they really saw a moving object. Perceived size constancy was not perfect, however; a number of *Ss*, especially among the children, reported that the square changed its size a little during the motion. The student group was also asked to try to see the stimulus pattern as stationary and changing its size. Fifteen of 42 *Ss* admitted this possibility but gave it the lowest value on a rating scale, whereas the moving constant shape alternative was always given the highest scale value.

MARMOLIN'S EXPERIMENT

In the experiment described by Koffka, but also under the strictly controlled conditions in Johansson's experiment, we find that the size constancy is not perfect. More or less pronounced changes of size of the perceptual object can be seen together with the motion. Marmolin, (1973a, 1973b) has made this the subject of a special study. His experimental setup was similar to Johansson's, but by use of a minicomputer he caused the square to shrink and grow in a linear way between maximal size and 1.5% of this size. Applying scaling techniques, Marmolin was able to show that an equation including a factor standing for a certain amount of size change added to the motion gave a satisfactory fit over the whole range of change. Because of the relatively small numerical value of this change in perceptual size, it was less impressive for size changes less than 50%. There a simple power function with an exponent similar to that found in studies on static distance perception could be accepted.

A significant guess from Marmolin's research is that the small but consistent deviation from perfect constancy under motion is not due to lack of information but rather is a characteristic of the functioning of the visual system.

EFFECTS OF FURTHER REDUCTION OF STIMULUS INFORMATION

Holway and Boring introduced a systematic cue reduction in order to investigate perceptual size constancy as a function of the amount of distance information. Our review of experiments relevant to kinetic size constancy has made it clear that even under Holway and Boring's most reduced stimulus condition approximative kinetic size constancy is established. The next step is to investigate the perceptual outcome of still more severe reductions of stimulus information.

A way to such reductions, is found in the fact that neither a homogeneous surface nor an unbroken outline of a geometric figure is necessary for perceiving a stable form. It is well known that a static row of points can be perceived as a line and that a quadrangle, for instance, can be perceptually represented by four such point arrays. In an extreme case and for visual angles of less than 5° just the four corner points are sufficient for eliciting the perception of a corresponding quadrangle, three points represent a triangle, and so on. This remarkable capacity of the visual system has been taken advantage of in some experiments that have been studying the basic mechanisms underlying motion perception, experiments that are highly relevant also in our present context.

In connection with the experiments with a contracting-expanding homogeneous bright square, described above, Johansson instead of exposing the square caused four spots arranged as corners of a square to move simultaneously to and fro along its imaginary diagonals. The outcome of this experiment was very clear-cut. With few exceptions these diagonally moving spots are seen as the bright corners of a rigid square moving sagittally in depth. The coordinated motion of the spots ties them together perceptually to a strong spatial unit.

Geometrically four points represent the absolute minimum information about a quadrangle. No redundancy exists. Take away one of the points and we get a triangle. It seems evident that it is the coordinated motion of the elements far more than the figural closure that results in a rigid perceptual unit in motion. This conclusion is far more strengthened in the interesting experiments with three moving spots carried out by Börjesson and von Hofsten (1973) and which are commented on in a later section.

SUMMARY ON KINETIC SIZE CONSTANCY

The results of the experiments reviewed above are conclusive: when a proximal pattern continuously changes within a certain range of angular size and frequency, the concomitant percept is a sagittal motion of an object with an approximately constant size. This has been found to hold in general also under the most impoverished stimulus conditions.

This clearly implies that a continuous change over time of the visual angle of a stimulus pattern acts as a source of information which is absent in the corresponding static stimulation. Under these kinetic conditions no secondary information is needed for evoking perception of a constant object in motion. In Titchener's and Boring's terms: stimulation of sensory "core" type yields perceptual size constancy under kinetic conditions. No assumptions about influence of "context" interpretation are required.

Kinetic Shape Constancy, a Background

GEOMETRICAL CONSIDERATIONS

In the conceptual system of geometry and kinetics there exist just two categories of motion, namely *translations* and *rotations* together with their combinations. In our study of perceived sagittal motion induced from shrinking or growing frontal patterns, we have dealt exclusively with perceived translation in depth. The geometrical reason for this, of course, is that a perspective projection from a sagittal translation of an object is a stationary shrinking or growing pattern. A perspective projection into the eye of an object or surface in rotation, however, typically results in a form change of the projected figure rather than a change of its size. Therefore, as we now are going to study kinetic shape constancy (i.e., perceived constant shape from a retinal projection that continuously changes its shape) we will be dealing with patterns that perceptually and most often also geometrically represent rotations.

STATIC SHAPE CONSTANCY AS DEPENDENT ON DEPTH CUES

It is well known that there is a good parallel between static size constancy and static shape constancy when it is a question of the dependence of depth cues. Already Thouless (1931), one of the pioneers in research on shape constancy, has stated that under full cue conditions there is a good correspondence between real and perceived shape of slanted disks but that the degree of constancy is declining with reduction of depth cues. Later research in this field has confirmed Thouless' findings (e.g., Epstein & Park. 1963; Flock, 1964, 1965; Freeman, 1965, 1966) but also made probable that the cognitive attitude of the observer as manipulated by the instruction also can play a certain role (Epstein, Bontrager, & Park, 1962; Lichte & Borresen, 1967; Landauer, 1969).

SHAPE CONSTANCY DATA FROM RESEARCH ON MOTION PERCEPTION

Musatti. Still earlier than Thouless' findings on static shape constancy effects, Musatti's (1924) studies on "stereo-kinetic phenomena" established the existence of kinetic shape constancy under conditions excluding secondary depth cues. This study, however, was published only in Italian and therefore regrettably has had only limited influence on the discussion in this field.

METZGER'S EXPERIMENT

Ten years later Metzger published his highly original and to some extent epochmaking investigations on the perceptual effects from coordinated motions of a number of shadow lines moving over a screen (Metzger, 1934a, 1934b). Figure 11-1 illustrates the principle in Metzger's setup. A number of pegs are arranged in various patterns on a rotating turntable. The shadows from these pegs are projected onto a translucent window in an otherwise opaque screen and move

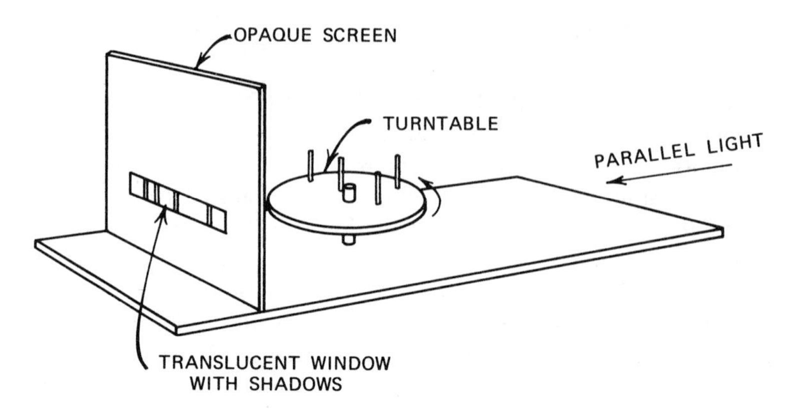

Figure 11-1 Metzger's shadow-caster arrangement. The shadows from a number of pegs on a slowly rotating turntable are projected onto a screen. The sinusoidally oscillating shadows are seen as belonging to one or several rigid units rotating in depth.

sinusoidally back and forth over the window with the same frequency but with different amplitudes and different phase relations.

In these experiments an observer usually does not perceive what is presented on the screen: a number of lines moving back and forth over a frontal screen while continuously changing their mutual distances. Instead he spontaneously sees one or more rigid units performing rotary motions in depth. For most peg constellations the grouping changes over the inspection period. In spite of this instability in perceptual organization, however, each perceived unit, when perceived as rotating, is characterized by constant distance between its elements. It is of great theoretical interest to observe that there exists on the screen no secondary information about depth and generally nothing corresponding to the perceived units, whereas percieved constant distance within each unit has a counterpart in the constant distance between the pegs on the turntable. Metzger's results have been verified by White and Mueser (1960).

WALLACH AND O'CONNELL'S EXPERIMENT

As a third milestone in the research about visual perception of motion in depth and as a contribution of essential importance to our discussion we mention Wallach and O'Connell's (1953) investigation of what they termed the kinetic depth effect. These authors raised in a very clear way the problem of perceptual shape constancy from changing shape of a proximal pattern. In their experiment they used a shadow-caster technique similar to Metzger's. Between a point source of light and a translucent screen they placed objects of different types which were slowly rotated about a central vertical axis. The shadows generated

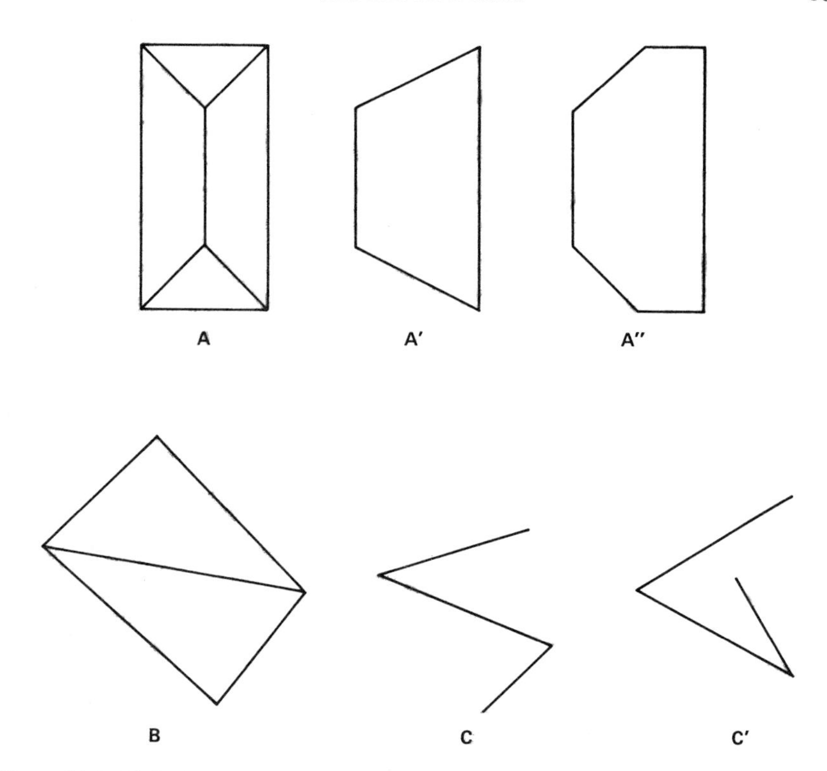

Figure 11-2 Wallach and O'Connell's shadow-casters. A is a solid form (a roof) shown with two of its shadows, A′ and A″; B, is a wire quadrilateral and C a wire "helix" a top view of which is shown in C′. (After Wallach and O'Connell, 1953.)

by these objects were projected on the screen and seen from behind by the *Ss*. Examples of these shadow-casters are shown in Figure 11-2.

A typical result from the experiments was that the continuously changing shadows were perceived as rigid objects rotating in depth. The subject could also easily identify the shape of the shadow-casters. An important and restricting condition for perceiving rigid objects in rotation, however, was that the shadows had to change simultaneously in two dimensions. Static presentations of the objects in different slants resulted in perception of frontoparallel surfaces of different shapes.

Thus again we find that continuously changing proximal stimuli, in which all possibilities of secondary information about motion in depth is excluded, elicit vivid perception of rigid rotation in depth. This result is obtained in spite of the fact that the arrangement with shadows on a visible screen yields some good secondary information about changes on a frontal surface. It is of interest to read what the authors say about the vividness of object perception from stimulus A in Figure 11-2:

> In experiments like this the impression of three-dimensional form is so natural that many *Ss* who are not psychologists are not astonished by their observations. They correctly assume that behind the screen is just an object as they see. (p. 207).

GIBSON AND GIBSON'S EXPERIMENT

The two Gibsons were interested in the question whether geometrically perfect perspective transformation in projections of rigid objects in oscillatory motion in depth brought about veridical perception of shape, orientation, and amplitude of oscillation (Gibson & Gibson, 1957). These authors used the shadow-caster technique, but in a more sophisticated way. Although the earlier technique described above had resulted in semiparallel projections, the arrangement used by these authors, and previously developed by J. J. Gibson (1957), produced perfect central projection on a visible screen. Rectangular as well as "amoeboid" forms in slow oscillatory motion about a vertical axis were studied and the data gave quantitative measures of perceived motion amplitude.

The outcome of the experiment was that such patterns changing in strict accordance with principles of perspective brought about not only veridical perception of the shadow-caster shape but also fairly accurate reports of slant and oscillation. It was also found, however, that for some shorter intervals during the session some *Ss* instead perceived a frontal surface changing its shape. In a control group presented with the same figures under static conditions the irregular patterns were generally seen as frontoparallel and the rectangular patterns as slanted but to a considerably smaller degree than the slant of the shadow-caster.

EPSTEIN'S EXPERIMENT

In 1964 Epstein carried out an experiment with 75 *Ss* that can be regarded as a check of the conclusions from Wallach and O'Connell's and Gibson and Gibson's investigations. Epstein's (1965) problems setting is very close to the main problem in this chapter: perceived spatial invariance under kinetic conditions.

The result is unequivocal. The static presentations of the pattern gave some kind of depth in about one-third of the cases, but this changed to 100% at the onset of motion. Also the estimated angles of the turning points were good (or rather good) approximations of the real angles.

Epstein in his discussion also stresses that investigating the shape-slant invariance hypothesis under static conditions perhaps is rather inadequate. "It may be that the invariance-hypothesis is more appropriately tested under conditions in which either 0 or the target is in motion" (Epstein, 1965, p. 303).

Shape Constancy of a Rotating Object Elicited from Two Moving Dots.

In our treatment of kinetic size constancy we found that a few dots in coordinated motion could bring about a perception of a rigid object moving in depth. *Mutatis mutandis,* the same has been found to hold true for rotary motion. In a recent paper Johansson (1974c) reported experiments with a few bright dots tracing a common path in the form of a conic section (ellipse) or of combination of conic sections.

In the experiments the stimuli were controlled by a minicomputer and displayed on a CRT screen. They were seen through a collimating device which eliminated conflicting information from a flat screen. The motion tracks and accelerations were computed as central projections from a circular motion with constant speed. Figure 11-3 shows some of the two-dot patterns and the typical percepts elicited by those stimuli.

The two dots usually had counterphase relation in their motions around the path. This means that they represented projections of the end points on a line slowly rotating about its center point.

The percepts from these motion constellations were invariably reported in terms of rotation of a "rod" on a more or less slanting plane according to the eccentricity of the elliptical track (stimulus of type B). Also stimuli of type C were seen as a rigid rod, rotating but at the same time describing a pendulum motion. This description is a mathematically correct interpretation of the stimulus regarded as a projection from a rigid motion. Even more impressive is that two points that move in the same way but in a rectangular track and with constant speed as diagrammed in Figure 11-3 are perceived as a rigid rod in a complex rotation.

Thus even this minimum information of coordinated rigid motion in depth is enough to evoke an unavoidable impression of spatial constancy plus motion in most cases.

Summary on Kinetic Shape Constancy

We can generalize hypothetically from these examples of research on perception of rotary motion in depth under impoverished stimulus conditions in the following way. Whenever it is possible to describe a continuous proximal change of a

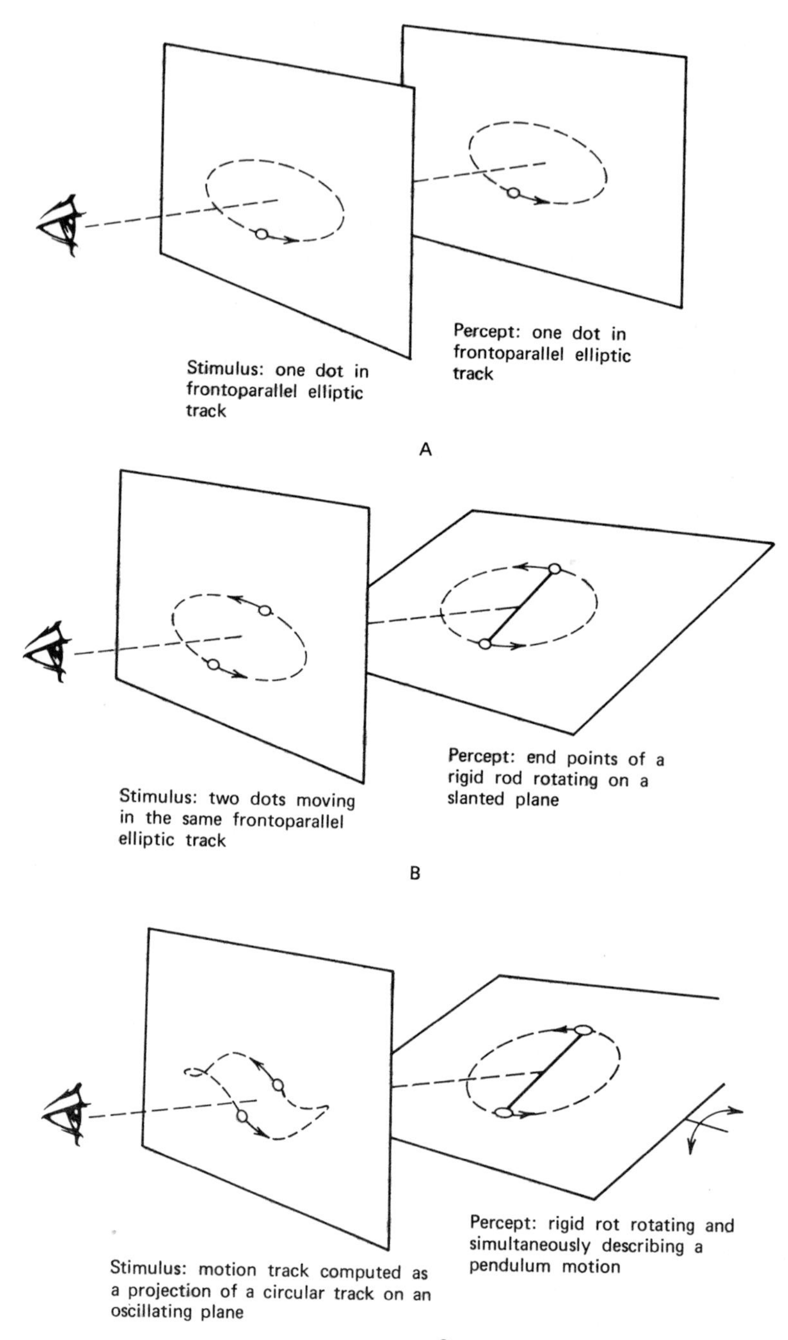
Stimulus: one dot in frontoparallel elliptic track
Percept: one dot in frontoparallel elliptic track
A
Stimulus: two dots moving in the same frontoparallel elliptic track
Percept: end points of a rigid rod rotating on a slanted plane
B
Stimulus: motion track computed as a projection of a circular track on an oscillating plane
Percept: rigid rot rotating and simultaneously describing a pendulum motion
C

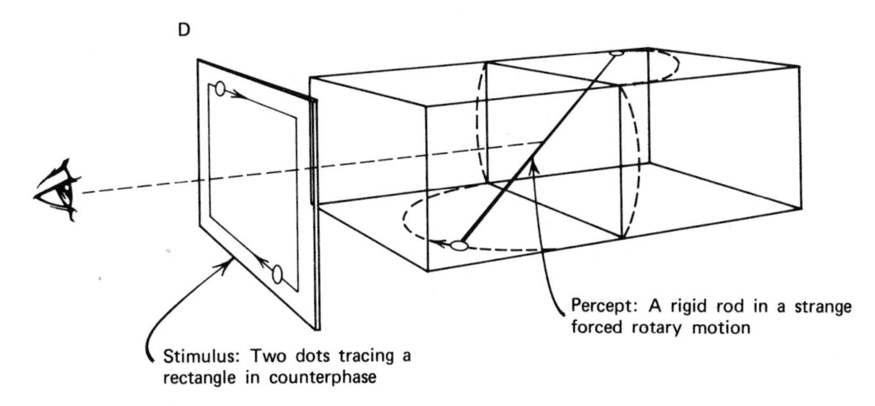

Figure 11-3 Examples of Johansson's (1974) demonstrations which show that two dots in coordinated motion along the same complex track evoke perception of a rigid unit ("a rod") that performs rotary motion in depth. A, control with only one spot; B, two dots trace a projection of a circular track on a slanted plane; C, projection of a combination of rotary and oscillatory motion; D, two dots trace a rectangle are perceived as a rigid motion in depth. (Reproduced from *Scientific American*, June 1975, with due permission.)

figure or a constellation of moving dots as a projection from a rigid structure, the perceptual apparatus responds with perceptual shape constancy.

Evidently constant shape plus motion in three-space is the highly preferred perceptual paradigm even under the most impoverished stimulus conditions, whereas seeing frontal figural deformation represents an artificial way of perceiving under these conditions.

Combined Size and Shape Constancy from Geometrically Complex Projections

From the point of view of kinetics most of the motions in our natural environment are described as combinations of translatory and rotary components. In the same way objects that are perceived as moving away from the observer in a curvilinear track or moving away and rotating at the same time represent cases of combined size and shape constancy. Good experimental evidence exists that under the most impoverished stimulus conditions projections from such more complex mechanical motions elicit perceptual object constancy. (Johansson, 1964, 1973, Börjesson & von Hofsten, 1973). We can also find a couple of examples on experiments that used dot patterns as stimuli which are especially convincing because of the complete lack of redundancy in the stimulus, and therefore I shall restrict my exemplification to them.

Börjesson and von Hofsten have studied the percepts evoked by three bright dots in various motion combinations over an invisible screen. They used a mini-

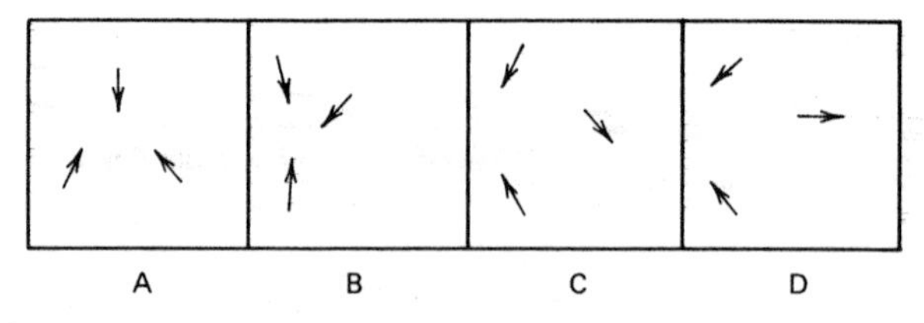

Figure 11-4 Combinations of motions of three dots (Börjesson & von Hofsten, 1973. (a) and (b) were generally perceived as form-constant triangles in simultaneous rotation and translation in depth, whereas (c) and (d) were most often seen as triangles rotating in depth and at the same time changing their shape. (After Börjesson & von Hofsten, 1973.)

computer and an optical device like Johansson's (1974c). Figure 11-4 illustrates some of the motion combinations studied. The general result was that their *Ss* reported seeing a triangle in complex motion in an empty space most often approximately rigid but sometimes also elastic. We may observe that geometrically a continuous motion of three points on a plane can always be described as projection from a rigid triangle freely moving in space. It seems as if the visual system in man is programmed for an interpretation of this kind and spontaneously produces perceptual object constancy even from the most scanty stimulus information when this is geometrically possible.

Our second example is found in an investigation of "biological motion," the complex motion pattern generated by the skeletal bones of man under various kinds of locomotion. Johansson (1973) attached small bright spots to the 10 main joints of a coworker and recorded with film and video technique the motion tracks of these spots seen against a totally dark background. 11-5, are unavoidably seen as representing a person walking, running, dancing etc. It seems as if the very complexity of the pattern is a highly favorable condition for perceptual efficiency even under nonredundant conditions of stimulus information. In a subsequent investigation Johansson (1976a) has shown that such a perceptual organization of a pattern of moving dots is a primary act. Exposure times down to a tenth of a second have proved to be sufficient for perceptual organization of the dot pattern, yielding a correct interpretation in fully naïve observers. The path lengths of the different dots under this time interval extends over only –1′–18′ of visual angle!

Perception of Constancy Under Bending and Deformation

In current texts on space perception the term object constancy implicitly denotes perception of *rigid* objects. In everyday perception of our environment, however, we are adequately dealing also with perceptual objects that are constant in far more limited meaning. As an example of such an object which

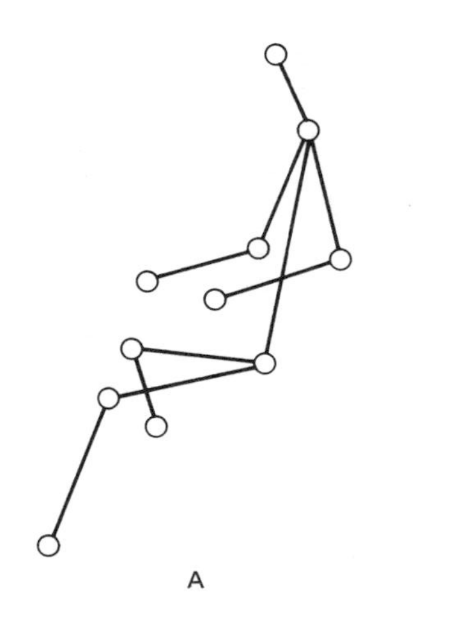

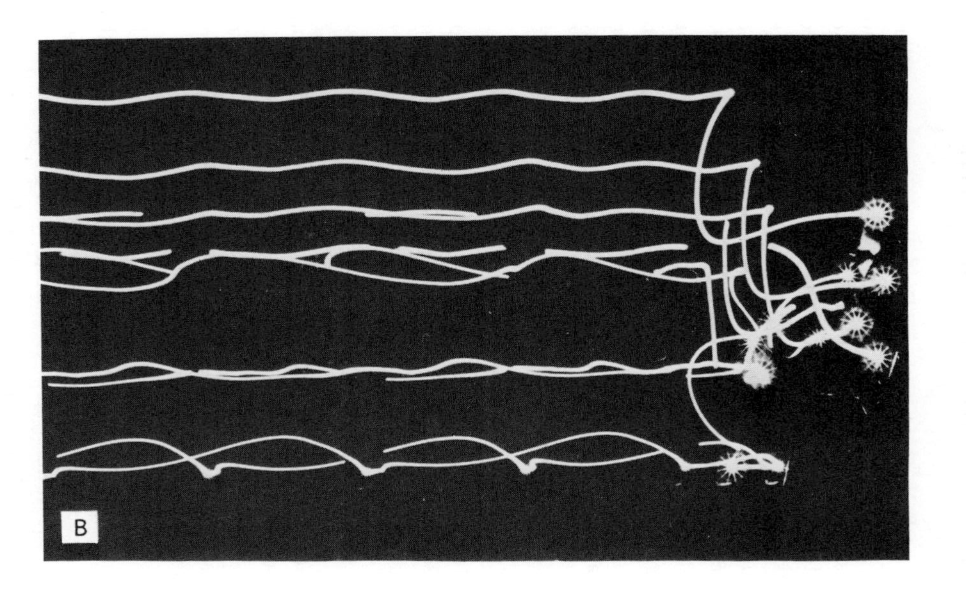

Figure 11-5 (a) The arrangement with light spots on the joints of a person sitting in a totally dark room; (b) The light tracts from this person recorded by an open camera when he stands up from a sitting position and starts walking. When the person is sitting motionless a meaningless constellation of bright spots is seen, but as soon as he starts moving an intricate combination of spot motions is immediately perceived as a person performing the activity described.

has permanent shape without being rigid in the ordinary meaning we can think of the pages of this book. A sheet of paper is flexible and bendable but cannot expand or contract. The same holds true for blades of grass and branches of trees We also perceive deformable permanent objects like a piece of dough or clay and elastic objects like rubber balls. Our experience also covers "objects" with sometimes limited permanence; for instance, wisps of smoke, flames of fire, and low drifting clouds. These examples of partial object constancy under change are of great interest from a theoretical point of view but they represent a field or research still characterized by a nearly complete lack of experimental investigation. There exists one investigation on the perception of bending together with a number of observations and data from experiments on motion and form change.

PERCEPTION OF BENDING

In a recent investigation Jansson and Johansson (1973) studied the perception of bending by changing a square in a systematic nonperspective way. The square was seen through a collimating device and was deformed in the ways shown in Figure 11-6.

When one of the corners was moving along the diagonal, as illustrated in A 29 of a total of 30 *Ss* reported seeing a flexible bending of a square surface. One *S* reported seeing an elastic shrinking-stretching. In the same way 26 of the 30 *Ss* saw the pattern in B as a bending passing the invisible diagonal line. Pattern C was finally seen as a bending by about two-thirds of the *Ss* and as a deformation of a frontal surface by the rest.

A study of Figure 11-6 shows that the various patterns of change can be regarded as approximations to projections from a flat surface, bending in the way described by the *Ss*. It is close at hand to interpret this outcome as indicating a strong tendency in the visual system to respond with the maximum of object constancy allowed by the principles of projection. In fact, the results from this investigation show that, opposite to the traditional assumption, frontoparallel change is in no way preferred, although such a change is easily seen under certain stimulus conditions.

These effects are demonstrated more informally in an educational film on motion perception from the Uppsala laboratory, edited by J. Maas.* In this film a.o. an initial outline square is submitted to a series of random but continuous, complex, and rapid changes, including twisting, deformation, and change of size. The film demonstrates in an impressive way how the most complex changes in a frontal quadrangle can yield a vivid perception of a moving object, the shape and size of which are approximately constant on most parts of the demonstration.

*James B. Maas, *Motion Perception II.* New York: Houghton Mifflin. 1, 1971.

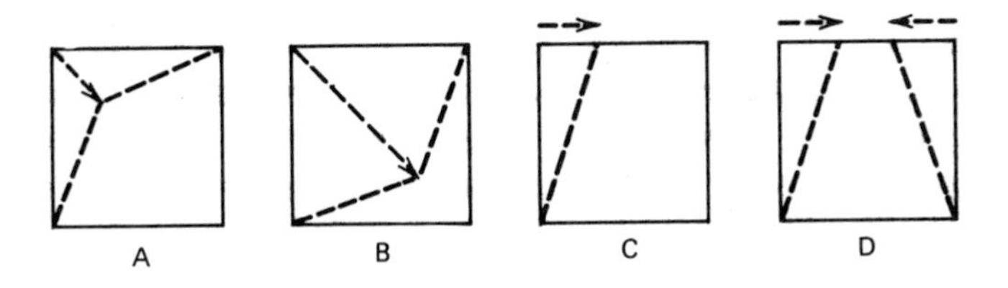

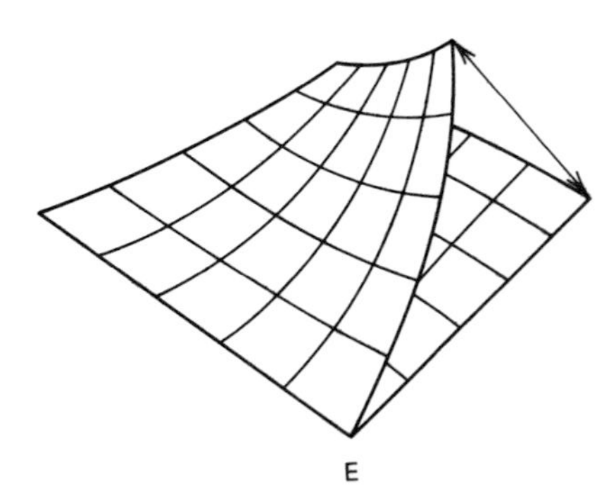

Figure 11-6 Examples of form changes of a frontal square (a to d) evoke perception of bending of a flexible surface; (e) illustrates the percept from (a), (a to d after Jansson & Johansson, 1973; e after Johansson, 1975.)

The object is seen as freely moving in three-space under bending and twisting. At some moments however, phases of surface deformation are also seen.

PERCEPTION OF DEFORMATION

In many experiments on motion perception under restricted cue conditions we find examples of perception of form change. It is interesting to note that in all these cases it is a frontal surface that is seen to change. Thus we meet with two-dimensional perception proper. It is never a surface slanting in three-dimensional space that is seen. This demonstrates a certain preference for the special case of frontoparallel orientation when other valid directional information is lacking.

Several of the experiments described as basic indications of the priority of perceptual object constancy also provide examples of perceived form change, albeit as nonpreferred alternatives. These often occur in cases in which the frontal screen carrying the stimulus pattern is visible and are most probably an effect of conflicting cues. However, also under better controlled conditions, we sometimes meet with two-dimensional form changes, which are of great interest because they can help us to a better understanding and a specification of the principles of perceptual stimulus decoding of changing patterns.

We may remember that Koffka in his experiment with a continously shrinking-growing circular area sometimes met with perception of a stationary disk changing its size. This result has never been confirmed by me under screen-free conditions. In the same way Ittelson, in similar experiments, always achieved perception of motions in depth. Most probably Koffka's result was in part an effect of his experimental conditions and there are good reasons to state that continuous size change is a perfect stimulus for perceiving rigid motion in depth.

Shrinking of a circle is characterized by a proportionally equal change in two dimensions. Introducing change of a figure in only one dimension has been found to often bring about perception of deformation of a frontal surface. See Wallach and O'Connell's experiments (p. 385) and Johansson's (p. 393).

Wallach and O'Connell even raised the hypothesis that simultaneous change in two dimensions of the proximal pattern is a necessary condition for perceiving rigid motion in depth. Braunstein (1972) took up this hypothesis to an experimental investigation, using only length changes in straight lines. He established that even under those conditions a majority of *S*s perceived motion in depth of lines with constant length. Especially efficient were changes simulating polar projections under large visual angles. We may also remember Metzger's and White and Mueser's results from one-dimensional changes in line patterns. Johansson (1964) also reports an experiment with one-dimensional change in a homogenous surface, which resulted in about 50% of motion in depth of an approximately constant surface. Perhaps Wallach and O'Connell's result was an effect of the mode of stimulus presentation on a visible screen, where conflicting information about a static frontoparallel surface dominated a comparatively weak tendency to perceive object contancy and motion in depth from one-dimensional changes. In summary, it appears that the visual apparatus is effectively programmed for transforming two-dimensional change in the stimulus to rigid motion in three dimensions but that its ability to go from an one-dimensional change to a three-dimensional percept is rather limited. The circle and ellipses are interesting exceptions to this principle, which is easily understandable from the point of view of projective geometry. Change of these figures in only one dimension regularly brings about perception of rotation of a circle about a diameter at right angles to the direction of change.

Eriksson together with some coworkers (Eriksson, 1970; Eriksson & von Hofsten, 1971; Eriksson & Hemström, 1973) has in a number of investigations studied perception of two-dimensional line or dot patterns under reduced stimulus conditions. These patterns were attached to a slanted plate which was slowly

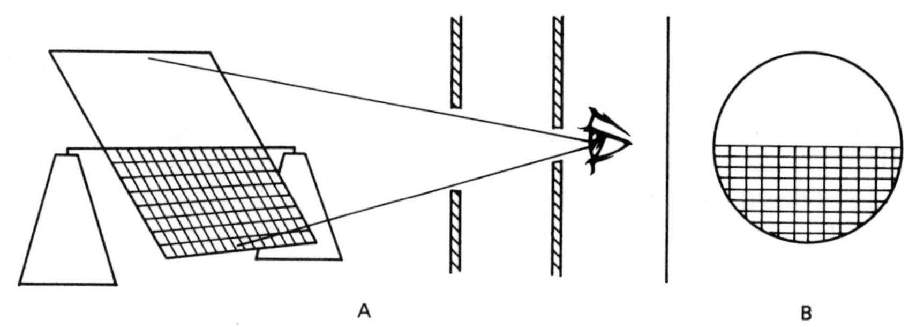

Figure 11-7 (a) Diagram of Eriksson's apparatus. (b) The apparance of stimuli seen through the reduction screen. (From Eriksson & von Hofsten, 1971.)

oscillating about its frontoparallel horizontal axis. The amplitude of oscillation was fixed (10° to 25°). For a diagram of the apparatus see Figure 11-7. In all but one experiment the oscillating stimulus pattern was seen through a hole in a reduction screen.

Eriksson's theoretical position in these papers is a programmatic empiristic one and the experiments were designed.for studying possible cognitive factors under these conditions of stimulus presentation. Therefore in most of the experimental designs data were obtained from *S*s initially naïve with regard to the real character of the distal stimulus and later from the same *S*s who were more sophisticated in this respect or who were under the influence of certain instructions. It is the percept descriptions from naïve *S*s that are of interest on our discussion. (For discussion of experience and cognition as explanatory constructs see p. 396 and p. 397).

Most of these spontaneous percept descriptions were in terms of elastic deformations of a frontal surface. In closely related experiments on projections from oscillation in depth of rigid distal structures Gibson and Gibson (1957) and Epstein (1965) received spontaneous descriptions of oscillating rigid objects from all *S*s. Thus it is evident that in Eriksson's experiments there must exist some special characteristics that effectively prevented seeing the patterns as constant objects in motion. It would be of theoretical interest to get these characteristics specified. Regrettably the experimenters systematically varied only the figural aspects of the oscillatory patterns but not the amplitude of oscillation or the perspective. This makes it hard to reach a conclusive interpretation. One possibility is that the outcome simply is an effect of the limited amplitude of oscillation together with the limited displacement information in the dot patterns. In these cases the proximal deviation from one-dimensional change could hypothetically have fallen below the perceptual threshold with consequences similar to those in Wallach and O'Connell's experiment. If so, the outcome of these experiments would be of limited interest to us.

Far more intriguing are the results from the experiments with line patterns like those illustrated in Figure 11-7, which were also seen as frontoparallel deformations. All these patterns were seen through a hole in the reduction screen. The results probably reveal that an induced static contour like the hole in the screen will be perceptually combined in a perceptual hole with the enclosed changing line or dot pattern. Such contour effects are well known from static configurations. If this interpretation is correct, it explains why the patterns were perceived as static deformations. Further research on effects of simultaneous but conflicting information from enclosed contours (static and moving) and enclosed changing patterns would be of considerable theoretical interest.

Also theoretical interesting is Börjesson and von Hofsten's (1973) finding that although three spots in rectilinear motion with more or less random directions and amplitudes are perceived most often as an approximately form constant triangle moving in space in certain combinations of element motions they are perceived as an elastic triangle. Evidently certain still unspecified motion constellations or figural factors in such dot patterns can prevent perception of a unitary object in motion.

ANALYSIS AND THEORY

Background

In the history of experimental psychology we meet with a number of theories about the mechanisms or processes that yield object constancy or, more generally, three-dimensionality in visual perception. We have paid some special attention to the traditional cue theory. In addition to or combined with this theory we meet some other theoretical paradigms in which still higher explanatory value is given to the organism's experience and cognitive activity for filling the gap between optical impingement and perception. As examples we mention Brunswik's ecological anchorage of the perceptual constancies and the transactionists appeal to experience as a necessary constituent of perceptual constancy.

Our review of experiments on three-dimensional motion perception, together with our comments, has made clear that the classical reference to secondary cues for explaining visual perception of a three-dimensional space has no relevance when it is a question of kinetic object constancy. We also find that explanations of these constancy effects in terms of experience and cognitive factors are superfluous and sometimes irrelevant. In a majority of the experiments described an explanation of this type must be regarded as highly artificial to say the least and in some of them it would clearly contradict the actual results. An impressive example of the inadequacy of the empiristic or cognitive account is the outcome of the experiment in which two bright spots trace a rectangular path. Common human experience and cognitive interpretation of this motion pattern would in all likelihood bring about a perception of just two spots moving in a common closed rectangular path. This is always the percept, *mutatis mutandis*, received when only one of the dots is visible.

It is not mere coincidence that the observations we have described do not lend themselves to interpretation in terms of cognitive factors. It is the effect of a deliberate choice of experimental material. The steering principle behind this choice has been to search for experiments in which the stimulus information about three-dimensionality and object constancy is drastically reduced but in which the percepts still reveal object constancy. This is just an application of the principle of parsimony. What can be excluded without destroying the perceptual effect studied cannot be a primary source of the effect. With this material in our hands and thus rejecting explanations mainly derived from studies in static space perception it is now one of the primary tasks for our analysis to abstract factors in the impoverished stimulus events that prove to be essential for the perception of kinetic space constancy.

In the traditional discussion of visual space perception we often find a strange and unhappy tendency to theoretical polarization: innate mechanisms *or* learning. Even in the contemporary debate we sometimes meet trends of this oversimplification or singlemindedness. In my theoretical treatment I shall be dealing with the physiological background to perception of object constancy and with perceptual mechanisms that are automatic in character. Some readers perhaps will regard this as a sign of such singlemindedness. Therefore I should like to take particular care in stating my theoretical position in this respect. In addition to the fundamental basis given by the anatomy of the eye and the neural structure of the visual system, modifications from adaptive mechanisms, perceptual learning, and cognitive attitudes in some situations can strongly influence the actual perception of space. This is experimentally verified in a large number of experiments. As an example related to the problem of kinetic space constancy we can recall numerous experiments with distorting glasses. Furthermore, it is legitimate to stress that it is hard to deal with the problems of veridical perception in regard to size, shape, and absolute distance without going into the problems of such adapting and modifying mechanisms. Also, if this aspect falls outside our central theme, I hope that my readers will find that the relativistic theory, advanced in the following pages, presupposes adapting mechanisms.

An Image Model or a Flow Model?

Our first task is to specify a model for description of kinetic visual stimulation that is as efficient as possible for our present purpose. In the history of vision research we can discern two different categories or types of model for the description of proximal stimulus. For the moment, neglecting some differences within these two categories, I term them *the image* and *the flow model.*

THE IMAGE MODEL

The image model is the traditional one and too well known to need a more thorough description. It was outlined clearly as early as in Berkeley's famous "A new theory of vision" (1709) but its history is older. This type of model is still by far the most common one today and is generally accepted in the framework of the cue theory. In fact most of our present knowledge about visual space perception has been gathered within this theoretical framework. The image model is founded on the fact that the optics of the eye focuses the light onto the retina and thus can project a sharp image. It is this "retinal image" that in the image model has been accepted as an appropriate description of the proximal stimulus. A background knowledge of the basic principles of image projection through a positive lens, together with knowledge of the orientation of the distal object in relation to the eye, makes it possible to specify the proximal stimulus in accordance with this model.

As mentioned above the image model is the commonly accepted one. It is in fact so dominant that it is often regarded and treated as a factual statement rather than as a constructed model and simplification, perhaps appropriate only for a certain class of studies. In our present context it is of special interest to observe that questions about the spatial constancies were initially formulated with this model as a background and generally still are. The questions in this respect were roughly of the following type: how can different retinal images evoke similar percepts?

A serious limitation of the image model is that the dimension of time is not represented and consequently changes over time are hard to deal with in this framework. This limitation may be tolerable when it is a question of studies of static percepts. In the domain of motion perception, however it has brought about many theoretical difficulties. Many of the early students of visual perception were lead to think about motion perception as the outcome of some kind of higher mental activity resulting in an integration of a hypothetical series of static images. Typical in this respect is Wundt (e.g., see Wundt, 1910, p.611) and also Titchener.

THE FLOW MODEL TYPE

The difficulty in dealing with motion perception within the framework of the image model rather early inspired the development of more or less advanced flow models. In these models the optical change over time rather than the optical image was chosen as the basis for description of proximal stimulus. Nowadays we know that there never exists anything like a sharp image on the retinae and also that some slight amount of optical change is a necessary condition for static perception (e.g. Riggs et al., 1953). Furthermore, from a physical point of view the optical stimulus in everyday perception essentially has the character of a flow. This is due to various kinds of motion of the eye in relation to the environment: motions of the whole body, head movements, and eye movements as well as motions of objects in the environment. In primitive organisms motion seems to be a necessary condition for perceptual response. (For further discussion see Johansson, 1975). Let us pinpoint some main contributions in the development of the flow model.

Ernst Mach, the Pioneer. Ernst Mach (1838-1916), the famous physicist, psychologist and philosopher, was a highly influential pioneer in the structuring of the flow model. In his studies on motion perception he regarded optical change *per se* as the primary stimulus for perceived motion, equally primary for the sensory system as the stimuli for brightness and colour. Therefore he coined the term "Bewegungsempfindung" (motion sensation) for the visual response to moving stimuli. Mach's position was probably a consequence of several interacting factors. Mach was a physicist and a mathematician by profession and therefore accustomed to dealing with spatial change over time. He specialized early in the perception of motion and carried out thorough and highly important investigations on perception of rotation, stemming from the vestibular systems as well as from the visual one. Being of the same pioneer generation as Wundt and Hering, he was relatively free from the pressure of earlier schools and traditions.

The Gestaltists. In their treatment of visual perception the Gestaltists generally applied a model of the image type. This is evident from their treatment of Gestalt laws (e.g., Wertheimer, 1923; Koffka, 1935) and their approach to the problems of spatial constancies (Koffka, 1935). However, the Gestaltists have also made important contributions to the study of motion perception and, as is well known, the Gestalt school was born from an investigation on stroboscopic motion. In this branch of research the Gestaltists follow the line from Mach and deliberately regard the optical change *per se* as the effective stimulus quality. We may remember Wertheimer's hypothetical physiological explanation of perception of stroboscopic motion as an outcome of a neural shortcircuiting process (Wertheimer, 1912), and Koffka has in a pronounced way declared his position as belonging to the Mach group (Koffka, 1931).

J. J. Gibson's Contributions. More than any other investigator since Mach, J. J. Gibson deserves credit for fashioning the flow model into a useful tool. I regard his consequent endeavor in this respect as one of his most essential contributions to the theory of visual perception and he has profoundly influenced my own thinking in this respect. In Gibson's work we meet for the first time with a coherent analysis of visual information under locomotion. Already in his book *The Perception of the Visual World* (1950) Gibson discussed the visual stimulus for perception of locomotion as a continuous flow of radiant energy over the retina (pp. 118-144). In this approach Gibson introduces a new, highly fruitful method of analysis. The analysis of the stimuli for locomotion was further elaborated and given in strict mathematical form in Gibson, Olum, and Rosenblatt (1955).

Gibson's flow model probably grew from his research on aviation problems in connection with aircraft landing during World War II and it has formed the theoretical background for a number of applied studies on aviation and in car driving (Calvert, 1954; Crawford and Briggs, 1962; Havron, 1962; Gordon, 1963, 1965; Llewellyn, 1971). In Gibson (1957) a flow model was introduced into the framework of projective geometry by proposing an analysis of proximal stimulus from moving stimuli in terms of continuous perspective transformations. This approach was further advanced in Gibson and Gibson (1957) in which the two Gibsons presented an experimental study and theoretical analysis of the stimulus for rigid rotary motion. The stimuli were described in terms of continuous perspective transformations in the retinal projection and in this way the Gibson flow model (1950) was elaborated in a significant way. Later Hay (1966) further expanded Gibson's analysis and Lee (1974) gave an elegant and simple reformulation of Gibson's flow model by applying cylindrical coordinates. In the later part of this chapter we shall find, of decisive importance, Gibson's proposal that the proximal stimulus in vision be considered in terms of continuous perspective transformation in the optical flow.

Johansson's Vector Model. Following the trend of Mach, but also of Karl Duncker and Edgar Rubin, Johansson (1950) structured a flow model by applying vector analysis to motion perception. The proximal stimulus was described in terms of optical motion vectors rather than Euclidean images, and the vector field was analyzed in terms of relative and common vector components.

During the last decades flow models have been applied by many students of motion perception. As typical examples we can mention Green (1961), von Fieandt and Gibson (1959), von Fieandt (1961), Braunstein (1972), Johansson (1958, 1964, 1973, 1974a-c, 1975), Gibson (1966), Börjesson and von Hofsten (1972, 1973).

Application of a flow model rather than the traditional image model in the stimulus description is crucial to a better understanding of kinetic space constancy. In the following we search for correlations between the optical flow generated by the object motion and corresponding percept. Our problem is, what invariance exists in the stimulus flow that corresponds to the shape and/or size invariance in the perceived object?

Choice of Geometry for Stimulus Analysis

Traditionally the description of the proximal stimulus in the study of spatial constancies is given (a) in terms of images projected onto the retina and (b) by means of an Euclidean analysis of these images. In this way the strong Euclidean demands for congruence have been the determining criteria for characterizing abstracted "images" as equal or unequal.

Perfectly equal figures according to Euclidean geometry mean congruent figures—and only those—and this definition of equality has formed the basis for the problems of spatial constancies. Two retinal images projected from the same distal object but not fulfilling the Euclidean demands for figural invariance because of differences in absolute size and/or shape are consequently classified as different. When two such noncongruent projections are perceived as the same object at different distances and/or in different orientation, we say that we are dealing with the problem of size or form constancy.

Let us now be unconventional enough to ask for the rationale behind this traditional approach. First, is Euclidean congruence really the only possible model for the study of proximal projections and their relations? The answer to this question is easy to give. It is no. We know that in the geometry of today the Euclidean system is just one among a number of geometric systems with different sets of axioms. For our purpose it is expedient to pay attention to Felix Klein's famous "Erlangen Program" of 1872. In this program Klein introduces a classification of the geometry in groups in accordance with the principles of figural invariance under transformation. In this context a geometry is defined as "the study of those properties of figures which are not affected by transformations belonging to a particular group." The Euclidean (or congruency) geometry forms one pole in the hierarchical series of groups. On a lower level we find the affine geometry. This is the geometry of parallel projections, which projects parallel lines to parallel lines and preserves the ratio of collinear and parallel line segments but not their absolute lengths. Still fewer invariants characterize the next important level, the projective geometry, which is the geometry of the central or polar projection. In this geometry the parallel axiom is rejected and the affine plane is augmented with points and lines at infinity. The topology represents a still "lower" level.

Our next question, which is a consequence of the answer to the first one, has far more the character of a real problem. It says: does Euclidean congruency really afford an adequate framework for the study of spatial constancy in visual perception in general but particularly for constancy effects in motion perception or can geometry offer better alternatives?

Two basic facts imply that the Euclidean metric with its restrictive conditions for figural invariance under transformation is badly suited to the study of perceptual spatial invariance under stimulus change. The first, and from a theoretical point of view the most important, is that is has been considered as something like a general principle for sensory functioning, underlying all types of perceptual comparison, that it is *relations* and not absolute measures that are compared and reacted to. The second contraindication is represented by the findings that brought about the problem of spatial constancy: the fact that the percept hardly ever corresponds to the Euclidean description of the proximal stimulus. (It is only under frontoparallel motion that we get a correspondence between perceptive and Euclidean equality.)

As we have already observed, the tradition of visual stimulus description was established long before the explicit formulation of the visual constancy problems and this determined the setting of these problems. Now that we propose to start anew in the search for an adequate geometric model for stimulus analysis it is essential to maintain some distance from the traditional type. For this purpose the following viewpoints may be of some help.

Certainly Euclidean geometry is one of the most magnificient mathematical tools ever constructed by man. Still in Berkeley's time it was considered as a set of laws of nature. Berkeley knew of no other type of geometry, and therefore it is understandable that his theory presupposes that the visual apparatus in its stimulus analysis must work in accordance with this set of laws. In this way, and because of the brilliance of Berkeley's theory and his longlasting authority, we received as a cultural inheritance the Euclidean description of retinal images (which forms the basis of today's cue theory). Many excellent theorists still work as if there were no alternatives to this outdated problem setting and this is disturbing, especially in the study of kinetic space perception.

As we try to free ourselves from this tradition and to make our choice of geometric model on more rational considerations, the following criteria seem to be of special importance.

(1). The geometric system chosen must be well suited for comparison of spatial relations.

(2). Its compatibility with the anatomical and physiological structure of the visual systems must be taken into consideration.

(3). Experimentally found perceptual invariance ought, if possible, to have some kind of counterpart in the geometric principles of invariance, characteristic of the system.

Our criticism of the Euclidean congruency model was founded on the fact that it seldom describes the percept. In its turn the rationale behind this criticism is the opinion that, in the study of perception, the mathematical description must be preferred which best and most directly covers the percepts reported in experiments. When we look at the experiments that use a shadow-caster technique, as reviewed above, we find a common denominator between percept description and a special type of geometric description. The percepts evoked by these stimulus events generally represent rigid or semirigid objects and, because of the experimental technique, the optical projections can always be described geometrically as continuous perspective transformations. All this material taken together indicates that Gibson (1957) pointed to a promising way for us when he called attention to perspective transformations in the proximal stimulus, hypothesizing them as representative stimulus correlates of critical importance for perception of objects in three-dimensional motion.

In our search for an alternative to the Euclidean description of stimulus we ought to consider and seriously investigate the possibility of accepting the geometry that deals with perspectivity, thus projective geometry, as a possible alternative.

Strong arguments for accepting the principles of projective geometry as an alternative model to the Euclidean is in my opinion the fact that optical projection into the eye obeys the laws of the projective geometry and that perspective transformations generate perceptual shape and size constancy.

In order to avoid misunderstandings perhaps it ought to be underlined from the beginning that accepting the hypothesis that the visual system in stimulus analysis works in accordance with some relativity principles from projective geometry in no way implies that a Euclidean analysis of these relations, for example, in terms of trigonometry, should not be applied in research on space and motion perception. This is a question of choice of technique and not of model. As made clear in the following, it is the analogy between projective invariance under perspective transformation and perceptual constancy from changing optical stimulation that makes the principles of central projection, as established in projective geometry, an appropriate model for visual space perception. The technical methods for demonstrating this invariance present another question. This computation can be done appropriately with applications of trigonometry and analytical Euclidean geometry. We may remember, however, that projective geometry has also developed a system for relative coordinates, homogeneous coordinates, which specify directions rather than absolute localization, thus affording an elegant substitute for absolute cartesian coordinates.

A classical and well-known application of quantitative methods for constructing projective relations is found in the techniques of perspective drawing with the establishment of vanishing points (projective points at infinity) and horizons (lines at infinity). More closely related to the study of kinetic constancies are the on-line applications of trigonometry in computer science for the purpose of simulating object motion or locomotion by means of displays in which continuous perspective transformations represent these objects or environments. This fascinating technique is already in use in advanced architectural planning and in the training of space pilots. In short, we can apply various quantitative mathematical techniques to the establishment of projective invariance and therefore it is essential to distinguish between choice of model and techniques for investigating the validity of the model.

Further reasons for accepting projective geometry as a model for visual stimulus analysis and arguments for including parallel projection in the paradigm are given in some of the following sections. We now proceed with a brief characterization of projective geometry as a possible basic model.

PROJECTIVE GEOMETRY AS AN ALTERNATIVE

Projective geometry is a young branch of mathematics. It was developed as an independent system during the nineteenth century and has as its basis the principle of central projection, which is the principle behind image generation by means of a pinhole or positive lens. It has in fact developed from a scientific study of perspective drawing and image projections through lenses.

Projective geometry has its own set of axioms that does not include the parallel axiom. While Euclidean geometry is a system for an absolute metric, dealing with measurements of length of line and size of angle projective geometry deals exclusively with spatial *relations.* It compares relations between distances and angles, not their absolute measures. In central projection as in vision parallel lines intersect at a point on the horizon of their plane (the "line at infinity"). Consequently, when it is a question of comparing figures, projectively equivalent, with invariant projective relations. Any rigid figure in the figure but will attach no special meaning to invariant absolute size and shape.

Because of its metric the Euclidean geometry accepts transformations only in form of the two types of rigid motion: translation and rotation. Only under these transformations will a figure retain its shape and size so that it can be superimposed on a congruent one. Under central projection, in contrast, the shape of a figure can change and still the various projections will remain projectively equivalent, with invariant projective relations, for any rigid figure in motion in relation to the center of perspectivity, the station point, in a three-dimensional projective system always generates a set of equivalent projections on the picture plane. This characteristic of projective geometry depends on the fact that there a number of figural properties of relations still exist which stay invariant under such transformation. Straightness in a line is such a property; the property of a tangent to a curve remaining as a tangent under projective transformation and the remarkable property of a conic to result in a conic and of a cross ratio to stay invariant under central projection are other examples.

Therefore a set of perspective transformations generated as shadows on a screen by a shadow-caster in motion is always projectively equivalent. Just as in Euclidean geometry a figure can be means of rigid motion be brought to cover a congruent one, so can a figure by a perspective transformation be brought to cover an equivalent one. Thus perspective transformations on a picture plane can be thought of as a projective equivalent to Euclidean congruence and rigid motions in our environment. It is this parallelism that makes projective geometry such an interesting alternative to the Euclidean model for stimulus analysis in the study of proximal stimulus-to-percept relations subsumed under the term spatial constancy perception. In the traditional formulation of these problems it is implicity assumed that the visual system abstracts and evaluates absolute measures in the stimulus flow. This is an a priori assumption not supported by facts. Instead, let us assume that it is projectively invariant relations that are spontaneously abstracted. Much of the work reviewed in the earlier sections lends empirical support to this latter hypothesis. Even the cue theory admits that this relational aspect plays some role. Linear perspective is accepted as a secondary cue for depth and this is certainly a description of proximal image in terms of projective invariance. Acceptance of the relativity model, however, means that we in fact must disclaim the traditional setting of the problems of spatial constancies. I think many readers after all will regard this as a rather hazardous step. Therefore we may leave the decision open until we have far more thoroughly scrutinized the arguments for a shift to the relativity model. I have in some papers explicitly argued for a geometric model of the projective type (Johansson, 1973, 1974a, 1974c, 1975, 1976b). In the following I will develop these arguments further.

Evolution and Visual Geometry

My first argument in favor of the projective model is taken from well-known facts about the anatomy of the visual organs. It is a fascinating fact that at all stages of animal life straight through the evolution of species there are organisms that are sensitive to light and make use of this sensitivity for receiving directional information about their environments and events in them. Already among the protozoa we meet a diffuse light sensitivity demonstrated by selective directional responses in form of motion. At somewhat higher levels we find animals with light-sensitive neurons on the surface of their bodies. This arrangement provides crude directional information of type above-below. Still a little higher in the series we find that neurons responding to light are sunk in a depression, or a little chamber, on the animal's surface. This anatomical structure is regarded as an important step toward development of the more advanced visual organs found for example in vertebrates. It introduced the principle of screening that made possible, in a successively more and more efficient way, simultaneous recording of narrow pencils of light reflected from structures in the environment, stationary or moving.

Technically we know this principle for the *camera obscura* or the pinhole camera. Good examples of pinhole camera eyes are found among the mollusks. In terms of geometry the optical pinhole principle represents an application of the principle of central projection with its one-to-one mapping from one plane to another via straight lines, or "rays", crossing in a distinct point not on either plane. The principle of central projection once applied in the first crude forms of eyes became decisive for the development of the eyes of human type thus the vertebrate eyes and for the other major advanced type, the arthropod.

The rather wide opening for light entrance in the first primitive eyes of the chambered retina type of course is far from the ideal pinhole. As compared with the neural system of the vertebrate eyes, the neural organization of the light-sensitive inner surface of the chamber is rather uncomplicated. Nevertheless, it is functionally efficient. However, we must hypothesize that parallel with the successive geometric-optic refinement of the visual organs there developed a more and more intricate neural network which brought about a signal processing on a comparable level of sophistication. Without this instrument for data processing, which was developing in pace with the optical system, the optical advances were meaningless and thus in conflict with a basic principle in the biological evolution.

In an important respect the pinhole principle has a serious drawback. When the size of the hole approaches the character of a real pinhole (which is necessary for improving the resolving power of the eye), the amount of light allowed to enter becomes too small for exciting even highly sensitive receptors. A positive lens at a focal distance from the sensitive area was the evolutionary solution to this dilemma, and early in the evolutionary process we meet organisms with a lens device. A positive lens instead of a pinhole represents another application of the principle of central projection, geometrically more complex but optically more efficient because of the enormous gain in the amount of light that hit the retina. Adhering to the principle of central projection as a basis for neural processing means that we meet basically the same principle for data processing from the first crude forms of visual organs up to the fully developed vertebrate eye. Evidently nature found early but efficient methods of neural processing of directional optical information and has never discarded it. Also the arthropod eye is an instrument for recording ray directions, although the optical solution differs from that found in the vertebrate eye. For further related viewpoints on the evolution of vision see Gibson (1966).

In summary, we find that organisms from the most primitive stages of evolution of visual organs have been able to respond to the direction of light and in more or less advanced form utilized principles of central projection. Therefore it is highly probable that also the neural network developed for recording the incoming optical information is efficiently constructed for handling changing central projections. From the point of view of spatial information it is not the projected image as an image that is of special importance but instead the directional information available at the geometrical point of ray crossing, the nodal point in the optical system.

With this background it would hardly be rational in the study of the evolution of the visual systems to stick to a two-step type of theory for a geometrical analysis such as the cue theory with its constructs concerning an inadequate primary data treatment followed by secondary correction to achieve three-dimensionality and object constancy. Here Occam's razor certainly is needed. The primary optical information, as developed above, is of directional character and specifies relations between changes of directions. Therefore it is logical to think about the neural network in primitive visual systems as initially adapted to a direct and straightforward abstraction of certain constant relations in the optical flow.

It would also be strange from a biological point of view to hypothetize that in man (and perhaps some other vertebrates) a principle for optical analysis, found effective in more primitive species, for some reason had been scrapped and replaced by a totally different and less efficient one. I like to think of the visual system in man as working on the same principles as those of the lower species but that because of the advanced cortical organization in man his system is far more flexible and adaptable.

Experimental Support for a Relational Model

In order to get a stable basis, let us stress a fact already alluded to. A complete set of changing shadow images projected from a moving shadow-caster under the experimental conditions described in the first part of this chapter represents a family of perspective transformations. This means that all members of the family are projectively equivalent because, as noted above, the projective properties stay invariant under a perspective transformation. We found as a general result of all these shadow-caster experiments that the continuously changing shadow patterns brought about perception of objects with constant shape, moving in three-dimensional space. This was found to hold true even when all spatial information other than change of shape itself was eliminated and when under static conditions a frontoparallel figure was seen. Combining these two facts, the geometrical characteristics of the stimulus and the geometrical characteristics of the percept in a series of experiments brought about as a first tentative conclusion the statement that any continuous sequence of perspective transformations projected onto the retina of the human eye evokes perception of rigid motion, that is, motion of an object with constant shape.

As a next step we examine the experiments that use as stimuli a few bright dots in motion. We have there no physical contours that can determine the perception of shape geometrically and therefore with this type of stimuli we have a good chance to get in touch with some organizing principles built into the visual system.

Let us look first at some of my own experiments especially designed for this purpose, the experiments in which two dots trace a conic section (Johansson, 1974c). The notable result of these experiments is that in all the motion combinations studied the perceptual distance between the dots stayed constant and the continuous geometric change of distance between them on the picture plane was perceptually represented by a motion in depth.

This perceptual effect must be a strong manifestation of a basic principle in the perceptual decoding of changing patterns because all information provided by contours is lacking. When, in spite of this lack of information, fixed end points on an imaginary line are perceived, must this be regarded as an effect of a sensory intergration. Evidently, projective invariance in the spatial relation between the two dots is continuously abstracted and spontaneously yields for each moment of time information about a specific direction in three-space of a line with constant length. In the experiments in which the two dots trace an ellipse a plane of rotation with constant slant is perceived and this slant has been found to represent a good approximation to the simulated slant. The same effect was also found when more complex motion paths were generated. Patterns with combinations of conic sections and with two dots tracing rectangles were always seen as complex rotation patterns of a rod of constant length in three-dimensional motion.

It hardly seems possible to interpret these experimental results in the classical way as a primary perception of the frontal motions of two dots corrected by some kind of secondary interpretation stemming from experience and cognitive processes. Instead, the most plausible interpretation is that the neural part of the visual apparatus is constructed for a direct and automatic analysis of the optical message in a way analogous to filtering out projective invariance. This conclusion received additional support from the experiments with three spots in simultaneous motion (Börjesson & von Hofsten, 1973) together with a number of unpublished experiments of my own. From a geometric point of view three spots in simultaneous and continuous motion on a frontal screen can always represent a central projection of a triangle which is moving in space and the apexes of which are represented by the spots. In fact, this is also the typical percept from such a combination of moving dots. Thus again the percept represents an approximately correct projective interpretation of the stimulus pattern.

A noteworthy aspect of this sensory achievement is perhaps the fact that under most impoverished conditions of stimulus information the slant (or rather the direction in space) of the plane on which the rod or triangle is seen moving at a given moment is determined with such high certaintly and geometric perfection. In such experiments the produced value of this angle was a good approximation of the computed one. In Johansson (1973, 1974a) this is tentatively explained as a possible effect of determining the points of concurrency of pencils of projected parallel lines, thus specifying the projected horizon of the plane.

Further Elaboration of the Relational Model

A common theme in this chapter is the search for geometric-mathematical analogs to kinetic constancies. So far we have found that a description of the proximal stimulus in accordance with the principles of perspective transformation under central projection with their anchorage in the relational invariances of projective geometry affords a good mathematical correlate to these constancies. Consequently I have proposed to exchange the traditional Euclidean model for a projective one. In this situation, however, some further clarification is needed. Many observations have shown what could be expected a priori: although the visual system works according to certain principles of projective geometry, it by no means does so in any orthodox way. The analogy is found essentially in the abstraction of invariant relations in the stimulus flux. Let us look a little closer into this. We found rather decisive arguments in favor of a projective model in the review of experiments that dealt with continuously changing projections. In some of these experiments, however, we also found a somewhat puzzling trend: these results are a little too good for the theory. This trend appears in some of the experiments on rotation. There the visual system responds with form constancy in spite of the fact that from the point of view of central projection adequate information about invariance is lacking. This holds true for Metzger's experiment (1934a, 1934b) with moving rod shadows and for Wallach and O'Connell's (1953) experiments on "the kinetic depth effect," in which approximately parallel projection was used. Even Braunstein's (1975) experiments with one-dimensional change of a line under perfect parallel projection demonstrate the same effect (cf. Fisichelli, 1946; Johansson, 1964; Braunstein, 1966; Hershberger, Carpenter, Starzec, & Laughlin, 1974).

These examples, taken together, clearly show that changes in accordance with central projection is not a necessary condition for perception of rotation in depth. This perception is readily elicited from continuous transformations under *parallel* projection. Therefore a relational model bound exclusively to the principles of central projection would be too restrictive. The model must be extended to encompass the principles of parallel projection. This statement, concerns only the perception of rotation. For perception of translatory motion in depth transformation under central projection is a necessary condition. This is perhaps a rather trivial statement because under these condition parallel projection would imply constant size. In sagittal motion the stimulus would not change at all, and for translatory motion in other directions the stimulus would geometrically specify frontoparallel motion.

Sometimes geometers treat parallel projection as a special limiting case of projective geometry, namely, as a central projection with its center (the station point) at infinity (for discussion see Felix Klein, 1925). This generalization seemingley has little relevance to our discussion. The argument is that it would imply that displays with transformation under parallel projection would represent extremely distant objects, but this is never the perceptual effect. Usually the objects are seen rather close to the observer, 1 to 3 m. However, some other characteristics of visual decoding of perspective transformation discussed below, make it relevant to accept this generalization and to hypothesize that when it is a question of rotation the visual system will treat parallel and central projection in accordance with a generalized paradigm.

Before arguing further along this line I must remind the reader that the parallel projections form their own transformation group—*the affine geometry.* In contrast to projective geometry affine geometry lacks the points and lines at infinity, but like the projective geometry it deals exclusively with relations. As a link between the orthogonal (Euclidean) and the projective geometry, however, it is characterized by a higher number of invariant relations under transformation than the projective geometry; for example, parallel lines stay parallel under parallel projection and the ratio between points on a line is also invariant (the cross-ratio is invariant under central projection).

Applying these facts in our discussion leads us to assume that the visual system takes advantage of the higher number of invariant relations under continuous parallel transformation and consequently interprets them as representing figural constancy under rotary motion. In this way it is comprehensible that parallel projection brings about perception of object constancy under rotation in depth. In another well-known respect, however, parallel projection (as well as central projection under small angles) is perceptually unspecific. Although it informs about three-dimensional form constancy and rotation in depth, it leaves perceived direction of rotation ambiguous and the perspective reversible. It is only under rather large angles of central projection (simulated or real) that this ambiguity disappears. It is of interest to observe that the natural environment of man and other terrestrial animals is filled with examples of parallel projections. The sun and the moon are at optical infinity and consequently all shadows and pencils of light under sunshine and moonlight represent perfect examples of parallel projections.

We observed earlier that, as opposed to theoretical expectations, parallel projections under impoverished experimental conditions could bring about perception of objects moving rather close to the observer. This implies that perspective, at least under these conditions, does not determine perceived distance to the object. It is easily found from daily experience that under full cue conditions the visual system is tolerant in regard to geometrically inadequate perspective. Consider first the perception of the static perspective in paintings, drawings, and photographs. In these pictures the perspective is fixed and does not change with the motion of the observer, in spite of which the perspective is considered adequate over a wide range of viewing distances picture-observer. Geometrically there is just one localization of the observer that is projectively adequate. This point in space is determined by the drawer's construction of the perspective or on the refracting power of the camera lens. Still more relevant in the present context are the analogous effects encountered when we look at movies or television pictures. Seen from all but one distance, the moving objects should appear as undergoing more or less pronounced deformation during motion. It is only in extreme wide angle presentation or photographing, however, that we actually percieve this deformation. Again we meet a high degree of tolerance with respect to distance in simulated perspectives, paired with rather perfect information about form constancy and motion from the same perspective transformation.

Finally we return to the outcome of some of the reviewed experiments on perceived translation in depth from shrinking or growing frontal patterns. As the reader will remember, in some of these experiments *S*s reported seeing a slight change of size in the moving object. Marmolin's systematic and thorough investigation of this side effect is especially informative because he was able to separate in a quantitative way the component of size change from the component of motion in depth. If it holds true, as assumed above, that the perceived distance from observer to object in such situations is determined partly by factors other than the simulated perspective, we can probably expect to meet something like a cue conflict in these experiments. Hypothetically the perceived elasticity in these experiments stems from a conflict between the perspective transformation and extraneous distance information like oculomotor distance registering and the specific distance tendency (Gogel, 1969). In my experience the latter effect seems to have a strong influence on perceived distance in such experiments when convergence and accommodation have been well controlled. The importance of oculomotor registering of distance for object constancy during motion has been clearly demonstrated by Wallach and his coworkers in many ingenious investigations in this field. Modification of stereoscopic depth perception by means of the telesteroscope and by glasses that simultaneously change both convergence and accommodation resulted perceptually in initial elasticity of rotating objects and rapid (< 10 minute) adaptation toward shape constancy (e.g., see Wallach, Moore, & Davidson, 1963; Wallach, Frey, & Bode, 1972; Wallach & Frey, 1972). Wallach and his coworkers were able to establish that the adaptive processes took place mainly in the oculomotor distance recording.

Visual Vector Analysis of the Stimulus Flow

Accepting projective invariances as fundamental determinants for perceptual object constancy under kinetic conditions implies accepting a model for space perception founded on abstraction of invariant spatial *relations*. This is an important step toward a better understanding of the kinetic constancies but not a sufficient one. As we have found, it can nicely explain the perception of constant objects under motion in the experiments with shadow-casters, but it does not cover the outcome from the experiments with perceived combination of form change and rigid motion (Johansson, 1964) or the experiments on bending (Jansson & Johansson, 1973) and the experiments on biological motion (Johansson, 1973) as reviewed above. In fact, accepting projective geometry as a model is not sufficient for a better understanding of visual decoding of changing proximal stimuli. It must be regarded mainly as an important first step toward a more comprehensive theory of relativity in vision.

The second major component in this theory is the perceptual vector analysis of the stimulus flow. This principle indicates that the proximal flow on the receptive surface is often sensorially analyzed in a number of independent flow components. Let us go a little deeper into this. By definition motion is displacement in a given reference system. Only when a perceptual theory programmatically considers motion in relation to moving reference systems in hierarchiacal series will it afford a satisfactory theoretical framework for the study of kinetic spatial constancies or, in the long run, for all aspects of visual space perception.

The study of relative motion in visual perception was introduced by Rubin (1927) and Duncker (1929) who coined the term "Bewegungsanalyse" (motion analysis). This study was further extended by Johansson (1950, 1958, 1964, 1973, 1974, 1975) who borrowed his theoretical framework from vector algebra and introduced the term *perceptual vector analysis.* The experimental studies in this field and their theoretical consequences have recently been reviewed by the present author (Johansson, 1976b). They are also thoroughly treated in (Johansson, 1964, 1975). Therefore only some basic facts from this field are given here.

It is easy to determine that our everyday perception is replete with examples of perceptual analysis of the proximal flow in a number of isolated "streams." First, consider a classical example borrowed from Rubin (1927). When you see a friend waving to you from a slowly moving train, his hand describes a wavy motion in relation to you rather like a sinusoidal curve. However, you perceive a horizontal translatory motion of the person and only an up and down motion of the hand. The sinusoidal stimulus motion has been perceptually split up in two components in a mathematically correct way. In the same way you cannot perceive the cycloidal motion of the valve on a rolling bicycle wheel but you can see it as describing a circular track with the hub in its center as the hub moves linearly along the road. Again the stimulus motion, the cycloid, has been split up into two components in which one represents a common linear motion of the whole system and the other a rotary motion in relation to this moving reference frame. For an experimental analysis see Johansson (1974b).

As a third example recall the experiment with biological motion described above. When a person is walking, we perceive a translatory motion of the whole body, whereas the upper legs are seen as describing pendulum motions in relation to the hips, the lower legs as describing pendulum motions about the knee joints, and so on; thus we encounter complex hierarchies of relative motions. In the experiment described it was found that this perceptual analysis was carried out in a fully compelling way even when only the motions of the main joints of the person in relation to the static background were presented. Furthermore, it was found that this analysis had the character of a spontaneous perceptual act, accomplished in a fraction of a second (Johansson, 1976a).

The perceptual separation of motion components into those common to the whole stimulus flow under consideration and those components of motion relative to this common component in hierarchical order have been found to form a good approximation to mathematical vector analysis in accordance with the principles sketched. This has been thoroughly investigated in the special case of frontoparallel motion (Johansson, 1950, 1973, 1974b) and has been found to hold true also for projections from complex motion in depth (Johansson, 1964, 1973, 1974a, 1974c). In the latter case we are, of course, dealing with proportionally equal rather than metrically equal vectors. The following experiment adapted from Johansson (1964) is informative and instructive. Let a bright surface of any shape, seen against a homogeneous background, continuously shrink and expand in the vertical dimension as illustrated in Figure 11-8A. Some observers will see this surface as oscillating about a horizontal axis but in many cases a frontal surface changing its shape will be seen (cf, the discussion of effects of one-dimensional changes on p. 393). Let us restrict our comments to this latter group of observers because it is of special interest in our analysis.

We now introduce a change also in the horizontal dimension of the figure with about half the amplitude of the vertical and with the same frequency and phase as shown in B. Thus we still have a continuous form change in the surface but a somewhat more complex one. The perceptual effect, however, is significantly different. All our observers see the same object as before, although its shape now changes to a smaller degree. Simultaneously with the form change the "object" moves back and forth in depth as shown in C. What has happened is a perceptual vector analysis of the optical flow through the pupil. A common component of vertical and horizontal change in relation to a projected point at infinity has been abstracted. In our case this point at infinity is localized in the center of the figure, due to the symmetry of change. Therefore by means of a simple vector diagram we can mathematically abstract this component as shown in D.

Our diagram demonstrates that this component in all points on the contours of the figure represents a proportionally equal motion toward the projected point at infinity. Thus it represents a projectively invariant component in the pattern of change or, in other words, the projection of a rigid motion in depth. After abstracting this common component we have as residual vectors the remaining vertical change. In this case it is seen as an elastic deformation. Therefore we must conclude that the visual system automatically abstracts a maximum of projective invariance in the transformation when no substantial projectively invariant carrier like the contours of a figure exists. Rigid motion is what is basically sought.

I sum up the outcome of this discussion of perceptual vector analysis of relations by describing space perception in our everyday environment. When we move around in our environment, light reflected from static and rigid structures like the ground, houses, walls, floors, and furniture will, because of the principles of central projection, generate a unitary continuous perspective transformation in our eyes. All spatial relations in this changing array stay projectively invariant

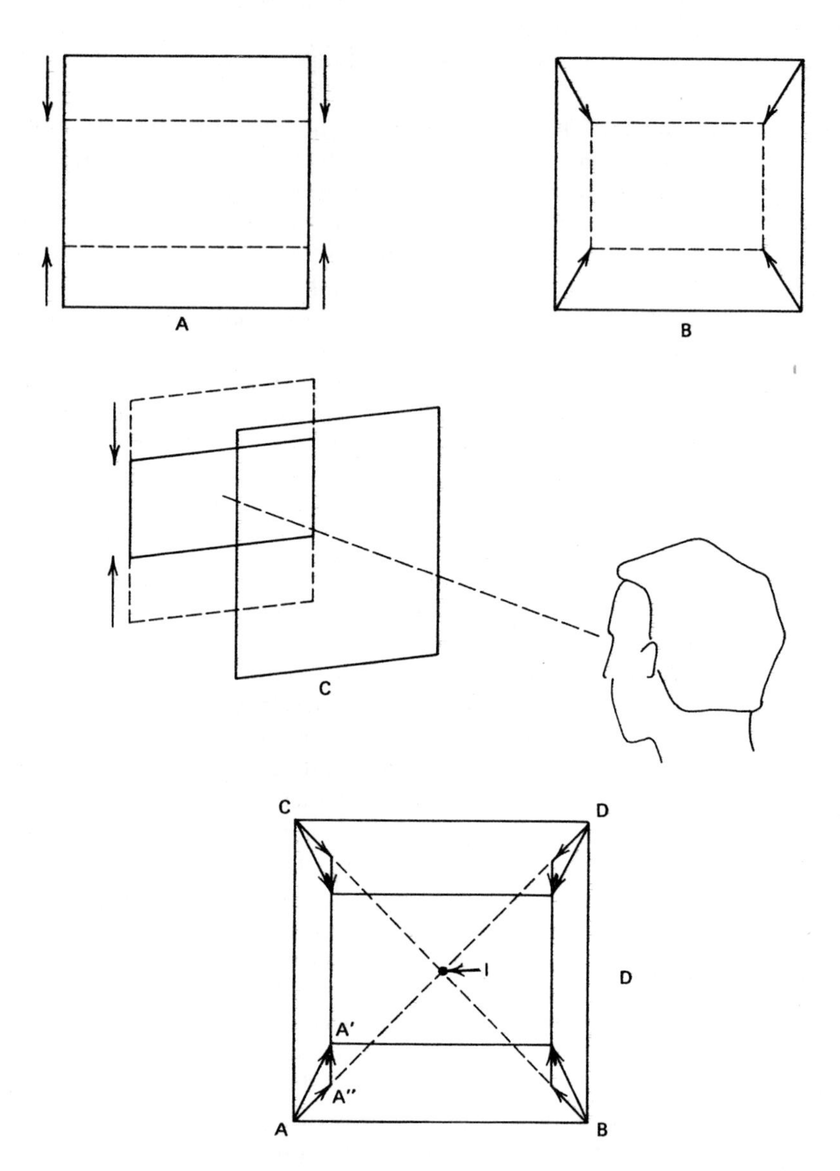

Figure 11-8 (a) A homogeneous bright square (about 3° vis. angle) shrinking-growing in the vertical dimension (about .5 p sec^{-1}); about 50% of the *S*s saw this pattern as a frontal surface changing its form while another group of Ss reported seeing it as a square oscillating about a central axis; (b) the figure (a) with simultaneous change also in the horizontal dimension; (c) the percept from (b) in about 50% of *S*s: a figural change of the same type as evoked by (a) but taking place in a surface simultaneously moving back and forth in depth; (d) a vector analysis of (b) analogous to the percept described in (c). The vectors AA″ – DD″ depict the proximal motion of the four corners A – D. These vectors are treated as sums of the corresponding components AA′ and A′A″, etc.; AA′ – DD′ represent a perspective transformation of the four corners of the square under sagittal motion in depth. The corners move in projectively parallel tracks because the lines AI, BI, CI, and DI are projections of parallel lines intersecting in the projection (I) of their common point of infinity. The vectors A′A″ – D′D″ represent the component of deformation in the percept (alternatively they can represent an oscillation in depth under parallel projection).

under the continuous change and consequently, in accordance with the theory of perceptual relativity brought forward, these structures are perceived as constant in shape and size. This holds true for objects or structures close to us as well those far away. The whole rigid space forms a rigid unit because its projection is invariant.

In this environment rigid objects in motion generate specific components in the total flow which geometrically represent specific invariant projective relations when described with the common total flow as a frame of reference. In accordance with the principle of perceptual vector analysis, these flow components are perceived as units in motion in relation to the perceptual motion state of the environment. In an analogous way motion in relation to moving objects, like the legs of a walking person or the wings of a bird, is perceived. "Objects" without constant shape, like cigarette smoke, are, as we have analyzed them, seen as moving and changing shape at the same time.

SUMMING UP

In our analysis of the problem of how constant objects can be perceived in continuously changing optical patterns we have been forced to depart from the traditional formulation of the problems of spatial constancy. This theoretical structure has been found to be an effect of the initial choice of a geometric model for description of proximal stimuli. It is founded on the implicit assumption, never specified, that the visual apparatus in its neural treatment of the optical pattern must apply the metric of Euclidean geometry. Therefore the traditional problem can be given in the following setting: "how can a sequence of images which are not congruent in Euclidean meaning be visually interpreted as representing the same object?"

In this chapter an attempt has been made to show that—especially in the study of motion perception—this problem will vanish when we accept the geometry of central projection (which is the geometry for projection through a positive lens like that of the eye) as the mathematical framework for stimulus description. Instead of anchoring the analysis in Euclidean congruence, or lack of congruence, in proximal images, the question of perceived spatial relations under projective pattern transformation has been stressed as decisive. This means that the traditional theory founded on an absolute metric has been replaced by a study of invariant relations under figural form changes.

The theoretical stress on relativity as a key to a better understanding of visual space perception and of object constancy under motion has been further accentuated by demonstrating that the visual analysis of complex patterns of proximal changes represents a separation of components of relative and common displacements. In this analysis a common motion component acts as a frame of reference for the deviating components. This has been discussed in terms of visual vector analysis.

The proximal patterns of change have been discussed in terms of continuous flow rather than as a temporal series of images. Therefore is has been stressed that it is often advantageous to describe the proximal event in terms of vector calculus.

A great number of experiments of perception of motion in three dimensions support a theory of the type described. My final conclusion therefore is that the traditional problems about perceptual size and shape constancy are false problems in the meaning that they vanish as soon as the irrelevant metric type of stimulus analysis is discarded. The closest counterpart to these problems in the framework of a relational flow theory are problems about perceptual specifications of relative distance and slant.

REFERENCES

Berkeley, G. *An essay towards a new theory of vision.* 1709. In M. W. Calkins (Ed), *Berkeley: Essays, principles, dialogues.* New York; Scribner's, 1929.

Boring, E. G. *Sensation and perception in the history of experimental psychology.* New York; Appleton Century, 1942.

Börjesson, E. & von Hofsten, C. Spatial determinant in depth perception in two-dot motion patterns. *Perception & Psychophysics*, 1972, 11, 263-268.

Börjesson, E. & von Hofsten, C. Visual perception of motion in depth. Applications of a vector model to three-dot motion patterns. *Perception & Psychophysics*, 1973, 13, 169-179.

Braunstein, M. L. Sensitivity of the observer to transformations of the visual field. *Journal of Experimental psychology*, 1966, 72, 683-689.

Braunstein, M. Perception of rotation in depth: a process model. *Psychological Review*, 1972, 79, 6, 510-524.

Calvert, E. S. Visual judgments in motion. *Journal of the Institute of Navigation*, 1954, 7, 233-251.

Crawford, A. & Briggs, N. L. A theoretical approach to the study of driving. *Note LN/192) ACNLB*, Road Research Laboratory, United Kingdom, 1962.

Day, R. H. *Human perception.* Sydney: Wiley Australasia, 1969.

Duncker, K. Ueber induzierte Bewegung. *Psychologische Forschung*, 1929, 12.

Epstein, W. Perceptual invariance in the kinetic depth-effect. *American Journal of Psychology*, 1965, 78, 301-303.

Epstein, W., Bontrager, H., & Park, J. N. The induction of nonvertical slant and the perception of shape. *Journal of Experimental Psychology*, 1962, 63, 472-479.

Epstein, W. & Park, J. N. Shape constancy: Functional relationships and theoretical formulations. *Psychological Bulletin*, 1963, 60, 265-288.

Epstein, W., Park, J., & Casey, A. The current status of the size-distance hypothesis. *Psychological Bulletin*, 1961, 58, 6, 491-514.

Eriksson, E. S. A cognitive theory of three-dimensional motion perception. *Department of Psychology, University of Uppsala, Report 75*, 1970.

Eriksson, E. S. & Hemström, B. The perceptual effect of two-and three-dot patterns in distal oscillation. *Department of Psychology, University of Uppsala, Report 134*, 1973.

Eriksson, E. S. & von Hofsten, C. A study of perceptual systems in the perception of oscillating surfaces. *Department of Psychology, University of Uppsala, Report 96*, 1971.

Fieandt, K. von. The perception of a visual shape and of its frame-surface during continuous transformation. *Scandinavian Journal of Psychology*, 1961, 2.

Fieandt, K. von and Gibson, J. J. The sensitivity of the eye to two kinds of continuous transfomations of a shadow-pattern. *Journal of Experimental Psychology*, 1959, 57, 344-347.

Fisichelli, V. R. Effects of rotational axis and dimensional variations on the reversal of apparent movements in Lissajous figures. *American Journal of Psychology*, 1946, 59, 669-675.

Flock, H. R. Optical texture and linear perspective as stimuli for slant perception. *Psychological Review*, 1965, 72, 505-514.

Flock, H. R. A possible optical basis for monocular slant perception. *Psychological Review*, 1964, 71, 380-391.

Freeman, R. B. Function of cues in the perceptual learning of visual slant. An experimental and theoretical analysis. *Psychological Monographs: General and applied*, 1966, 80, (Whole 610).

Freeman, R. B. Jr. Ecological optics and visual slant. *Psychological Review*, 1965, 72, 501-504.

Gibson, J. J. *The perception of the visual world*. Boston: Houghton Mifflin, 1950.

Gibson, J. J. *The senses considered as perceptual systems*. New York: Houghton Mifflin, 1966.

Gibson, J. J. Optical motions and transformations as stimuli for visual perception. *Psychological Review*, 1957, 64, 288-295.

Gibson, J. J., & Gibson, E. J. Continuous perspective transformation and the perception of rigid motion. *Journal of Experimental Psychology*, 1957, 54, 129-138.

Gibson, J. J., Olum, P., & Rosenblatt, F. Parallax and perspective during aircraft landings. *American Journal of Psychology*, 1955, 68, 372-385.

Gogel, W. C. The sensing of retinal size. *Vision Research*, 1969, 9, 1079-1094.

Gordon, D. A. Static and dynamic visual fields in human space perception. *Journal of optical Society of America*, 1965, 55, 1296-1303.

Gordon, D. A. & Michaels, R. M. Static and dynamic visual fields in vehicular guidance. *Highway Research Record*, 1963, 84, 1-15.

Graham, C. H. Visual space perception. In C. H. Graham (Ed.), *Vision and Visual perception*, New York: Wiley, 1965, pp. 504-547.

Green, B. F. Figure coherence in the kinetic depth effect. *Journal of Experimental Psychology*, 1961, 62, 272-282.

Hastorf, A. H. & Way, K. S. Apparent size with and without distance cues. *Journal of General Psychology*, 1952, 47, 181-188.

Havron, M. D. Information available from natural cues during final approach and landing. Human Science. *Research Incorporated, Report HSR-RR-62/3-MK-X*, 1962.

Hay, J. G. Optical Motions and Space Perception. *Psychological Review*, 1966, 73, 6, 550-565.

Hershberger, W. A., Carpenter, D. L., Starzec, J. & Laughlin, N. K. Simulation of an object rotating in depth: constant and reversed projection ratios. *Journal of Experimental Psychology*, 1974, 103, 5, 844-853.

Hillebrand, F. Das Verhältnis von Accomodation und Konvergenz zur Tiefenlokalisation. *Zeitschrift für Psychologie und Physiologie der Sinnesorgane*, 1894, 7, 77-151.

Holaday, B. E. Die Grössenkonstanz der Sehdinge bei Variation der inneren und äusseren Wahrnehmungsbedingungen. *Archiv für die gesamte Psychologie*, 1933, 88, 419-486.

Holway, A. H. & Boring, E. G. Determinants of apparent visual size with distance variant. *American Journal of Psychology*, 1941, 54, 21-37.

Ittelson, W. H. Size as a cue to distance: radial motion. *American Journal of Psychology*, 1951, 64, 188-202.

Jansson, G. & Johansson, G. Visual pepception of bending motion. *Perception*, 1974c, 2, 321-326.

Johansson, G. & Jansson, G. Perceived rotary motion from changes in a straight line. *Perception & Psychophysics*, 1968, 4, 165-170.

Johansson, G. Spatio-temporal differentiation and integration in visual motion perception. *Psychological Research*, 1976a. (in press)

Johansson, G. Visual event perception. In *Handbook of Sensory Physiology*. Berlin: Springer-Verlag, 1976b. (in press)

Johansson, G. Visual motion perception. *Scientific American*, June 1975, 76-88.

Johansson, G. Projective transformations as determining visual space perception. In R. B. MacLeod and H. L. Pick Jr. (Eds.), *Perception: Essays in Honor of J. J. Gibson*, Ithaca, New York: Cornell University Press, 1974a.

Johansson, G. Vector analysis in visual perception of rolling motion. *Psychologische Forschung*, 1974b, 36, 311-319.

Johansson, G. Visual perception of rotary motion as transformations of conic sections. *Psychologia*, 1974, 17, 226-237.

Johansson, G. Visual percpetion of biological motion and a model for its analysis, *Perception & Psychophysics*, 1973, 14, 201-211.

Johansson, G. Perception of motion and changing form. *Scandinavian Journal of Psychology*, 1964, 5, 181-208.

Johansson, G. Rigidity, stability and motion in perceptual space. *Acta Psychologica*, 1958, 14, 359-370.

Johansson, G. *Configurations in event perception*. Uppsala: Almqvist & Wiksell, 1950.

Kilpatrick, F. P. & Ittelson, W. H. The size-distance invariance hypothesis. *Psychological Review*, 1953, 60, 223-231.

Klein, F. *Elementarmathematik vom höheren Standpunkte aus II.* Dritte Auflage, Berlin: Julius Springer, 1925, p. 145f.

Klein, F. *Vergleichende Betrachtungen über neuere geometrische Forschungen*. Erlangen, 1872.

Koffka, K. *Principles of Gestalt Psychology*. London: Routlege & Kegan, 1935.

Koffka, K. Die Wahrnehmung von Bewegung. *Bethes Handbuch der normalen und pathologischen Physiologie XII, Berlin:* Band, 2 Häfte, 1931, pp. 1166-1214.

Landauer, A. A. Influence of instructions on judgments of unfamiliar shapes. *Journal of Experimental Psychology*, 1969, 79, 129-132.

Lee, D. N. Visual information during locomotion. In R. B. MacLeod and H. L. Pick Jr. *Perception: Essays in honor of J. J. Gibson*, Ithaca, New York: Cornell University Press, 1974.

Leibowitz, H. W. *Visual perception*. New York: Macmillan, 1965.

Lichte, W. H. & Borresen, C. R. Influence of instructions on degree of shape constancy. *Journal of Experimental Psychology*, 1967, **74**, 538-572.

Llewellyn, K. R. Visual guidance of locomotion. *Journal of Experimental Psychology*, 1971, **91**, 245-261.

Marmolin, H. Visually perceived motion in depth resulting from proximal changes. I. *Perception & Psychophysics*, 1973a, **14**, 133-142.

Marmolin, H. Visually perceived motion in depth resulting from proximal changes. II. *Perception & Psychophysics*, 1973b, **14**, 1, 143-148.

Metzger, W. Betrachtungen über phänomenale Identität. *Psychologische Forschung*, 1934a, **19**, 1-60.

Metzger, W. Tiefen-Erscheinungen in Optischen Bewegungsfeldern. *Psychologische Forschung*, 1934b, **20**, 195-206.

Murch, G. M. *Visual and auditory perception.* New York: Bobbs-Merrill, 1973.

Musatti, C. L. Sui fenomeni stereocinetic. *Archiva Ital, Psicologia*, 1924, **3**, 105-120.

Renshaw, S. Object perceived size as a function of distance. *Optometry Weekly*, 1953, **44**, 2037-2040.

Riggs, L. A., Ratliff, F., Cornsweet, J. C. & Cornsweet, T. N. The disappearance of steadily fixated visual test objects. *Journal of Optical Society of America*, 1953, 495-501.

Rubin, E. Visuell wahrgenommene wirklische Bewegungen. *Zeitschrift für Psychologie*, 1927, **I**, 103.

Stavrianos, B. K. The relation of shape perception to explicit judgments of inclination. *Arch. Psychol.*, 1945, 296.

Thouless, R. H. Phenomenal regression to the real object. *British Journal of Psychology*, 1931, **22**, 11-30.

Wallach, H. & Frey, K. J. On counteradaptation. *Perception & Psychophysics*, 1972, **11**, 161-165.

Wallach, H., Frey, K. J. & Bode, K. A. The nature of adaptation in distance perception based on oculomotor cues. *Perception & Psychophysics*, 1972, **11**, 110-116.

Wallach, H., Moore, M. E., & Davidson, L. Modification of stereoscopic depth-perception. *American Journal of Psychology*, 1963, **76**, 191-204.

Wallach, H., O'Connell, D. N. The kinetic depth effect. *Journal of Experimental Psychology*, 1953, **45**, 4, 205-217.

Wertheimer, M. Untersuchungen zur Lehre von der Gestalt II. *Psychologische Forschung*, 1923, **4**, 301-350.

Wertheimer, M. Experimentelle Studien über das Sehen von Bewegung. *Zeitschrift für Psychologie, 1912*, **61**, 161-265.

White, B. J. & Mueser, G. E. Accuracy in reconstructing the arrangements of elements generating kinetic depth displays. *Journal of Experimental Psychology*, 1960, **60**, 1-11.

Woodworth, R. S. *Experimental psychology.* New York: Holt, 1938.

Woodworth, R. S. & Schlosberg, H. *Experimental psychology.* Revised ed. New York: Holt, 1954.

Wundt, W. *Grundzüge der physiologischen Psychologie.* II. Heidelberg: 1874.

Zeigler, P. & Leibowitz, H. Apparent visual size as a function of distance for children and adults. *American Journal of Psychology*, 1957, **70**, 106-109.

CHAPTER

12

LESSONS IN CONSTANCY FROM NEUROPHYSIOLOGY

WHITMAN RICHARDS

Massachusetts Institute of Technology

In the last 15 years rapid strides have been made toward understanding the feature encoding characteristics of the brain. This advance is due almost entirely to neurophysiological techniques for recording from single cells or fibers in the visual pathways (Hartline, 1940; Granite, 1947; Kuffler, 1953). As a result, we now know that neurons respond selectively to stimuli such as directional movement of dark objects (Lettvin, Maturana, McCulloch, & Pitts, 1959) or spots (Barlow & Hill, 1963) or even may require a bar of specific width and orientation precisely located in the visual field (Hubel & Wiesel, 1959; Bishop et al., 1971). As more data have been accumulated, a long list of specific trigger features

Supported by grant #EY 00742 from NIH and AFOSR Contract #F44620-72-0076 from ARPA. Drs. D. Pollen, P. Schiller, D. Berson, and A. Witkin very kindly provided helpful criticisms on early versions of this manuscript.

has been compiled (Jung, 1973), many of which appear at birth or require little developmental experience (Hubel & Wiesel, 1963, 1974; Grobstein, et al., 1973; Pettigrew, 1974; Blakemore & Van Sluyters, 1975; Sherk & Stryker, 1975). With this rapid explosion in our knowledge of the features extracted by the brain, we might expect a concomitant growth in our understanding of brain function. In particular, what insights into constancy mechanisms have been provided by these neurophysiological findings?

Unfortunately many constancy phenomena , such as perception itself, require interactive processing by the observer. The perception of size, for example, can be grossly distorted by the observer's conception of the frame of reference (see Ittelson's summary of the Ames' demonstrations, 1952), by instructions (Gilinsky, 1955; Foley, 1972), or by many other factors illustrated throughout this book (see especially Epstein, Chapter 13). Perception is clearly not a passive process. Yet almost all the findings of neurophysiology are addressed to the passive filtering properties of perceptual mechanisms. In spite of this severe constraint on available data, however, some insight into constancy mechanisms can be extracted from the neurophysiological results. These insights take the form of three principles that become most obvious on examining two primitive constancies—brightness and color. From our understanding of the (passive) neurophysiological mechanisms underlying these two low-level constancy phenomena some further insight into higher-level constancies such as size and shape can be obtained, but we must not lose sight of the fact that a complete understanding of the constancies must at some future date also incorporate the active components of the perceptual process. At this time the relevant neurophysiological results are too premature to provide a strong basis for active constancy mechanisms at work. Therefore the only active mechanism incorporated into this chapter is the selective attention known to occur between a variety of passive feature-extracting operations.

BRIGHTNESS CONSTANCY

The most primitive constancy is probably that of brightness. Over a wide range luminance levels, for example, as noon proceeds to dusk, the brightness of objects does not change noticeably. Many years of psychophysical research have shown that this constancy phenomenon is due principally to the fact that the brightness of objects is judged among them (Wallach, 1963; Heinemann, 1972). Thus reflectance rather than luminance is the closer correlate of brightness, for relations between reflectance and brightness will remain invariant with changes in the level of ambient illumination, whereas the luminance value will not (Land & McCann, 1971). How does the visual system extract such "brightness" relations, discounting almost completely the true level of illumination or luminance?

Thanks to the explorations of the neurophysiologists, the mechanism under-

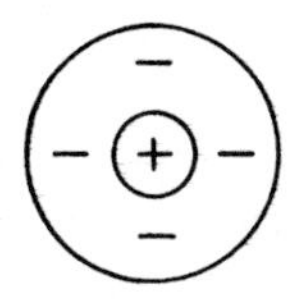

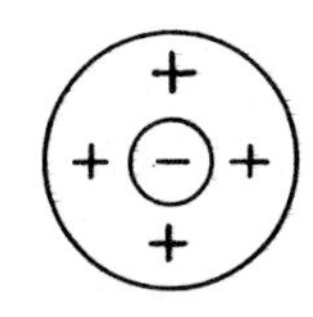

Figure 12-1 "Kuffler Units." Center and surrounding regions of the retina act antagonistically on the activity of a neuron. Diffuse illumination does not lead to a response.

lying this aspect of brightness constancy is now fairly well understood (Baumgartner, 1964; Horn, 1975) and is the result of early information processing in the visual sytem based on the comparison of local regions of activity via lateral inhibition. Almost all known visual systems begin their feature extraction by creating "Kuffler units," as shown in Figure 12-1, in which a small central region of the retina excites (or inhibits) the activity of a neuron as the surrounding region acts in an antagonistic manner. These center-surround interactions are usually exquisitely balanced so that uniform illumination over the entire receptive field causes no change in neural activity. Thus the "Kuffler" unit filters out ambient illumination and reports changes in luminance (or reflectance), especially at boundaries. The essence of brightness constancy thus lies in the mechanism of brightness perception itself.

Although this basic mechanism can be interpreted in other ways (Ratliff, 1965; Cornsweet, 1970; Marr, 1974), the other formalisms are all roughly equivalent. All require a network of lateral interactions that leave the activity of the net unchanged when illumination is diffuse but magnify neural activity when there is a step in luminance or reflectance. Thus the absolute level of illumination is irrevocably lost and only changes in luminance are passed on to higher levels of processing. Brightness constancy is thus the result of a filtering out or discarding of information by a visual pathway at a very early stage.

The insight into brightness perception in terms of the primitive "Kuffler unit" the neurophysiologist has given us shows quite clearly how a simplification in information processing may lead to a specific sensory loss which is interpreted as a "constancy" phenomenon. In the case of brightness there are two important factors that are critical to the creation of this information loss which leads to the constancy. These factors are summarized as rules for future generalization:

Rule 1. A constancy phenomenon is the result of the relative insensitivity of a "channel" to selected veridical* cues.

Rule 2. The filtering operation of the "channel" is accomplished by an antagonistic comparison between like analyzers which have previously extracted the veridical cues to be discarded (note that it is necessary only to discard the veridical cues previously featured in the pathway. Other veridical cues never encoded have already been discarded by default).

*A veridical cue is a genuine or real cue as opposed to a derived or implicit cue. *Luminance* is a veridical cue to object intensity, whereas *brightness* is the derived sensation, hence must always remain a nonveridical cue (e.g., to object distance).

We shall also see from subsequent discussion that a third rule is necessary:

Rule 3. The observer must be able to attend selectively to the pathway containing the "channel" that has filtered out the veridical cues (especially when the filtering is performed in one pathway but not another).

If we apply these rules to brightness constancy, we will see that Rule 1 requires that absolute luminance be discarded, for this is the primary veridical cue that is ignored during brightness constancy. Rule 2 suggests that the filtering of luminance is accomplished by comparing the activities of two analyzers that are, in fact, responding to absolute luminance. Thus the Kuffler-type unit compares neural activities in neighboring regions and reports only the difference (or more correctly the ratio) between the center and surround illumination. Because this filtering is accomplished by spatial differentiation, the response of the higher level feature detector will be subject to spatial dependencies, and although this higher level feature detector now has discounted luminance and instead reports only brightness relations or reflectance it has accomplished this task by subjecting itself to possible errors in the interpretation of luminance gradients or boundaries. (These errors in interpretation, however, may still be beneficial to the behavior of the organism.)

Finally, Rule 3 requires that for the brightness constancy mechanism to be effective in perception this channel must be the principal source of information about object brightness. If not, then the observer must at least pay attention to this channel over others carrying discordant information. Clearly the visual system does not discard information about absolute luminance entirely, for pupil size is strongly dependent on absolute luminance rather than on brightness contrast (LeGrand, 1968) and thus may possibly be the pathway tapped by experienced photographers who guage ambient illumination quite well. Thus at least one pathway is still available to carry information about the absolute level of illumination. But this midbrain pathway mediating pupil size is normally completely subservient to the geniculostriate pathway that is presumed to contain the dominant channel for brightness on which most of us base our perceptions.

COLOR CONSTANCY

Will the same three rules gleaned from neurophysiology also be successful in the interpretation of color constancy? Here also in this field are data on neural encoding which are obtained at several levels in the visual pathway (Wagner et al, 1960; Wiesel & Hubel, 1966; Daw, 1967; DeValois & Jacobs, 1968; Gouras, 1972; Zeki, 1973). How do these data bear on the problem of color constancy?

Like brightness contancy, psychophysical results suggest that color constancy is based on a filtering operation that leaves the visual system relatively insensitive to the properites of the ambient illumination. In color, however, the spectral

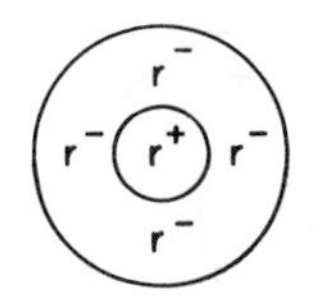

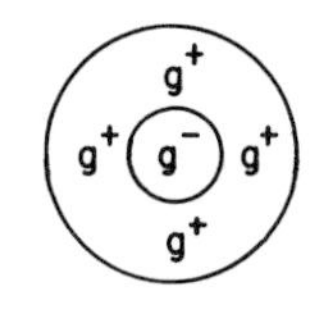

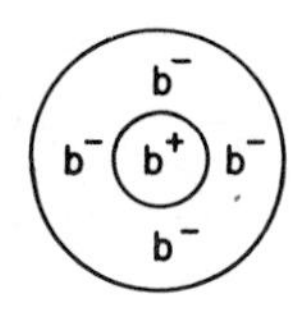

Figure 12-2 Possible color analyzers that satisfy constraints for a color-constancy channel. The r, g, and b symbols indicate excitatory (+) or (-) inputs derived from different cones.

composition of the source as well as its illuminance, is discarded (Helmholtz, 1866). Regardless of the illuminant, the color appearances of objects tend to remain the same even though their spectral properties must change as the composition of the illuminant changes. The effect is as if we were selectively "blind" to the spectral character of the light source (Rule 1). Clearly the objective of the channel or pathway of interest is to convey some but not all of the information about the spectral composition of the light reflected from the illuminated objects.

To discount the color of the illuminant, we can follow the leads suggested by the preceding analysis of brightness constancy and compare the activities of like analyzers (Rule 2) to filter out the global effects of the illumination; for example, a possible solution might be to compare red-cone activity in one retinal region with adjacent red-cone activity in a neighboring region (McDougall, 1901; Alpern & Rushton, 1965). Thus the "Kuffler units" reporting to the "color channel" would have "red" centers and "red" surrounds or "green" centers and "green" surrounds, and so on, as illustrated in Figure 12-2. No neurophysiological evidence is available at present, however, to support such an encoding scheme.*

Before proposing another basis for color constancy, the relation between brightness constancy and brightness contrast should once again be noted. These mechanisms are inextricably intertwined, for it is the extraction of brightness contrast that leads to brightness constancy. Similarly, the psychophysical evidence suggests that the encoding of color contrast occurs simultaneously with color constancy (Graham, 1965). Color contrast lacks the spatial dependencies of brightness contrast, which suggests different encoding mechanisms (Horst et al., 1967). Brightness contrast requires and flourishes in the presence of borders. This is consistent with the spatial organization of the Kuffler unit. Color contrast, on the other hand, is maximal when luminance steps are minimized (Kirschmann, 1891), which suggests that the encoding of color has a spatial encoding between centers and surrounds that is different from that of the brightness analyzing mechanism.

*Neurons with "red" centers and "yellow" surrounds are common, as are "green" versus "yellow" Kuffler units (Wiesel & Hubel, 1966). Because "yellow" represents a convergence of "red" plus "green" cone input, such R-Y or G-Y units would partially discount the illuminant (Richards, 1972).

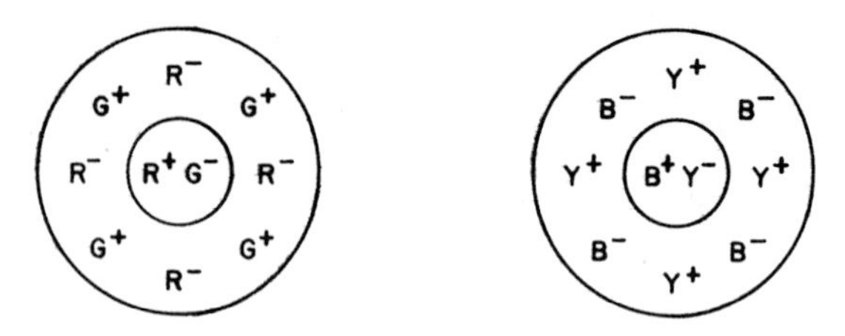

Figure 12-3 Double-opponent color analyzers. The R, G, Y, and B, symbols indicate excitatory (+) or inhibitory (-) inputs with different spectral sensitivities.

Daw (1967) who was the first to recognize the significance of this difference, proceeded to show that the goldfish retina contained a new type of receptive field which he identified as a double-opponent feature analyzer. Figure 12-3 illustrates the composition of these receptive fields. The center of the field is excitatory to red but inhibiting to green, whereas the surround acts antagonistically (following Rule 2). If red or green (or any wavelength) is shown uniformly across the entire field, there is no unit response because of a balanced antagonism between centers and surrounds. Thus ambient illumination of any color is discounted according to Rule 1. On the other hand, if a red-green border is placed across the edge of field so that it just grazes the central region, the central region will be excited by the red and this response will be reinforced by the excitation of green in the surround. Thus the color difference enhanced. It will be strongest when the red and green illuminations are roughly equal because subtractive and additive neural mechanisms are nonlinear (pseudologarithmic) functions of illuminance.* Provided that the brain can selectively tune in to activity from these types of units (Rule 3), we now have a good neurophysiological mechanism for color constancy as well as simultaneous color contrast.

MOTION CONSTANCY

As we drive down a road, objects flow across the retina at different rates, yet the moving environment appears as a whole. We tend to ignore the flow gradients composed by the passing trees and concentrate instead on the point of origin of the flow. When a bird suddenly breaks up the normal gradient of flow by flying across our path, our attention is immediately captured. Again we have a constancy phenomenon at work, one quite analogous to color constancy, in which the more global ambient conditions are generally discounted and in

*Note that these color opponent units require two levels of processing. They must discount the illuminant and compare "red-green" activities.

which local differences or novelty are emphasized. In the case of motion the continuous stream of background flow is ignored and instead only the local perturbation of movement introduced by the interposing bird is significant. Do our rules for constancy mechanisms again help us build a bridge to the relevant neurophysiology?

The simplest motion detector is that described by Barlow, et al. (1964) in the rabbit retina. These neural units are vigorously activated by a spot or bar moving in one direction (the "preferred" direction) across their receptive fields but are inhibited by motion in the opposite direction (the "null" direction). This directionally selective response is relatively independent of the type of stimulus, that is, its shape, color or contrast, or speed of movement, although the smaller objects are generally more effective than the large. This directional response is likely to be due to a wave of lateral inhibition that precedes the movement of stimuli in the null direction, cancelling or preventing a discharge from neural units to be subsequently activated by the stimulus (Barlow & Levick, 1965). The motion detector thus does not measure velocity directly, as a speedometer does, but infers instead the presence of movement by attempting to calculate the relation $\Delta x/\Delta t$ by using relatively fixed values of Δx (retinal distance over which inhibition occurs) and Δt (the time delay for inhibition to become effective). Although the direction of encoded movement is directly related to the direction of real movement, and hence is encoded veridically, the speed of motion is not. Thus the initial encoding of movement already sacrifices some information about absolute rate. Because some ordinal relations may still be preserved, however, differential rates and especially direction of movement could still be detected, provided that the outputs of two or more elemental be preserved, however, differential rates and especially direction of movement could still be detected, provided that the outputs of two or more elemental motion detectors can be compared. A mechanism for motion constancy built along the lines of one for color must then require a higher level type of motion analyzer constructed from the primary elemental detector of Barlow and Levick (1965).

Figure 12-4 shows two types of higher level motion analyzer. Each arrow represents a lower level elemental motion detector similar to that found by Barlow et al. (1964). To obtain motion constancy the higher level analyzer should be made up of antagonistic pairs of units, in which each of the subunits (arrows) responds optimally to movement in the same direction (Rule 2). By separating the antagonistic components spatially the average flow rate will be discarded (Rule 1) and only differential local movement will be reported. This analyzer would become remarkably effective for discriminating average flow if it were relatively insensitive to changes in speed and instead responded best to changes in the direction of movement.

Selective attention to motion channels is also required for a constancy mechanism (Rule 3). Evidence illustrating such selective channels includes the effectiveness of movement for triggering "unconscious" saccades to novel positions in the visual field before an analysis of object recognition is performed. Such "unconscious" acts most certainly are mediated by pathways different from the "conscious." Following the earlier proposals for dissociating the function of the visual pathways (Riddoch, 1917; Marquis, 1935) we now consider orienting behavior the primary responsibility of the midbrain mechanisms, whereas the analysis of form, brightness, and color is the responsibility of geniculo-striate mechanisms (Schneider, 1967, 1969; Ingle, 1967; Michael, 1969; Ewert, 1974). Thus strong behavioral as well as neurophysiological (Allman & Kaas, 1974) and anatomical evidence (Graybiel, 1972) supports the notion of separate visual pathways encoding separate visual functions, thus providing a good basis for the simultaneous encoding of different constancies in parallel pathways. To complete our insight into motion constancy, however, we lack a crucial test by the neurophysiologist. The requirement not yet tested is whether the responses of movement analyzers in the colliculus (or prestriate cortex receiving collicular input) are as good in a field of motion as they are when the background is stationary.

SIZE CONSTANCY

Although size perception has been the subject of intensive study for centuries, our understanding of its basis is meager. Even current neurophysiology has provided little insight into size perception, at least if we would hope to find neural units whose responses provide an accurate measure of object size (Marg & Adams, 1970). Considering that the representation of the visual field is locally

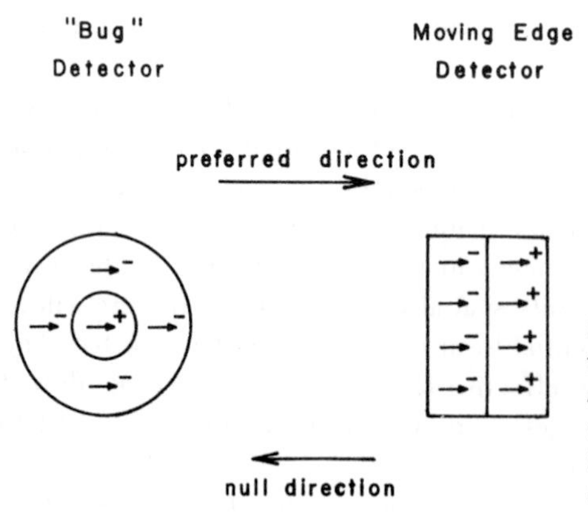

Figure 12-4 Motion analyzers suitable for extracting motion constancy. The arrows indicate symbolically the arrangement of excitatory (+) and inhibitory (—) inputs from low-level Barlow-Levick type directional motion detectors.

disjoint, at least in the cortex in which both orientation and position are encoded how can we even hope to find size-detecting neurons? Perhaps, however, the absence of this category of feature detector is in itself the essence of size constancy, for if absolute (angular) size is not encoded the strongest veridical cue to size must be ignored. The simplest basis for size constancy therefore may lie simply in the fact that the angular extent of objects is not encoded by our nervous system (i.e., Rule 1).

A failure to encode absolute angular size will still leave many other non-veridical cues to the size of objects. Two main categoried will be considered: dynamic and static cues. A dynamic cue to size occurs when an object approaches or recedes, thus generating a boundary of flow (or shear) on the retina; for example, fixate an object on the wall, and while holding fixation steady move one hand in and out. There is a dramatic change in size. The major cue to size change is probably flow (or shear), for if the flow rate is made very low by moving the hand extremely slowly the growth in size on approach is not nearly so obvious.

Of course, when fixation is held rigid, the in-out movement of the hand leads to a correct perception of size change. There is no size constancy of the hand. In contrast, if fixation moves in and out with the hand (as if the hand itself were fixated), the apparent size of the hand will show less change. Size constancy thus occurs to some degree in the presence of vergence movements but not without it.

If we apply our rules for constancy mechanisms, we can propose that in active vergence a selective attention mechanism (Rule 3) ignored the flow pattern, thus eliminating the basis for size judgment. This corollary-discharge mechanism, however, may not need invoking. If we refer again to Figure 12-4 we will see that higher level motion detectors of the kind described here will not respond if the flow on each side of the boundary is in the same direction. During vergence tracking the eyes are moving inward (or outward), thus causing the background field to move at the same time the approaching (or receding) object is moving. For each half-field the movement is in the same direction as the movement of the object's boundary. The flow cue to size change would thus be considerably weakened, depending on the breadth of tuning of the motion detectors to speed of movement. Without the flow cue, size changes would not be noted so easily.*

*Two types of motion detector are illustrated in Figure 12-4; one is circular without orientation filtering and the other is orientation dependent. Clearly the former is more suitable for triggering "unconscious" fixation, whereas the latter is more appropriate for detecting the "shear" required for the model of size constancy. Following this distinction further, the first detector should be more typical of midbrain mechanisms, whereas the second is more appropriate for cortical analysis (see Schiller & Stryker, 1972).

An analogous proposal may be made for stationary scenes in which object size is inferred. In the static case in which no dynamic cues to size change are available other cues must be noted. One obvious possibility is texture, or the grain of object and background. According to Rule 2, the sought-for neural analyzer should be one that compares the texture in one region of the field with the texture in a neighboring region. If the textures or grains are identical, the channel created by the texture analyzers should be inactive. Figure 12-5 illustrates possible receptive fields that have this property. When a uniform textured surface is presented to these Hubel-Wiesel "simple" cells, the units will not respond. On the other hand, if the boundary between two textures is moved across the edge of the receptive field (illustrated on the left of Figure 12-5), the unit will be strongly activated (Hammond & MacKay, 1975; Orban et al., 1975). The unit shown in the right-hand portion of Figure 12-5 will respond best when a line element matching the central excitatory region is presented without a similar line in the neighboring field. Channels built from these analyzers could provide a suitable basis for size constancy mechanisms for two reasons: (a) they are optimally stimulated by textures or grains that differ in neighboring regions, and (b) the length of the stimulus is not the critical stimulus variable, hence have discarded an important dimension of size and extent.* These units report the boundaries or regions of texture change and say little about the actual size of the object that creates the boundary.

Because there will be many analyzers in any one region of the visual field, each with a different width, judgments about the actual grain size could be quite poor (Richards & Polit, 1974). Instead, activity of these units would indicate only the presence of a boundary created by a texture difference. As long as this difference (or ratio) in texture or grain is maintained, the population should continue to report the presence of a boundary until the grain of the field exceeds the range of widths detectable by the analyzers. This invariance, which is based on analyzers built to detect change, is quite analogous to that proposed for the dynamic cues to size change. Both mechanisms, the static texture and dynamic flow analyzers (shear), receive further assistance toward maintaining size constancy by the increasing coarseness of the analyzing mechanisms as their receptive fields become more eccentrically located in the visual field (Hubel & Wiesel, 1974; Fisher, 1973). This trend toward increasing coarseness with retinal eccentricity nicely compensates for the increase in size

*A still more sophisticated analysis of texture could be made utilizing "periodic" cells best responsive to gratings. Pollen et al (1971) also comment upon the potential these units have as encoders of spatially invariant information about size.

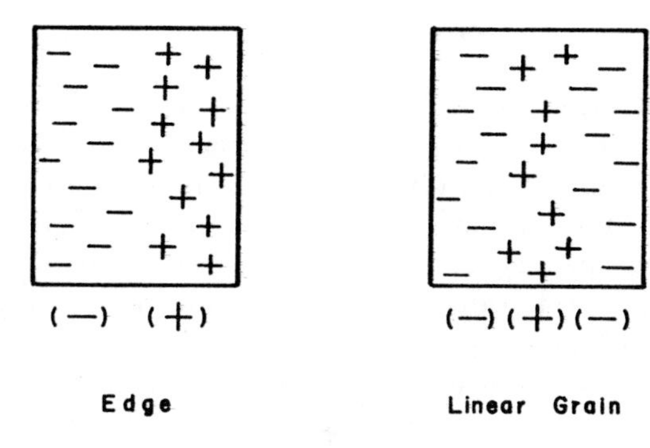

Figure 12-5 Texture-respective fields that might provide a suitable basis for size constancy. The plusses and minuses are symbolic representations of inputs derived from low level "Kuffler" units that are excited (+) or inhibited (-) by an increase in contrast.

of approaching objects and will tend to continue to highlight the dominant boundary of interest.

In summary, our rules suggest that the neurophysiological correlate for size constancy may be based on the fact that simple Hubel-Wiesel cortical cells are designed to detect boundaries or texture elements that stand out against a rich background. Most of these units do not appear to respond to extent (at least length), hence have removed an important veridical cue to absolute size. *But such filtering operations are necessary at some stage if size constancy is to occur. Already at the cortex, although a global topography of the visual field is intact, local topography has been lost. The encoding of the continuity of a boundary or line is sacrificed at the expense of retaining the detection of orientation. This kind of information loss is extremely beneficial to a size-constancy mechanism because the only cue to size extracted is texture or orientation of a boundary. Thus a channel might base its size perceptions on the simple hypothesis that as long as the texture (ratio of background) and orientation of the boundary did not change object size must remain constant.

*In this respect some of the hypercomplex units of Hubel and Wiesel would not be a suitable basis for size constancy. Instead, hypercomplex units, particularly those that seek stopped ends of intersections, may well underlie depth constancy in which interposition would seem to play a key role. If so, such hypercomplex units should be more sensitive to intersecting lines (or surfaces) with different disparities rather than to boundaries in the same plane. The disparity analyzers currently proposed by the neurophysiologists for stereopsis seem to me to provide an inadequate basis for depth constancy; hence this phenomenon has been omitted from discussion elsewhere.

SHAPE CONSTANCY

If simple cortical analyzers have discarded the veridical cues to absolute size, what cues remain to assess shape? The three most obvious are probably (a) orientation, (b) disparity, and (c) the angle of intersection of two (or three) lines or surfaces. Of these three remaining cues only orientation and disparity have been documented by the neurophysiologist (see footnote on p. 431), but if orientation and disparity are to be used for shape perception consider the complex calculation that must be performed to determine whether the shape of an object will remain constant under rotation. In particular, the relations between the lengths of two edges and their relative disparity must be calculated to ascertain whether the new object is a projective transformation of the original (or its schema). Is the visual system really capable of such a calculation, even for the simple case in which a square is rotated in three-dimensional space to create two separate trapezoidal images on each retina?

According to our derived rules, the answer to the question why the trapezoidal shapes of a rotated square still appear "square" does *not* lie in trying to comprehend how a complex calculation is made. Rather, it depends on an understanding how the shape of an object may be assessed *without* utilizing the orientation and disparity cues.

Consider now what happens when a square is rotated in space out of the frontal plane. As the square rotates, the leading and trailing edges, as well as the inclined edges and surface move at different rates. Johansson (1975) has shown that these differential movements are powerful stimuli to shape perception, even from two-dimensional projections of rather complex shapes. In some manner we can make inferences about object shape solely from these differential flow patterns, in complete ignorance of the disparity or orientation cues. Perhaps in the static case also a similar mechanism is at work, in which the orientation of the sides, their relative disparity, and their geometrical relations (other than ordinal relations) are ignored and shape is assessed by another mechanism, perhaps texture gradients rather than flow. Clearly, more work in both psychophysics and neurophysiology is needed to answer these questions.

Finally, at still higher levels of visual processing other "constancies" occur which are often cast in terms of equivalences. A report has appeared on one of these "equivalences," that is, on mirror-image symmetry, which nicely illustrates the spirit of the arguments I have proposed. In this study (Gross et al., (1975) monkeys with inferior-temporal lesions solved a mirror-image problem more readily than normal monkeys when both groups were equated on ability to learn a nonmirror image discrimination. Thus we have a paradoxical result in which ablating a cortical area improved a learning discrimination! If inferior-temporal cortex were the "channel"-mediating mirror-image constancy, then, by our argument, this cortical regions should be "blind" to mirror-image equivalences (i.e., it might subserve symmetry perception). When this cortex is removed, the constancy would then also be eliminated because the animal would be forced to use remaining channels not "blind" to the mirror-image transformation.

CONCLUSION

The psychophysical findings of Johansson and the behavioral results of Gross illustrate constancy perception at high levels. Yet these phenomenon can be understood in terms of some simple principles of neural encoding, often by rather simple feature detection. We must not be misled into looking further for more and more complex elements of visual perception (such as detectors specific to certain sizes and shapes). Rather the basic building blocks are already apparent and may be sufficient for encoding various symbolic operations in each of the many higher level visual, auditory, or other sensory areas (Marr, 1974). If constancies are viewed not as successes in encoding but rather as selectively attended failures in cue processing, the study of constancies should help to build bridges from psychophysics to neurophysiology by indicating the conditions under which feature-encoding fails as well as succeeds. This essence underlies the three rules that summarize the insights provided by the neurophysiologists.

REFERENCES

Allman, J. M. & Kaas, J. H. A crescent-shaped cortical visual area surrounding the middle temporal area (MT) in the owl monkey. *Brain Research*, 1974, **81**, 199-213.

Alpern, M. & Rushton, W. A. H. The specificity of the cone interaction in the after-flash effect. *Journal of Physiology* (London), 1965, **176**, 473-482.

Barlow, H. B. & Hill, R. M. Selective sensitivity to direction of movement in ganglion cells of the rabbit retina. *Science*, 1963, **139**, 412-414.

Barlow, H. B., Hill, R. M., & Levick, W. R. Retinal ganglion cells responding selectively to direction and speed of image motion in the rabbit. *Journal of Physiology. (London)*, 1964, **173**, 377-407.

Barlow, H. B. & Levick, W. R. The mechanism of directionally selective units in rabbit's retina. *Journal of Physiology (London)*, 1965, **477-504**.

Baumgartner, G. Neuronal mechanisms des Kontrast- und Bewegungsseheus. *Berichte cleutscher Ophthal Gesillschaft*, 1964, **66**, 111-125.

Bishop, P. O., Coombs, J. S., & Henry, G. M. Interaction effects of visual contours on the discharge frequency of simple striate neurons. *Journal of Physiology*, (London) 1971, **219**, 659-687.

Blakemore, C. & Van Sluyters, R. C. Innate and environmental factors in the development of the kitten's visual cortex. *Journal of Physiology* (London), 1975, **248**, 663-716.

Cornsweet, T. N. *Visual Perception.* New York: Academic; 1970.

Daw, N. W. Goldfish retina: Organization for simultaneous contrast. *Science*, 1967, **158**, 942-944.

DeValois, R. & Jacobs, G. H. Primate color vision. *Science*, 1968, **162**, 533-540.

Ewert, P. J. The neural bases of visually guided behavior. *Scientific American*, 1974, **230** #3, 34-49.

Fisher, B. Overlap of receptive field centers and representation of the visual field in the cat's optic tract. *Vision Research*, 1973, **13**, 2113-2120.

Foley, J. M. The size-distance relation and intrensic geometry of visual space: implications for processing. *Vision Research*, 1972, **12**, 323-332.

Gilinsky, A. S. The effect of attitude upon the perception of size. *American Journal of Psychology*, 1955, **68**. 173-192.

Gouras, P. Color oppency from fovea to striate cortex. *Invest. Ophthal.*, 1972, **11**, 427-434.

Graham, C. H. *Vision and Visual Perception.* New York: Wiley New York, 1965.

Granite, R. *Sensory Mechanisms of the Retina.* London: Oxford University Press: 1947.

Graybiel, A. M. Some extrageniculate visual pathways in the cat. *Invest. Ophthal.*, 1972, **11**, 322-332.

Grobstein, P., Chow, K. L., Spear, P. D., & Mathers, L. H. Development of rabbit visual cortex. Late appearance of a class of receptive fields. *Science*, **180**, 1973, 1185-1187.

Gross, C. G., Lewis, M. & Plaisier, D. Inferior temporal cortex lesions do not impair discrimination of lateral mirror images. Annual Meeting, Neurosciences Society, New York, November 2-6, 1975, p. 74.

Hammond, P. & MacKay, D. M. Differential responses of cat visual cortical cells to textured stimuli. *Experimental Brain Research*, 1975, **22**, 427-430.

Hartline, H. K. The effects of spatial summation in the retina on the excitation of the fibers of the optic nerve. *American Journal of Physiology*, 1940, **130**, 700-711.

Heinemann, E. G. Simultaneous brightness induction. D. Jameson & L. M. Hurvich (Eds.), *Visual Psychophysics (Vol. VII/4), Handbook of Sensory Physiology.* Berlin: Springer Verlag; 1972.

Helmholtz, H. L. F., von (1866). *Treatise on physiological optics* (J. P. C. Southall, Ed. and Trans.). New York: Dover; 1962.

Horn, B. K. P. (1975). On lightness. (See reference in Marr, 1974).

Horst, G. J. C. van der, Weert, C. M. M. de, & Bouman, M. A. Transfer of chromaticity-contrast at threshold in the human eye. *Journal of the Optical Society America*, 1967, **57**, 1260-1266.

Hubel, D. H. & Wiesel, T. N. Receptive fields of single neurones in the cat's striate cortex. *Journal of Physiology* (London), 1959, **148**, 574-591.

Hubel, D. H. & Wiesel, T. N. Receptive fields of cells in striate cortex of very young, visually inexperienced kittens. *Journal of Neurophysioloy*, 1963, **26**, 994-1002.

Hubel, D. H. & Wiesel, T. N. Uniformity of monkey striate cortex: a parallel relationship between field size, scatter, and magnification factor. *Journal of Comparative Neurology*, 1974a, **158**, 295-306.

Hubel, D. H. & Wiesel, T. N. Ordered arrangement of orientation columns in monkeys lacking visual experience. *Journal of Comparative Neurology*, 1974b, **158**, 307-318.

Ingle, D. Two visual mechanisms underlying the behavior of fish. *Psychologische Forschang*, 1967, **31**, 44-51.

Ittelson, W. H. The Ames demonstration in perception. Princeton, New Jersey: Princeton Universtiy Press, 1952.

Johansson, G. Visual motion perception. *Scientific American*, 1975, 232, #6, 76-88.

Jung, R. Visual Centers in the Brain. *Handbook of Sensory Physiology* Vol VII/3). Berlin: Springer Verlag, 1973.

Kirschmann, A. Ueber die quantitativen Verhaltnisse des simultanen Helligkeits- und Farben-contrastes. *Philosophische Studiern*, 1891, **6**, 417-491.

Kuffler, S. W. Discharge patterns and functional organization of mammalian retina. *Journal of Neurophysiology*, 1953 **16**, 37-68.

Land, E. H. & McCann, J. S. Lightness and retinex theory. *Journal of the Optceal Society of America*, 1971, **61**, 1-11.

LeGrand, Y. *Light, Color and Vision*. London: Chapman & Hall, 1968.

Lettvin, J. Y., Maturana, H. R., McCulloch, W. S. and Pitts, W. H. (1959). What the frog's eye tells the frog's brain. *Proceedings of the Institute of Radio Engineers*, 1959, **47**, 1940-1951.

Marg, E. & Adams, J. E. Evidence for a neurological zoom system in vision from angular changes in some receptive fields of single neurons with changes in fixation distance in the human visual cortex. *Experientia*, 1970, **26**, 270-272.

Marquis, D. G. Phylogenetic interpretation of the functions of the visual cortex. *Archives of Neurology and Psychiatry*, 1935, **33**, 807-815.

Marr, D. The computation of lightness by the primate retina. *Vision Research*, 1974, **14**, 1377-1388.

McDougall, W. Some new observations in support of Thomas Young's theory of light and color vision. *Mind*, 1901, **10**, N. S. 210-245.

Michael, C. R. Retinal processing of visual images. *Scientific American*, 1969, 220, #5, 104.

Orban, G. A., Callens, M., & Cole, J. M. Unit responses to moving stimuli in area 18 of the cat. *Brain Research*, 1975, **90**, 205-219.

Pettigrew, J. D. The effect of visual experience on the development of stimulus specificity by kitten cortical neurons. *Journal of Physiology*, (London), 1974, 237, 49-74.

Pollen, D. A., Lee, J. R., & Taylor, S. H. How does the striate cortex begin the reconstruction of the visual world? *Science*, 1971, **173**, 74-77.

Ratliff, F. *Mach bands: Quantitative studies on neural networks in the retina*. San Francisco: Holden Day, 1965.

Richards, W. One-stage model for color conversion. *Journal of the Optical Society of America*, **62**, 697-698.

Richards, W. & Polit, A. Texture matching. *Kybernetik*, 1974, **16**, 155-162.

Riddoch, G. Dissociation of visual perceptions due to occipital injuries, with especial reference to appreciation of movement. *Brain*, 1917, 4, 14-57.

Schiller, P. H. Stryker, M. Single-unit recording and stimulation in superior colliculus of the alert rhesus monkey. *Journal of Neurophysiology*, 1972, 35, 915-924;

Schneider, G. E. Two visual systems. *Science*, 1969, 163, 895-902.

Schneider, G. E. Contrasting visuomotor functions of tectum and cortex in the Golden Hamster. *Psychologische Forschung*, 1967, 31, 52-62.

Sherk, H. & Stryker, M. P. Quantitative study of cortical orientation-selectivity in the visually-inexperienced kitten. *Journal of Neurophysiology*, 1975, 39.

Wagner, H. G., MacNichol., E. F., & Wolbarsht, M. L. The response properties of single ganglion cells in the goldfish retina. *Journal of General Physiology*, 1960, 43, 45-62.

Wallach, H. The perception of neutral colors. *Scientific American*, 1963, 208, 107.

Wiesel, T. N. & Hubel, D. H. Spatial and chromatic interactions in the lateral geniculate body of the rhesus monkey. *Journal of Neurophysiology*, 1966, 29, 1115-1156.

Zeki, S. M. Colour coding in rhesus monkey prestriate cortex. *Brain Research*, 1973, 53, 422-427.

CHAPTER

13

OBSERVATIONS CONCERNING THE CONTEMPORARY ANALYSIS OF THE PERCEPTUAL CONSTANCIES

WILLIAM EPSTEIN

University of Wisconsin

The aim of this concluding chapter is to offer a set of observations concerning various methodological and theoretical aspects of the contemporary analysis of the perceptual constancies as it is reflected in the preceding chapters. The hope is that these observations may suggest a number of useful directions for future research.

METHODOLOGICAL NOTES

Sampling of Stimuli

Long ago Brunswik admonished investigators to sample situations as well as subjects. In some fields of human experimental psychology this advice was not necessary, inasmuch as sampling of stimulus materials is a commonplace feature of experimental design. A familiar example is the field of memory and verbal learning. In the field of perception, however, not only was the admonition warranted when Brunswik wrote but it remains so today. As a rule, there is little sampling of stimuli; for example, in an interesting study of the effect of exposure duration on shape constancy (Leibowitz & Bourne, 1956) a single shape was presented at a single orientation in depth to the same subject for an extended series of trials with exposure duration variant. In an important study of the relation between perceived shape and perceived slant-in-depth (Kaiser, 1967) all objective shape-objective slant combinations generated the same projective shape, but only one projective shape was tested. In studies of size constancy it is not uncommon to present only one or two sizes at the various distances.

The obvious risk introduced by inadequate sampling of stimuli is the potential for misleading overgeneralization. It is simply that one cannot say whether the findings may be peculiar to the stimulus selected. Considering the prevalence of the "oblique effect," that is, inferior performance on a large variety of perceptual tasks for stimuli oriented in the oblique direction (Appelle, 1972), there is plainly the prospect of an interaction between an independent variable in an orientation constancy experiment (e.g., body position) and the orientation of the stimulus to be judged.

In addition, there may be a less obvious effect of the failure to sample stimuli. Consider a case of the investigation of the relation between perceived size and perceived distance in which many object size-objective distance combinations are sampled, all of which subtend the same visual angle. There is the possibility that the subject will become aware that projective size is invariant and the further possibility that the size-distance relationship manifest in his judgments is conditioned by this awareness.

Pictures as Stimuli

Having urged the sampling of stimuli, a few words are in order concerning the use of pictures as stimuli in experiments intended to study constancy mechanisms. A useful distinction is to be made between cases in which pictures are presented in a format that conceals the fact that the displays are pictorial (e.g., Johansson's procedures) and others in which the pictures are plainly presented as pictures. It is about the latter case that I wish to comment; for example, in an experiment designed to study shape constancy (Lappin & Preble, 1975) subjects were asked to identify angles on random polygons that were displayed in freely inspected meaningful photographic slides projected on a frontal parallel screen. There are two ways in which a subject may construe a photographic image: as an object in itself or as a surrogate for the palpable absent model. It is only in the event that the former interpretation is adopted that the subject's response bears on the question of perceived shape. If the photograph is construed as a surrogate, the test probes the cognitive rules that govern the inferences an observer is willing to make about the absent model on the basis of pictorial signs. The language of pictures is so well learned by adult western observers that it is doubtful that even explicit instructions to adopt the object attitude can ensure that inferences reflecting the treatment of pictures as surrogates will not be imported. In any event, interpretation of responses to pictorial stimuli, when the stimuli are presented as pictures, will be uncertain unless the interpretive mode adopted by the subject is known.

Temporal Aspects of Design.

A common test of the algorithm approach to the constancies is to assess the correlation between the two percepts in the invariance relationship; for example, between perceived illuminance and perceived whiteness or between judged body position and judged object orientation. In designing the test the investigator confronts a dilemma. If the two judgments for each stimulus situation are secured in immediate succession, there is the risk of engendering a cognitive bias favoring the sought-after correlation. If the two judgments are separated by judgments of other intervening stimuli, there is the risk that the intervening experiences will modify the perceptual or response system, thereby affecting the prospects of correlation between the two judgments. An additional question in both arrangements is whether probing for a single attribute of a set of attributes that is normally packaged, for example, testing for perceived depth interval when it is an interval at a distance that is usually experienced, alters the task of information processing to a degree that makes the report of questionable value.

Obviously both concerns may be evaluated empirically, although I am not aware that such evaluations have been conducted. The first concern can be approached in a straightforward manner by comparing the two arrangements in a single experiment with the same subjects and stimuli. Resolution of the second question is a bit more challenging. The following is one approach worth examining. The aim is to vary the processing load in order to determine whether the obtained relationship between the two perceptual judgments is affected by the load. The processing load is controlled by varying the demand imposed by the test and timing the test. The proposal is illustrated in the context of the size-distance relationship.

Over a long series of trials presents all possible combinations of *n* sizes and *n* distances for the desired number of replications judgments,are secured under four conditions: (a) preexposure notice, size only, (b) preexposure notice, distance only, (c) preexposure notice, both (size and distance), (d) preexposure notice, either (size or distance); probe signal immediately after exposure solicits report of size or distance. Comparing the "only" conditions (a and b) with the "both" condition (c) will help determine whether focusing attention on a single attribute modifies the relationship between perceived size and perceived distance. The inclusion of the "either" condition (d) is intended to determine whether any difference between the only and both trials is to be attributed to events during initial encoding (perceptual processing) or to factors affecting the response system.

A "Pure" Measure of Perception.

A number of writers (e.g., Epstein, 1967; Garner, Hake, & Eriksen, 1956; Natsoulas, 1967, 1968) have discussed the problem of distinguishing between perceptual and cognitive components of a subject's response. The question appears occasionally in experiments designed to study constancy. The following are a number of examples: does knowledge of the objective distal properties of the target affect perceived constancy or only what a subject is likely to report? Do age-related changes in measured constancy reflect differences in perception or differences in interpretation of the task? Under reduced-cue conditions, if a subject reports that all targets, whatever their objective distance, appear to be at the same nearby location, does this report simply reflect a best guess or do the targets really look that way?

One procedure for coping with this problem is to draw on percept-percept dependencies that are not familiar experiences for the subject and are not suggested to the subject by the experimental procedure. As an illustration, consider the interpretation of distance reports under reduced-cue conditions. Analysis of the algorithm for position constancy reveals that constancy depends not only on accurate registration of retinal displacement and eye movement but also on accurate registration of absolute (egocentric) distance. If the latter is misperceived, apparent motion of a stationary target will accompany movement of the head. Perception of stationarity depends on perception of distance. It is doubtful that this dependency is common knowledge to subjects or that it would be suggested by the typical operations for securing distance estimates. Accordingly, carefully solicited motion reports may be used to resolve interpretation of the distance reports. Conceivably, this general approach could also be applied in developmental studies to determine whether observed differences are perceptual.

Instructions

As Carlson has made clear in his chapter in this volume, attitudes of observation contribute greatly to variance in responding. Plainly, it is not possible to assess the findings of a study without detailed knowledge of the instructions administered to the subject. Here I wish to stress a point that merits reiteration. Instructions, and the attitudes of observation elicited by instructions, not only have main effects on the subject's report they also interact with other variables, making it all the more important to specify the instructions clearly. An example is provided by Epstein and Broota's (1975) study of the time it takes to make veridical size estimates under full cue conditions with viewing distance variant. A clear cut interaction between instructions (objective and phenomenal) and viewing distance was found. Under objective instructions reaction time for veridical size estimates was a positive linear function of viewing distance; under phenomenal instructions, however, the function relating reaction time to distance had a zero slope. Obviously this interaction significantly affects interpretation.

NEXT STEPS

Need for More Data

A number of our authors have commented specifically on the need for more data. Day and McKenzie have noted the paucity of data bearing on the questions of the perceptual constancies in the infant. Ono and Comerford were able to report only a few studies of the affect of viewing distance on the perception of depth based on disparity. In what follows I wish to single out three types of datum that I believe can contribute significantly to our understanding.

RELATING TO THE ALGORITHM APPROACH.

Two types of equivalence have generally been assumed to hold in tests of the approach: intersubstitutability of cues and intersubstitutability of paradigms. The first refers to the assumption that all cues with the same direct perceptual effect, for example, that determine identical perceived distances, enter into the algorithm in the same way. Elsewhere (Epstein, 1973) I have referred to this assumption as the claim that the invariance algorithm is indifferent to the source of information. This is an untested assumption and it may be incorrect. It is not inconceivable that the operation of the algorithm is specialized for certain selected inputs, nor is it hard to accept that a cue may specify a percept in a one-to-one manner, although the percept is not transmitted to a higher processing stage as information for the algorithm. Precisely, this contingency is implied by Ono and Comerford in their discussion of the role of oculomotor adjustment in calibrated disparity depth. It would be desirable to have systematic studies of the intersubstitutability of cues.

A number of paradigms for the assessment of the constancy algorithm may be distinquished. One procedure holds one of the variables in the presumed invariance relationship constant and determines whether the two remaining variables covary to maintain an invariant relationship. Logically, it should not matter which of the variables is held constant; for example, to assess the hypothesis that perceived velocity equals angular velocity times perceived distance (Epstein, 1973) we could hold either angular velocity constant, assessing the relationship between perceived velocity and perceived distance, or perceived distance constant, assessing the relationship between angular velocity and perceived velocity. Another paradigm allows all three variables to vary, as in the standard size constancy experiment in which a target of fixed size is presented at various distances. The assumption of intersubstitutability of paradigms is the premise that these paradigms are functionally equivalent procedures for evaluating the constancy algorithm. This is an untested premise, insofar as there has been no direct test of any of the algorithms using the three paradigms under otherwise comparable circumstances. Although such comparison might be accomplished by a retrospective examination of the literature, it is best accomplished by experiments explicitly designed for the purpose. My own impression, based on a comparison in memory, is that the algorithm does not hold up equally under the various paradigms: the best agreement is secured in the paradigm that holds the retinal counterpart constant.

THE DYNAMIC FOCUS

Johansson, in this volume, has argued persuasively that we ought to turn our attention to dynamic stimulus situations. I cannot improve on Johansson's arguments. I would only add that we need to consider the moving observer as well as the moving display. There can be no doubt that movement permeates everyday conditions of viewing. We cannot believe that we have adequately described the phenomena of constancies if we have neglected to study these common conditions of perception.

NEUROPHYSIOLOGICAL DATA

The last decade has witnessed a flood of results obtained by recording from single neurons in sensory pathways. Most insistent have been reports of feature detectors, single neurons that are optimally responsive to specific features of retinal stimulation. Orientation detectors, movement detectors, line detectors, and disparity detectors are only several examples of a constantly expanding list of specifically tuned neurons. Unquestionably these discoveries tell us something significant about the basis of the perceptual world. What precisely are we being told? Is it likely that further discoveries of detectors specialized for other features, together with what is already known, will suffice to explain perceptual experience? Gogel (1973) and Rock (1975, and in this volume) have answered in the negative. For an affirmative answer see Barlow, 1972.

There are two reasons for doubting the sufficiency of the single neuron approach to perceptual explanation: (a) the constancy phenomena show that perception is independent of local specific retinal features; (b) frequently, in order to predict the percept that will accompany a particular retinal input and the inferred neuronal output, it is necessary to know other aspects of the observer's perceptual experience; for example, to predict the effect of the angular displacement of a frame in relation to a surrounded point (induced motion) the *perceived* relative distances of the point and the frame (Gogel & Koslow, 1971) must be known. To predict the perceived whiteness of a disk surrounded by a ring it is not sufficient to take into account the luminance relationships and the inferred lateral inhibition. We must also know the perceived depth relationships between the surfaces contributing the luminances (Rock, in this volume; Gogel & Mershon, 1969).

The discontinuity in the chain of events from retinal input through neuronal output to perceptual reports of distal properties emerges clearly in a pair of studies aimed at identifying the site of size constancy (Blakemore, Garner, & Sweet, 1972) and orientation constancy (Mitchell & Blakemore, 1972). Using the selective adaptation procedure for identifying feature detectors in human observers, they found that adaptation was specific to the retinal properties of the stimulus; for example, with an adapting gradient at three times the distance of the test grating, the maximum elevation of threshold occurred exactly for the same angular spatial frequency as that of the adapting pattern, even though size constancy prevailed, so that the adapting and test pattern appeared to be totally different in bar width. Similarly, in the study of orientation selective adaptation was optimal for the pattern in the identical retinal orientation even though the constancy mechanism ensured that this pattern appeared to be in a different orientation despite the head tilt. The authors conclude that the mechanisms for size and orientation constancies are located in a site after the visual cortex. Our reason for referring to these studies is to call attention to the fact that although retinal input may predict neuronal output, as assessed by selective adaptation, neuronal output obviously could not have predicted perceptual output nor could perceptual output have predicted neuronal output.

The preceding is an argument that the mere compilation of lists of specific neuronal detectors is not likely to contribute decisively to out understanding of constancy mechanisms. It should not be construed as an argument against the relevance of the electrophysiological enterprise. What is needed is more electrophysiological work, but it must be work addressed to pertinent questions. A study of the psychophysical investigations of the perceptual constancies can supply such questions.

Need for Analysis

Considering the long history of the algorithm approach to the constancies, there is a remarkable lack of attention to the details of the process. Oyama's work, reported in Chapter 6, is a step in this direction. Identifying the causal links in the invariance relationship is an aspect of the analysis of the constancy algorithm as a process. Equally noteworthy is Rock's (Chapter 10) explication of the ratiomorphic process of unconscious inference. I know of no other writer who has approached the task with such directness. Rock's essay may act as a spur to encourage others to be more explicit about the process. Shebilske's model of the process of visual direction perception, presented in Chapter 2, illustrates the possibilities of the information-processing approach.

The work of Oyama, Rock, and Shebilske does not exhaust the field; for instance, an interesting application of the stage analysis approach to shape constancy was presented in a recent paper by Massaro (1973). This approach is also being used in a study of size perception by Epstein and Broota (in progress). The study is aimed at resolving a question surrounding the common finding that in the absence of distance cues size percepts conform to the law of visual angle. There are two interpretations of this finding:

1. In the absence of distance cues an observer tendency (Gogel, Chapter 5) operates to determine that targets presented at various distances will all appear to be located at the same distance. If this occurs, then, following the size-distance algorithm, perceived size will be directly proportional to visual angle.
2. In the absence of distance cues perceived size is directly determined by retinal size. The size-distance algorithm does not come into play (Rock, Chapter 10).

One way to describe the difference between these two interpretations is by comparing the processes implied by the competing interpretations to the process underlying size perception under normal viewing conditions when distance cues are available. Setting aside differences in how perceived distance is governed, the first interpretation claims that the process underlying full-cue and zero-cue size perception is the same. In both cases distances are taken into account. The second interpretation, on the other hand, claims that the processes under full- and reduced-cue viewing differ. In the latter case the distance-processing and combinatorial stages are absent. Figure 13-1 summarizes the differences graphically. We are using reaction times and the additive factors method (Sternberg, 1969) in an effort to distinguish between these two models of the perception of size under reduced viewing conditions.

Whatever the heuristic device or methatheory that may be employed, for example, perceptual processing as information transformation, the need for more explicit and self-conscious attention to testing statements about process is urgent.

COEXISTENCE OF THEORETICAL APPROACHES

In the introductory chapter I described three approaches to the explanation of perceptual constancy: an associative learning approach, an algorithm approach,

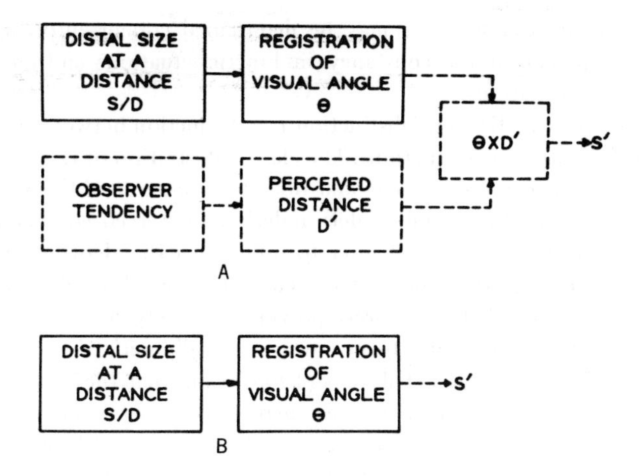

Figure 13-1 A. Under reduced cue conditions, perceived size S' is determined by the size-distance algorithm ($\theta \times D' = S'$). The observable events are represented by continuous line frames and arrows, the inferred events by broken line frames and arrows. B. Under reduced conditions perceived size S' is directly determined by visual angle θ.

and a proximal stimulus approach. The chapters that followed show that the latter two approaches are alive and well. Of course, coexistence of alternative theoretical accounts is not unusual in experimental psychology. Often the rival formulations account for the basic facts equally well, and it is difficult to exclude any of the contenders. This would appear to be the case in some of the instances of perceptual constancy.

Although we have learned to live comfortably with the prevailing theoretical pluralism, the algorithm and the proximal stimulus explanations need not be treated as complete competitors. To a degree the fact of their coexistence may be due to their potential for serving as complementary accounts.

Prospects for Complementary Application

Johansson (Chapter 11) has advocated the thesis that in analyzing precepts associated with kinetic stimulation, information sufficient to ensure constancy may be found in proximal stimulation. Johansson would contend that in kinetic stimulation invocation of the cognitivelike processing implied by the algorithm approach is superfluous. Undoubtedly the reader will have his own opinion of Johansson's thesis. Whatever that assessment may be, Johansson's proximal approach does not apply to static displays. We can agree with Johansson's criticisms of exclusive reliance on static displays in the analysis of constancies, while nevertheless accepting the often repeated laboratory findings of constancy in static displays. An explanation of these experimental facts is required and the algorithm approach seems well suited for the purpose. Accordingly, two complementary theoretical analyses may be maintained concurrently: a proximal stimulation explanation for constancy in kinetic situations and an algorithm analysis of statis situations.

Related to the kinetic-static distinction is a distinction between information-poor and information-rich situations. The classic illustration of this distinction is the difference between reduced-cue and full-cue viewing conditions. In his correlational analysis of the causal models underlying observed conformity to the invariance hypothesis Oyama (Chapter 6) found that the algorithm model was more likely to fit the data under reduced-cue conditions. Indeed, under full-cue conditions the algorithm model rarely provided a fit. Oyama's findings for the reduced-cue condition conform to my own informally derived conclusion. The findings also call to mind J. J. Gibson's (1966) claim that the minimal displays and reduced viewing conditions of many perceptual experiments create special situations that elicit cognitivelike operations. According to Gibson, these situations are so peculiar that there is no reason to believe that the operations exhibited under these conditions occur in normal information-rich situations. Clearly, the implication is that although the algorithm approach applies to information-poor cases the proximal stimulation approach applies to information-rich cases.

The distinction between the discrimination of relative and absolute distal properties provides another illustration of the sense in which the two explanations may complement each other. Consider perceived velocity. Two rival accounts have been formulated: in the proximal stimulation account perceived velocity is governed by relative angular velocity (Wallach, 1939). In the algorithm approach perceived velocity is governed by an algorithm that takes into account the perceived distance between the observer and the moving object (Epstein, 1973). Both explanations predict constancy of perceived velocity under the conditions in which it is obtained. Although it would appear that one of these accounts is superfluous, both may be necessary: one to explain discrimination of absolute velocity, the other to explain discrimination of relative velocity.

Perceptual experience of velocity includes discrimination of the specific velocity of a moving object, not merely whether it is moving faster or slower than another moving object. It is difficult to see how the proximal stimulation hypothesis can explain the discrimination of absolute velocity. The algorithm approach has no difficulty whatsoever in explaining this discrimination. Indeed, if constancy of perceived absolute velocity occurred in the absence of any visual context, there would seem to be no recourse but to a mechanism that takes distance into account. On the other hand, to explain relative velocity discrimination the proximal account seems entirely sufficient; taking distance into account would be redundant.

If the foregoing remarks about the potential complementariness of the alternative accounts has merit, it will be more profitable to press forward independently with each of the accounts rather than to attempt experiments that pit the alternatives against one another. Once we achieve a detailed specification of the possibilities and limitations of each account we will also have determined whether both are necessary.

REFERENCES

Appelle, S. Perception and discrimination as a function of stimulus orientation: The oblique 'effect' in man and animals. *Psychological Bulletin*, 1972, **78**, 226-278.

Barlow, H. B. Single units and sensations: A neuron doctrine for perceptual psychology. *Perception*, 1972, **1**, 371-394.

Blakemore, C. Garner, E. T. & Sweet, J. A. The site of size constancy. *Perception*, 1972, **1**, 111-120.

Epstein, W. The process of 'taking' into' account' in visual perception. *Perception*, 1973, **2**, 267-285.

Epstein, W. *Varieties of perceptual learning.* New York: McGraw-Hill, 1967.

Epstein, W. & Broota, K. D. Attitude of judgment and reaction time in estimation of size at a distance. *Perception & Psycholphysics*, 1975, **18**, 201-204.

Garner, W., Hake, H., & Eriksen, C. Operationism and the concept of perception. *Psychological Review*, 1956, **63**, 149-159.

Gibson, J. J. *The preception of the visual world.* Boston: Houghton-Mifflin, 1950.

Gogel, W. C. The organization of perceived space. I. Perceptual interactions. *Psychologische Forschung*, 1973, **36**, 195-221.

Gogel, W. C. and Koslow, M. The Effect of Perceived Distance on Induced Movement. *Perception and Psychophysics*, 1971, **10**, 142-146.

Gogel, W. C. & Mershon, D. H. Depth adjacency in simultaneous contrast. *Perception & Psychophysics*, 1969, **5**, 13-17.

Kaiser, P. J. Perceived shape and its dependency on perceived slant. *Journal of Experimental Psychology*, 1956, **51**, 277-281.

Lappin, J. S., & Preble, L. D. A demonstration of shape constancy. *Perception & Psychophysics*, 1975, **17**, 439-444.

Leibowitz. H. & Bourne, L. Time and intensity as determiners of perceived shape. *Journal of Experimental Psychology*, 1956, **51**, 277-281.

Massaro, D. W. The perception of rotated shapes: A process analysis of shape constancy. *Perception & Psychophysics*, 1973, **13**, 413-420.

Mitchell, D. E. & Blakemore, C. The site of orientational constancy. *Perception*, 1972, **1**, 315-320.

Natsoulas. T. What are perceptual reports about? *Psychological Bulletin*, 1967, **67**, 249-272.

Natsoulas, T. Interpreting perceptual reports. *Psychological Bulletin*, 1967, **67**, 249-272.

Rock, I. *An introduction to perception.* New York: Macmillan, 1975.

Wallach, H. On constancy of visually perceived speed. *Psychological Review*, 1939, **46**, 541-552.

AUTHOR INDEX

Numbers in *italics* indicate the pages on which full references appear.

SUBJECT INDEX